SCHAUM'S OUTLINE OF

THEORY AND PROBLEMS

of

TECHNICAL
MATHEMATICS

•

by

PAUL CALTER
Professor of Mathematics
Vermont Technical College

SCHAUM'S OUTLINE SERIES
McGRAW-HILL BOOK COMPANY

New York   St. Louis   San Francisco   Auckland   Bogotá   Düsseldorf   Johannesburg
London   Madrid   Mexico   Montreal   New Delhi   Panama   Paris
São Paulo   Singapore   Sydney   Tokyo   Toronto

0-07-009651-1

2 3 4 5 6 7 8 9 10 11 12 13 14 15 SH SH 8 7 6 5 4 3 2 1 0

Library of Congress Cataloging in Publication Data

Calter, Paul.
  Schaum's outline of theory and problems of technical mathematics.

  (Schaum's outline series)
  Includes index.
  1. Engineering mathematics. I. Title.
II. Title: Outline of theory and problems of technical mathematics.
TA330.C34   510'.2'46   78-8170
ISBN 0-07-009651-1

Preface

Perhaps the best way to learn practical mathematics is to work a large number of problems, and that's what this book is all about. It provides the student with an outline of the theory, as well as numerous practice problems.

The theory and problems have been selected to cover all or most of the mathematics presented in the usual Technical Mathematics course. It consists mainly of algebra, but there is also much trigonometry, plane and solid geometry, and the analytic geometry of the straight line.

These branches of mathematics are not kept separate in this book, but are intermixed as in a typical classroom presentation. However, each branch has been made as independent as possible so that the order of study may be changed.

Within each chapter, the material is divided into three distinct sections: Theory and Examples, Solved Problems, and Supplementary Problems.

Each Theory section contains a concise summary of the definitions, theorems, formulas, principles, and methods needed for that chapter. Each new point is illustrated by one or more fully worked examples. *Common Errors* are indicated. These are places where students often make mistakes. Each Common Error is enclosed in a box. You may think of this box as the outline of a trapdoor through which you will fall if you ignore the warning.

The *Solved Problem* section of each chapter contains fully worked solutions to typical problems. They are presented in the same order as in the Theory section, and are graded in difficulty within each group. As this is a book on applied mathematics, the problems, where possible, are of a practical nature, chosen primarily from the fields of Science; Civil, Mechanical, and Electrical Technology; and Finance.

Work the problems from your own field, to be sure, but also some from other fields if you have the time. At least read some of the other problems to see how a single mathematical idea can be applied to a bewildering range of problems that seem to have nothing in common.

Left out are many of those old favorites of math teachers, such as: "John is now twice as old as Mary, but in six years he will be eight years older than she: in how many years . . . etc."

The *Supplementary Problems* also follow the order of the Theory section. Answers to all Supplementary Problems are given.

After studying a particular topic in the Theory section and the Examples which accompany it, find a similar problem in the Solved Problems section. Try to work each problem without looking at the solution unless you get stuck. Even if you work the problem successfully by yourself, you should inspect the given solution. You may see some shortcuts, or a better method. Then try some of the Supplementary Problems.

Calculators are rapidly replacing tables for finding trigonometric and logarithmic functions. Accordingly, instructions for calculator use are given in this Outline, along with the usual instructions for the tables.

If you wish to buy a calculator for use in technical mathematics, it should have, in addition to the four basic operations:

the trigonometric functions	sin	cos	tan
the inverse trig functions	arc	or	inv
common logarithms	log		
natural logarithms	ln		
common antilogs	10^x		
natural antilogs	e^x		
exponents	y^x		

It should be able to work in either degrees or radians.

The above features will enable you to work any problem in this book without the use of tables.

SI metric units are used in about half the problems in this book.

The *Summary of Facts and Formulas* is intended not only as a working tool while using this book, but as a valuable reference afterward. Every mathematical formula used in this book is given in the Summary of Formulas, as well as a sampling of formulas from the various technologies. The latter are not intended to be in any way complete, but to give only the simpler formulas that are used in this Outline.

I would like to thank my colleagues at Vermont Technical College who reviewed portions of the manuscript and made valuable suggestions: Walt Granter, Charles Cooley, Don Nevin, and John Knox. Many of my students helped with the laborious job of checking all the numerical work: Tim Leonard, Lori Dorgan, Celia Lopez, Susan Clason, Paul Pietryka, Henry Lee, Gary Simon, Nancy Davis, and Nancy Brown. I am indebted to John Aliano and Ellen LaBarbera of McGraw-Hill for their expert and thorough editing of the manuscript.

PAUL CALTER

Contents

Summary of Facts and Formulas

No.				Page
PERCENTAGE	**1**		Amount = base × rate $\qquad A = BP$	8
	2		Percent difference = $\dfrac{\text{difference of the two numbers}}{\text{base}} \times 100$	9
	3		Percent change = $\dfrac{\text{new value} - \text{original value}}{\text{original value}} \times 100$	9
	4		Percent error = $\dfrac{\text{measured value} - \text{known value}}{\text{known value}} \times 100$	9
	5		Percent concentration of ingredient $A = \dfrac{\text{amount of } A}{\text{amount of mixture}} \times 100$	10 112
	6		Percent efficiency = $\dfrac{\text{output}}{\text{input}} \times 100$	10
EXPONENTS	**7**	Definition	$a^n = \underbrace{a \cdot a \cdot a \cdots a}_{n \text{ factors}}$	3 44
	8	Laws of Exponents	Products $\qquad x^a \cdot x^b = x^{a+b}$	44
	9		Quotients $\qquad \dfrac{x^a}{x^b} = x^{a-b}$	45
	10		Powers $\qquad (x^a)^b = x^{ab} = (x^b)^a$	45
	11		Product Raised to a Power $\qquad (xy)^a = x^a \cdot y^a$	45
	12		Quotient Raised to a Power $\qquad \left(\dfrac{x}{y}\right)^a = \dfrac{x^a}{y^a}$	45
	13		Zero Exponent $\qquad x^0 = 1$	46
	14		Negative Exponent $\qquad x^{-a} = \dfrac{1}{x^a}$	46
	15	Fractional Exponents	$a^{1/n} = \sqrt[n]{a}$	269
	16		$a^{m/n} = \sqrt[n]{a^m} = (\sqrt[n]{a})^m$	269
RADICALS	**17**	Rules of Radicals	Root of a Product $\qquad \sqrt[n]{ab} = \sqrt[n]{a}\,\sqrt[n]{b}$	269
	18		Root of a Quotient $\qquad \sqrt[n]{\dfrac{a}{b}} = \dfrac{\sqrt[n]{a}}{\sqrt[n]{b}}$	270
	19		Root of a Power $\qquad \sqrt[n]{a^m} = (\sqrt[n]{a})^m$	270

	No.				Page
OPERATIONS WITH ZERO	**20**	Addition and Subtraction		$a \pm 0 = a$	
	21	Multiplication		$a \times 0 = 0$	
	22	Zero Dividend		$\dfrac{0}{a} = 0$	
	23	Division by Zero		$\dfrac{a}{0}$ is not defined.	
	24	Zero Raised to a Power		$0^a = 0$	
	25	Zero Power		$a^0 = 1$	
	26	Zero Raised to the Zero Power		0^0 is not defined.	
	27	Root of Zero		$\sqrt[n]{0} = 0$	
	28	Zero Index		$\sqrt[0]{a}$ is not defined.	
	29	Zeroth Root of Zero		$\sqrt[0]{0}$ is not defined.	
SPECIAL PRODUCTS AND FACTORING	**30**	Common Factor		$ab + ac + ad = a(b + c + d)$	51
	31	*Binomials*	Difference of Two Squares	$a^2 - b^2 = (a - b)(a + b)$	51
	32		Sum of Two Cubes	$a^3 + b^3 = (a + b)(a^2 - ab + b^2)$	54
	33		Difference of Two Cubes	$a^3 - b^3 = (a - b)(a^2 + ab + b^2)$	54
	34	*Trinomials*	Test for Factorability	$ax^2 + bx + c$ is factorable if $b^2 - 4ac$ is a perfect square.	52
	35		Leading Coefficient = 1	$x^2 + (a + b)x + ab = (x + a)(x + b)$	52
	36		General Quadratic Trinomial	$acx^2 + (ad + bc)x + bd = (ax + b)(cx + d)$	52
	37		Perfect Square Trinomials	$a^2 + 2ab + b^2 = (a + b)^2$	54
	38			$a^2 - 2ab + b^2 = (a - b)^2$	54
	39	Factoring by Grouping		$ac + ad + bc + bd = (a + b)(c + d)$	55

	No.				Page
FRACTIONS	**40**	Signs of a Fraction		$-\dfrac{x}{y} = \dfrac{-x}{y} = \dfrac{x}{-y} = -\dfrac{-x}{-y}$	49
	41	Simplifying		$\dfrac{ad}{bd} = \dfrac{a}{b}$	56
	42	Multiplication		$\dfrac{a}{b} \cdot \dfrac{c}{d} = \dfrac{ac}{bd}$	56
	43	Division		$\dfrac{a}{b} \div \dfrac{c}{d} = \dfrac{a}{b} \cdot \dfrac{d}{c} = \dfrac{ad}{bc}$	57
	44	Addition and Subtraction	Same Denominators	$\dfrac{a}{b} \pm \dfrac{c}{b} = \dfrac{a \pm c}{b}$	57
	45		Different Denominators	$\dfrac{a}{b} \pm \dfrac{c}{d} = \dfrac{ad}{bd} \pm \dfrac{bc}{bd} = \dfrac{ad \pm bc}{bd}$	57
PROPORTION	**46**	In the Proportion $a : b = c : d$	The product of the means equals the product of the extremes.	$ad = bc$	95
	47		The extremes may be interchanged.	$d : b = c : a$	95
	48		The means may be interchanged.	$a : c = b : d$	95
	49		The means may be interchanged with the extremes.	$b : a = d : c$	95
	50	Mean Proportional		$a : b = b : c$	89
	51		Geometric Mean	$b = \pm\sqrt{ac}$	89
VARIATION	**52**	k = Constant of Proportionality	Direct	$y \propto x$ or $y = kx$	134
	53		Inverse	$y \propto \dfrac{1}{x}$ or $y = \dfrac{k}{x}$	134
	54		Joint	$y \propto xw$ or $y = kxw$	135
INTERSECTING LINES	**55**	Opposite angles of two intersecting straight lines are equal.			161
	56	If two parallel straight lines are cut by a transversal, corresponding angles are equal and alternate interior angles are equal.			162
	57	If two lines are cut by a number of parallels, the corresponding segments are proportional.			162

	No.				Page
ANY TRIANGLE	**58**	Areas		Area $= \dfrac{1}{2} bh$	163
	59			Area $= \sqrt{s(s-a)(s-b)(s-c)}$ where $s = \dfrac{1}{2}(a+b+c)$	164
	60		Sum of the Angles	$A + B + C = 180°$	164 242
	61		Law of Sines	$\dfrac{a}{\sin A} = \dfrac{b}{\sin B} = \dfrac{c}{\sin C}$	421 424
	62		Law of Cosines	$a^2 = b^2 + c^2 - 2bc \cos A$ $b^2 = a^2 + c^2 - 2ac \cos B$ $c^2 = a^2 + b^2 - 2ab \cos C$	422 426
	63		Exterior Angle	$\theta = A + B$	164
	64	Medians	A median joins a vertex with the midpoint of the opposite side.		
	65		The three medians meet at the centroid of the triangle.		
	66		The centroid cuts each median at 1/3 its length.		
	67	An angle bisector of a triangle divides the opposite side in proportion to the other two sides.			165
	68	A line parallel to one side of a triangle divides the other two sides proportionately.			165
	69	A line which joins the midpoints of two sides of a triangle is parallel to the third side and is equal to one-half the third side.			166
SIMILAR TRIANGLES	**70**	If two angles of a triangle equal two angles of another triangle, the triangles are similar.			167
	71	Corresponding sides of similar triangles are in proportion.			167
RIGHT TRIANGLES	**72**		Pythagorean Theorem	$a^2 + b^2 = c^2$	167 243
	73	Trigonometric Ratios	Sine	$\sin A = \dfrac{y}{r} = \dfrac{\text{opposite side}}{\text{hypotenuse}}$	210
	74		Cosine	$\cos A = \dfrac{x}{r} = \dfrac{\text{adjacent side}}{\text{hypotenuse}}$	210
	75		Tangent	$\tan A = \dfrac{y}{x} = \dfrac{\text{opposite side}}{\text{adjacent side}}$	210
	76		Cotangent	$\cot A = \operatorname{ctn} A = \dfrac{x}{y} = \dfrac{\text{adjacent side}}{\text{opposite side}}$	210
	77		Secant	$\sec A = \dfrac{r}{x} = \dfrac{\text{hypotenuse}}{\text{adjacent side}}$	210
	78		Cosecant	$\csc A = \dfrac{r}{y} = \dfrac{\text{hypotenuse}}{\text{opposite side}}$	210

	No.				Page
RIGHT TRIANGLES (*Cont.*)	**79**	Reciprocal Relationships	$\sin A = \dfrac{1}{\csc A}$ $\quad \cos A = \dfrac{1}{\sec A}$ $\quad \tan A = \dfrac{1}{\cot A}$		211
	80	Other Trigonometric Functions	Exsecant	$\text{exsec } A = \sec A - 1$	
	81		Versine	$\text{vers } A = 1 - \cos A$	
	82		Coversine	$\text{covers } A = 1 - \sin A$	
	83		Haversine	$\text{hav } A = \frac{1}{2} \text{ vers } A$	
	84	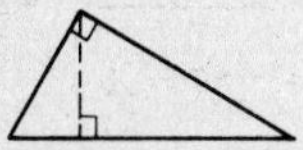	In a right triangle, the altitude to the hypotenuse forms two right triangles which are similar to each other and to the original triangle.		168
	85	*A* and *B* Are Complementary Angles 	Cofunctions	$\sin B = \cos A$ $\quad \cot B = \tan A$ $\cos B = \sin A$ $\quad \sec B = \csc A$ $\tan B = \cot A$ $\quad \csc B = \sec A$	245
CONGRUENT TRIANGLES	**86**	Two Triangles Are Congruent if	Two angles and a side of one are equal to two angles and the corresponding side of the other (ASA), (AAS).		169
	87		Two sides and the included angle of one are equal, respectively, to two sides and the included angle of the other (SAS).		169
	88		Three sides of one are equal to the three sides of the other (SSS).		169
QUADRILATERALS	**89**	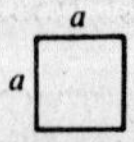	Square	Area $= a^2$	170
	90	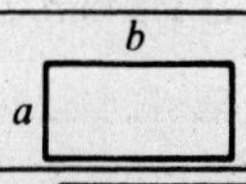	Rectangle	Area $= ab$	170
	91	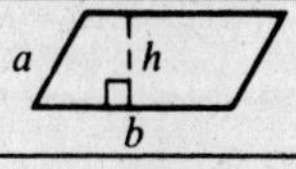	Parallelogram: Diagonals bisect each other.	Area $= bh$	171
	92	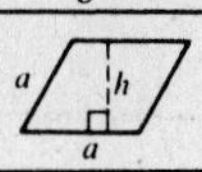	Rhombus: Diagonals intersect at right angles.	Area $= ah$	171
	93	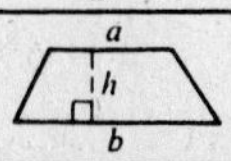	Trapezoid	Area $= \dfrac{(a+b)h}{2}$	171
CIRCLES	**94**	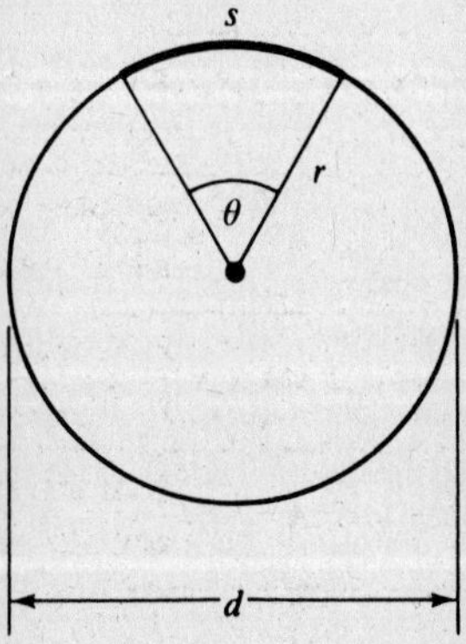	Circumference $= 2\pi r = \pi d$		171
	95		Area $= \pi r^2 = \dfrac{\pi d^2}{4}$		171
	96		Central angle θ (radians) $= \dfrac{s}{r}$		172 218
	97		Area of sector $= \dfrac{rs}{2} = \dfrac{r^2\theta}{2}$		172
	98		1 revolution $= 2\pi$ radians $= 360°$		172 206
	99		Any angle inscribed in a semicircle is a right angle.		174

	No.				Page
CIRCLES (*cont.*)	**100**		Tangents to a Circle	Tangent AP is perpendicular to radius OA.	174
	101			Tangent AP = tangent BP OP bisects angle APB.	175
	102		Intersecting Chords	$ab = cd$	173
	103			If two circles intersect, their line of centers is perpendicular to, and bisects, the common chord.	175
	104			If two circles are tangent to each other, the line of centers passes through the point of contact.	176
SOLIDS	**105**		Cube	Volume = a^3	177
	106			Surface area = $6a^2$	177
	107		Rectangular Parallelepiped	Volume = lwh	176
	108			Surface area = $2(lw + hw + lh)$	176
	109		Right Cylinder or Prism	Volume = (area of base)(altitude)	176 177
	110			Lateral area = (perimeter of base)(altitude) (not incl. bases)	176 177
	111		Sphere	Volume = $\dfrac{4}{3}\pi r^3$	179
	112			Surface area = $4\pi r^2$	179
	113		Cone	Volume = $\dfrac{1}{3}$ (area of base)(altitude)	178
	114			Lateral area = $\dfrac{1}{2}$ (perimeter of base) × (slant height)	178
	115		Frustum of Cone	Volume = $\dfrac{h}{3}(A_1 + A_2 + \sqrt{A_1 A_2})$	178
	116			Lateral area = $\dfrac{s}{2}$ (sum of base perimeters) = $\dfrac{s}{2}(P_1 + P_2)$	179
SIMILAR SOLIDS	**117**			Corresponding dimensions of plane or solid similar figures are in proportion.	179
	118			Areas of similar plane or solid figures are proportional to the squares of any two corresponding dimensions.	180
	119			Volumes of similar solid figures are proportional to the cubes of any two corresponding dimensions.	180

	No.					Page
DETERMINANTS	**120**	Second Order		$\begin{vmatrix} a_1 & b_1 \\ a_2 & b_2 \end{vmatrix} = a_1 b_2 - a_2 b_1$		294
	121	Third Order		$\begin{vmatrix} a_1 & b_1 & c_1 \\ a_2 & b_2 & c_2 \\ a_3 & b_3 & c_3 \end{vmatrix} = a_1 b_2 c_3 + a_3 b_1 c_2 + a_2 b_3 c_1 - (a_3 b_2 c_1 + a_1 b_3 c_2 + a_2 b_1 c_3)$		298
	122	Cramer's Rule	is	The solution to the set of equations $$a_1 x + b_1 y = c_1$$ $$a_2 x + b_2 y = c_2$$ $$x = \frac{c_1 b_2 - c_2 b_1}{a_1 b_2 - a_2 b_1} = \frac{\begin{vmatrix} c_1 & b_1 \\ c_2 & b_2 \end{vmatrix}}{\begin{vmatrix} a_1 & b_1 \\ a_2 & b_2 \end{vmatrix}} \quad \text{and} \quad y = \frac{a_1 c_2 - a_2 c_1}{a_1 b_2 - a_2 b_1} = \frac{\begin{vmatrix} a_1 & c_1 \\ a_2 & c_2 \end{vmatrix}}{\begin{vmatrix} a_1 & b_1 \\ a_2 & b_2 \end{vmatrix}}$$		293 296
COMPLEX NUMBERS	**123**	Powers of i		$i = \sqrt{-1}, \quad i^2 = -1, \quad i^3 = -i, \quad i^4 = 1, \quad i^5 = i, \quad \text{etc.}$		274
	124	Sums		$(a + bi) + (c + di) = (a + c) + (b + d)i$		275
	125	Differences		$(a + bi) - (c + di) = (a - c) + (b - d)i$		275
	126	Products		$(a + bi)(c + di) = (ac - bd) + (ad + bc)i$		275
	127	Quotients		$\dfrac{a + bi}{c + di} = \dfrac{(a + bi)(c - di)}{(c + di)(c - di)} = \dfrac{ac + bd}{c^2 + d^2} + \dfrac{bc - ad}{c^2 + d^2} i$		276
THE STRAIGHT LINE	**128**		Distance Formula	$d = \sqrt{(x_2 - x_1)^2 + (y_2 - y_1)^2}$		325
	129		Slope m	$m = \dfrac{\text{rise}}{\text{run}} = \dfrac{y_2 - y_1}{x_2 - x_1}$		325
	130			$m = \tan(\text{angle of inclination}) = \tan \theta$		327
	131		Equation of Straight Line	General Form	$Ax + By + C = 0$	328
	132			Slope-Intercept Form	$y = mx + b$	329
	133			Two-Point Form	$\dfrac{y - y_1}{x - x_1} = \dfrac{y_2 - y_1}{x_2 - x_1}$	330
	134			Point-Slope Form	$m = \dfrac{y - y_1}{x - x_1}$	329
	135			Intercept Form	$\dfrac{x}{a} + \dfrac{y}{b} = 1$	330

	No.				Page
THE STRAIGHT LINE (*cont.*)	**136**		If L_1 and L_2 are perpendicular, then	$m_1 = -\dfrac{1}{m_2}$	326
	137		Angle of Intersection	$A = \theta_2 - \theta_1$	333
QUADRATICS	**138**	General Form	$ax^2 + bx + c = 0$		328
	139	Quadratic Formula	$x = \dfrac{-b \pm \sqrt{b^2 - 4ac}}{2a}$		355 364
	140	Nature of the Roots	If a, b, and c are real, and	if $b^2 - 4ac > 0$ the roots are real and unequal if $b^2 - 4ac = 0$ the roots are real and equal if $b^2 - 4ac < 0$ the roots are complex	355
	141		Parabola, Vertex at Origin, Opening Upward	$x^2 = 4py$	
LOGARITHMS	**142**	Exponential to Logarithmic Form	If $b^x = N$, then $x = \log_b N$		381
	143		Products	$\log_b MN = \log_b M + \log_b N$	382
	144		Quotients	$\log_b \dfrac{M}{N} = \log_b M - \log_b N$	382
	145		Powers	$\log_b M^p = p \log_b M$	383
	146	Laws of Logarithms	Roots	$\log_b \sqrt[q]{M} = \dfrac{1}{q} \log_b M$	383
	147		Log of 1	$\log_b 1 = 0$	
	148		Log of the Base	$\log_b b = 1$	384
	149	Common Logarithm Notation	$\log_{10} N \equiv \log N$		384
	150	Natural Logarithm Notation	$\log_e = N \equiv \ln N$, where $e \cong 2.718$		384
	151	Log of the Base Raised to a Power	$\log_b b^n = n$		384
	152	Characteristic and Mantissa	$\log(1.85 \times 10^2) = \boxed{2}.\boxed{2672}$ Characteristic ——↑ ↑—— Mantissa		385
	153	Change of Base	$\log N = \dfrac{\ln N}{\ln 10} = \dfrac{\ln N}{2.3026}$		393 404

	No.				Page
SOME USEFUL FUNCTIONS	**154**	Exponential Growth		$y = ae^{nx}$	396
	155	Exponential Decay		$y = ae^{-nx}$	396
	156	Power Function		$y = ax^n$	394
MIXTURES	**157**	Mixture Containing Ingredients $A, B, C, \ldots$	Total amount of mixture = amount of A + amount of $B + \cdots$		112
	158		Final amount of each ingredient = initial amount + amount added − amount removed		112 116
	159	Combination of Two Mixtures	Final amount of A = amount of A from mixture 1 + amount of A from mixture 2		112
	160	Fluid Flow	Amount of flow = flow rate × elapsed time $A = QT$		116
WORK	**161**	If a job can be done in n days, then $1/n$ of a job can be done in one day.			115
	162	If m jobs can be done in n days, then m/n jobs can be done each day.			
	163	Total work done = amount done by A + amount done by $B + \cdots$			115
	164		Work = force × distance = Fd		
FINANCIAL	**165**	Unit Cost	Unit cost = $\dfrac{\text{total cost}}{\text{number of units}}$		
	166	Interest: Principal P Invested at Rate i for n years Accumulates to Amount S.	Simple	$S = P(1 + ni)$	100
	167		Compounded Annually	$S = P(1 + i)^n$	
	168		Compounded k times/yr	$S = P\left(1 + \dfrac{i}{k}\right)^{ki}$	

	No.				Page
FINANCIAL *(cont.)*	**169**	Straight-Line Depreciation	Total Depreciation	$D =$ purchase price $-$ salvage value $= P - S$	347
	170		Annual Depreciation	$A = \dfrac{\text{total depreciation}}{\text{number of years}} = \dfrac{D}{T}$	347
	171		Book Value y, after t Years	$y = P - At$	347
STATICS	**172**		Moment about Point a	$M_a = Fd$	113
	173	Equations of Equilibrium (Newton's First Law)	The sum of all horizontal forces $= 0$		114
	174		The sum of all vertical forces $= 0$		114
	175		The sum of all moments about any point $= 0$		114
	176		Coefficient of Friction	$\mu = \dfrac{f}{N}$	
MOTION	**177**	Linear Motion	Uniform Motion (Constant Speed)	Distance $=$ rate $\times$ time $\\ D = Rt$	87 110
	178		Uniformly Accelerated (Constant Acceleration a) — Displacement	$s = v_0 t + \dfrac{at^2}{2}$	99 372
	179		Velocity	$v = v_0 + at$	
	180		Newton's Second Law	$F = ma$	
	181		Average Speed	Average speed $= \dfrac{\text{total distance traveled}}{\text{total time elapsed}}$	110
	182		Linear Speed of Point at Radius r	$v = \omega r$	220
	183			Angle $=$ rate of rotation $\times$ time $\\ \theta = \omega t$	219
MATERIAL PROPERTIES	**184**	Density		Density $= \dfrac{\text{weight}}{\text{volume}}$	
	185	Mass		Mass $= \dfrac{\text{weight}}{\text{acceleration due to gravity}}$	
	186	Specific Gravity		$SG = \dfrac{\text{density of substance}}{\text{density of water}}$	103
	187		Total Force on a Surface	Force $=$ pressure $\times$ area	

	No.				Page
TEMPERATURE	188	Conversions between Degrees Celsius (C) and Degrees Fahrenheit (F)		$C = \dfrac{5}{9}(F - 32)$	87
	189			$F = \dfrac{9}{5}C + 32$	87
STRENGTH OF MATERIALS	190	Tension or Compression	Normal Stress	$\sigma = \dfrac{P}{a}$	141
	191		Strain	$\epsilon = \dfrac{e}{L}$	141
	192		Modulus of Elasticity and	$E = \dfrac{PL}{ae}$	96 140
	193		Hooke's Law	$E = \dfrac{\sigma}{\epsilon}$	141 146
	194	Thermal Expansion	Elongation	$e = \alpha L\,\Delta t$	
	195		New Length	$L = L_0(1 + \alpha\,\Delta t)$	97
	196		Strain	$\epsilon = \dfrac{e}{L} = \alpha\,\Delta t$	
	197	Temperature change $= \Delta t$ Coefficient of thermal expansion $= \alpha$	Stress, if Restrained	$\sigma = E\epsilon = E\alpha\,\Delta t$	
	198		Force, if Restrained	$P = a\sigma = aE\alpha\,\Delta t$	
	199		Force Needed to Deform a Spring	$F = \text{spring constant} \times \text{distance} = kx$	146
ELECTRICAL TECHNOLOGY	200	Ohm's Law		$\text{Current} = \dfrac{\text{voltage}}{\text{resistance}} \qquad I = \dfrac{V}{R}$	86
	201	Combinations of Resistors	In Series	$R = R_1 + R_2 + R_3 + \cdots$	109
	202		In Parallel	$\dfrac{1}{R} = \dfrac{1}{R_1} + \dfrac{1}{R_2} + \dfrac{1}{R_3} + \cdots$	99
	203	Power Dissipated in a Resistor		$\text{Power} = P = VI$	
	204			$P = \dfrac{V^2}{R}$	140
	205			$P = I^2 R$	
	206	Kirchhoff's Laws	Loops	The sum of the voltage rises and drops around any closed loop is zero.	321 344
	207		Nodes	The sum of the currents entering and leaving any node is zero.	

No.				Page
208	Resistance Change with Temperature		$R = R_1[1 + \alpha(t - t_1)]$	98 145 339
209	Resistance of a Wire $A \rightarrow$ L		$R = \dfrac{\rho L}{A}$	140 151
210	Combinations of Capacitors	In Series	$\dfrac{1}{C} = \dfrac{1}{C_1} + \dfrac{1}{C_2} + \dfrac{1}{C_3} + \cdots$	
211		In Parallel	$C = C_1 + C_2 + C_3 + \cdots$	
212	Charge on a Capacitor at Voltage V		$Q = CV$	
213	Impedance of a Series RLC Circuit		$Z = \sqrt{R^2 + \left(\omega L - \dfrac{1}{\omega C}\right)^2}$	98
214		Reactance	$X = X_L - X_C$	247
215		Magnitude of Impedance	$|Z| = \sqrt{R^2 - X^2}$	247
216		Phase Angle	$\phi = \arctan \dfrac{X}{R}$	247
217	Power in an ac Circuit	Apparent Power	$P_a = VI$ (voltamperes)	248
218		Average Power	$P = VI \cos \phi$ (watts)	248
219		Reactive Power	$P_q = VI \sin \phi$ (vars)	248
220		Power Factor	$F_p = \cos \phi$	248
221	Decibels Gained or Lost		$dB = 10 \log_{10} \dfrac{P_2}{P_1}$	412

ELECTRICAL TECHNOLOGY (cont.)

Chapter 1

Review of Arithmetic

1.1 NUMBERS

(a) Definitions

1. The *integers* are the *whole numbers*, both positive and negative, including zero.

    Example 1.1:   2300, -9, 0, and -23 are integers.

2. *Rational numbers* are those numbers that can be written as fractions.

    Example 1.2:   5/2, 3, $-11/3$, $4\frac{2}{3}$ are rational numbers.

3. *Irrational numbers* cannot be expressed as fractions.

    Example 1.3: π, $\sqrt{7}$, and $\sqrt[3]{9}$ are irrational numbers.

4. The *real numbers* consist of the rational and irrational numbers taken together.

    Example 1.4:   5, $\sqrt{15}$, -9, 27/5, and $-\sqrt[3]{17}$ are real numbers.

5. *Imaginary numbers* result when an even root of a negative number is taken.

    Example 1.5: $\sqrt{-4}$, $\sqrt[4]{-9}$, and $\sqrt[8]{-3}$ are imaginary numbers.

6. *Complex numbers* are combinations of real and imaginary numbers.

    Example 1.6: $3 + \sqrt{-5}$ and $-8 - \sqrt{-7}$ are complex numbers.

7. In the *decimal system*, the value of each digit in a number depends upon its position relative to the *decimal point*. The place values for a decimal number are given in Fig. 1-1. The decimal point is omitted when writing integers.

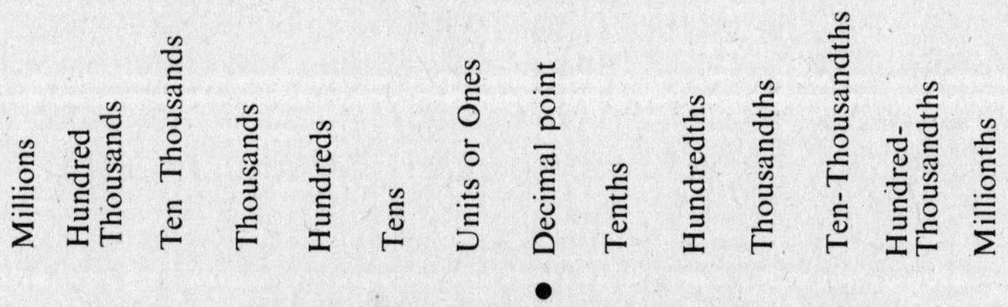

Fig. 1-1   Values of the positions in a decimal number.

Example 1.7:   The number 2375 has the value

$$(2 \times 1000) + (3 \times 100) + (7 \times 10) + (5 \times 1)$$

Example 1.8:   The number 0.123 has the value

$$(1 \times 0.1) + (2 \times 0.01) + (3 \times 0.001)$$

1

(b) The Number Line.   The *number line*, shown in Fig. 1-2, is a graphical representation of the real number system.   A point on the line, the *origin*, is chosen to represent the number zero.   The positive integers are plotted, equally spaced, to the right of the origin, and the negative integers to the left. Every other real number can then be plotted among the integers.

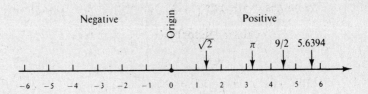

Fig. 1-2   **The number line.**

Example 1.9:   The number 9/2 is shown plotted halfway between 4 and 5 in Fig. 1-2.

(c) Signs of Equality and Inequality.   *Signs of equality and inequality* show the relative positions of two numbers on the number line.

   $a = b$ means *a equals b*, and that they occupy the same position on the number line.

   $a \neq b$ means that *a* and *b* are *not equal* and have different locations on the number line.

   $a > b$ means *a* is *greater than b*, and it lies to the right of *b* on the number line.

   $a < b$ means *a* is *less than b*, and it lies to the left of *b* on the number line.

Other symbols used are as follows:

   $a \geq b$ or $a \geqq b$ means *a* is *either greater than or equal to b.*

   $a \leq b$ or $a \leqq b$ means *a* is *either less than or equal to b.*

   $a \approx b$ or $a \cong b$ means *a* is *approximately equal to b.*

1.2   ARITHMETIC OPERATIONS WITH SIGNED NUMBERS

(a) Signed Numbers.   All numbers have an *algebraic sign*, which tells whether the number lies to the left or right of the origin on the number line.

Example 1.10:   The number 8 is 8 units to the right of the origin.   When no sign is written before a number it is assumed to be +.

Example 1.11:   The number −5 is 5 units to the left of the origin.

(b) Absolute Value.   The *absolute value* of a number *n*, written $|n|$, is its *magnitude*, regardless of algebraic sign.

Example 1.12

(a) $|3| = 3$ (c) $|4 - 7| = |-3| = 3$ (e) $-|9 - 11| = -|-2| = -2$

(b) $|-3| = 3$ (d) $-|6| = -6$

(c) Addition and Subtraction.   In order to add or subtract signed numbers, combine them according to their algebraic sign, as in the following examples.

Example 1.13

(a) $5 + 3 = 8$ (c) $-5 + 3 = -2$ (e) $-5 + (-3) - (-6) = -5 - 3 + 6 = -2$

(b) $5 + (-3) = 5 - 3 = 2$ (d) $5 - (-3) = 5 + 3 = 8$

(d) Multiplication.   The names of the parts of a multiplication problem are

$$(\text{factor})(\text{factor})(\text{factor})\cdots = \text{product}$$

Example 1.14

(a) $2(3) = 6$ (b) $2(-3) = -6$ (c) $-2(3) = -6$ (d) $-2(-3) = 6$ (e) $2(-3)(-4)(-1) = -24$

Note that the product of an *even* number of negative factors is always positive, while the product of an odd number of negative factors is negative.

(e) Division.   The names of the parts in a division problem are

$$\frac{\text{Dividend}}{\text{Divisor}} = \frac{\text{numerator}}{\text{denominator}} = \text{quotient}$$

The quotient will be positive if the numerator and denominator have the same sign, and it will be negative if they have opposite signs.

Example 1.15

(a) $\dfrac{6}{3} = 2$ (b) $\dfrac{-6}{3} = -2$ (c) $\dfrac{6}{-3} = -2$ (d) $\dfrac{-6}{-3} = 2$

(f) Reciprocals.   The *reciprocal* of a number n is $1/n$.

Example 1.16

(a) The reciprocal of 2 is 1/2.

(b) The reciprocal of 1/5 is 5.

(g) Powers.   In the expression

$$2^4$$

the number 2 is called the *base*, and the number 4 is called the *exponent*.   It is read "two to the fourth power."   Its value is

$$2^4 = 2 \cdot 2 \cdot 2 \cdot 2 = 16$$

Example 1.17

(a) $3^2 = 3 \cdot 3 = 9$ (b) $2^3 = 2 \cdot 2 \cdot 2 = 8$

In general,

$$\boxed{a^n = \underbrace{a \cdot a \cdot a \cdots \cdot a}_{n \text{ factors}}} \qquad \textbf{7*}$$

---

* All boxed and numbered formulas and theorems are tabulated in the Summary of Formulas in the front of this book.

A negative base raised to an *even* power gives a *positive* number. A negative base raised to an *odd* power gives a *negative* number.

Example 1.18

(a) $(-2)^2 = (-2)(-2) = 4$ (b) $(-2)^3 = (-2)(-2)(-2) = -8$ (c) $(-1)^{24} = 1$ (d) $(-1)^{25} = -1$

(h) Roots. If $a^n = b$, then

$$\sqrt[n]{b} = a$$

which is read "the *n*th root of *b* equals *a*." The symbol $\sqrt{}$ is a *radical sign*, *b* is the *radicand*, and *n* is the *index* of the radical.

Example 1.19

(a) $\sqrt{4} = 2$ because $2^2 = 4$ (d) $\sqrt{-4}$ imaginary (see Chapter 10)

(b) $\sqrt[3]{8} = 2$ because $2^3 = 8$ (e) $\sqrt[3]{-8} = -2$ because $(-2)^3 = -8$

(c) $\sqrt[4]{81} = 3$ because $3^4 = 81$

Common Error	An even root of a positive number is positive. Thus, $\sqrt{4} = +2$, *not* ± 2. This is called the *principal root*.

Example 1.20: Evaluate the following radicals: $\sqrt{9}, \pm\sqrt{16}, -\sqrt{25}, \sqrt{-25}, \sqrt[3]{-27}, -\sqrt[3]{-27}, \sqrt[3]{27}$.

Solution

(a) $\sqrt{9} = 3$ (c) $-\sqrt{25} = -5$ (e) $\sqrt[3]{-27} = -3$ (g) $\sqrt[3]{27} = 3$

(b) $\pm\sqrt{16} = \pm 4$ (d) $\sqrt{-25}$ is imaginary.   (f) $-\sqrt[3]{-27} = -(-3) = 3$

(i) Operations with Zero

Example 1.21: Addition and subtraction:

(a) $3 + 0 = 3$ (b) $0 + 6 = 6$ (c) $7 - 0 = 7$ (d) $0 - 4 = -4$

Example 1.22: Multiplication:

(a) $9 \cdot 0 = 0$ (b) $0 \cdot 2 = 0$

Example 1.23: Division:

(a) $\dfrac{0}{8} = 0$ (b) $\dfrac{7}{0}$ is not defined.   (c) $\dfrac{0}{0}$ is not defined.

Example 1.24: Powers:

(a) $0^2 = 0$ (b) $6^0 = 1$ (c) 0^0 is not defined.

Example 1.25: Roots:

(a) $\sqrt{0} = 0$ (b) $\sqrt[0]{4}$ is not defined.

1.3 COMMON FRACTIONS

(a) Definitions

1. A fraction whose numerator is smaller than its denominator is called a *proper fraction*.

Example 1.26:   2/3, 5/7, and 8/23 are proper fractions.

2. An *improper fraction* is one whose numerator is larger than its denominator.

Example 1.27:   5/3, 9/2, and 15/7 are improper fractions.

3. Combinations of whole numbers and proper fractions are called *mixed numbers*, or numbers in *mixed form*.

Example 1.28: $1\frac{2}{3}$, $3\frac{1}{2}$, and $2\frac{2}{5}$ are mixed numbers.

(b) Reducing a Fraction to Lowest Terms.

Divide numerator and denominator of a fraction by the same number to reduce it to *lowest terms*.

Example 1.29

Numerator and denominator
are divided by:

(a) $\dfrac{4}{8} = \dfrac{1}{2}$ 4

(b) $\dfrac{9}{27} = \dfrac{1}{3}$ 9

(c) $\dfrac{12}{24} = \dfrac{1}{2}$ 12

(c) Improper Fractions and Mixed Numbers.

To write an improper fraction as a mixed number, divide the numerator by the denominator, and express the remainder as the numerator of a fraction whose denominator is the original denominator.

Example 1.30:   Write 39/5 as a mixed number.

Solution: Dividing 39 by 5, we get 7 with a remainder of 4, so

$$\frac{39}{5} = 7\frac{4}{5}$$

To write a mixed number as an improper fraction, multiply the whole-number part by the denominator of the fraction part and add the numerator of the fraction part to this product. This sum will be the numerator of the improper fraction, and the denominator will be the denominator of the fraction part of the original mixed number.

Example 1.31:   Write $3\frac{4}{5}$ as an improper fraction.

Solution: The numerator of the improper fraction will be $5(3) + 4 = 19$ and the denominator will be 5, so

$$3\frac{4}{5} = \frac{19}{5}$$

(d) Multiplication. The product of two fractions is a fraction whose numerator is the product of the numerators of the original fractions and whose denominator is the product of the denominators of the original fraction.

Example 1.32

(a) $\dfrac{1}{2} \times \dfrac{1}{3} = \dfrac{1}{6}$ (b) $\dfrac{2}{3} \times \dfrac{3}{5} = \dfrac{6}{15} = \dfrac{2}{5}$ (c) $\dfrac{5}{8} \times \dfrac{2}{5} = \dfrac{10}{40} = \dfrac{1}{4}$

(e) Division. Invert the divisor and multiply.

Example 1.33

(a) $\dfrac{1}{2} \div \dfrac{1}{3} = \dfrac{1}{2} \times \dfrac{3}{1} = \dfrac{3}{2} = 1\dfrac{1}{2}$ (b) $\dfrac{5}{8} \div \dfrac{1}{2} = \dfrac{5}{8} \times \dfrac{2}{1} = \dfrac{10}{8} = 1\dfrac{1}{4}$ (c) $\dfrac{9}{16} \div 3 = \dfrac{9}{16} \times \dfrac{1}{3} = \dfrac{9}{48} = \dfrac{3}{16}$

(f) Least Common Denominator. The *least common denominator* LCD (also called the *lowest* common denominator) is the smallest number that is exactly divisible by each of the denominators. Thus the LCD must contain all the factors of each of the denominators. The factors should not be divisible by any number except one. When multiplied together, the factors should equal the denominator. To find the LCD, form a product of all the factors of each denominator. If the same factor appears in every denominator, include it only once in the LCD.

Example 1.34:   Find the LCD for the two fractions

$$5/12 \text{ and } 2/15$$

Solution: The denominator 12 factors into

$$2, 2, 3 \quad (2 \times 2 \times 3 = 12)$$

and 15 factors into

$$3, 5 \quad (3 \times 5 = 15)$$

The LCD must contain all these factors. Since 3 is a factor of both denominators, we write it only once.

$$\text{LCD} = 2 \times 2 \times 3 \times 5 = 60$$

(g) Addition and Subtraction. If the fractions have the same denominator, their sum is a fraction whose numerator is the sum of the numerators and whose denominator is the same as the original denominator.

Example 1.35

(a) $\dfrac{2}{9} + \dfrac{3}{9} = \dfrac{5}{9}$ (b) $\dfrac{2}{5} + \dfrac{4}{5} - \dfrac{1}{5} = \dfrac{5}{5} = 1$ (c) $\dfrac{3}{8} - \dfrac{5}{8} + \dfrac{7}{8} + \dfrac{1}{8} = \dfrac{6}{8} = \dfrac{3}{4}$

If the denominators are different, express each fraction as an equivalent one having the LCD as denominator, and combine as shown above.

Example 1.36

(a) $\dfrac{1}{2} + \dfrac{1}{3} = \dfrac{3}{6} + \dfrac{2}{6} = \dfrac{5}{6}$ (b) $\dfrac{3}{4} - \dfrac{2}{3} = \dfrac{9}{12} - \dfrac{8}{12} = \dfrac{1}{12}$ (c) $\dfrac{3}{4} - \dfrac{2}{5} + \dfrac{1}{2} = \dfrac{15}{20} - \dfrac{8}{20} + \dfrac{10}{20} = \dfrac{17}{20}$

(d) $\dfrac{5}{42} + \dfrac{3}{20}$ The factors of 42 are 2, 3, and 7, and the factors of 20 are 2, 2, and 5. The LCD is

$$2 \times 2 \times 3 \times 5 \times 7 = 420$$

so,
$$\frac{5}{42} + \frac{3}{20} = \frac{5}{42} \times \frac{10}{10} + \frac{3}{20} \times \frac{21}{21} = \frac{50}{420} + \frac{63}{420} = \frac{113}{420}$$

Do *not* add fractions as follows:

Common Errors	$\frac{1}{2} + \frac{1}{3} \neq \frac{1}{6}$ and $\frac{1}{2} + \frac{1}{3} \neq \frac{2}{5}$

(h) Equivalent Decimals and Fractions. To write a fraction as a decimal, divide the numerator of the fraction by the denominator.

Example 1.37:   Write the decimal equivalent of 7/8.

Solution: Dividing 7 by 8,

$$\frac{7}{8} = 0.875$$

1.4 PERCENTAGE

Percent means *by the hundred*. A percent gives the number of parts in every hundred. It is also called the *rate*.

Example 1.38:   A rejection rate of 2 percent means that 2 parts out of every 100 are rejected.

Percent is another way of expressing a fraction having 100 as the denominator.

Example 1.39:   An income tax rate of 23 percent (also written 23%) means that 23/100 of the taxable income had to be paid in taxes.

(a) Converting Decimals to Percent. To convert decimals to percent, move the decimal point two places to the right and affix the percent symbol.

Example 1.40

(*a*) $0.15 = 15\%$ (*b*) $0.74 = 74\%$ (*c*) $2.50 = 250\%$

(b) Converting Common Fractions to Percent. To convert fractions to percent, express the fraction as a decimal, and proceed as in **(a).**

Example 1.41

(*a*) $\frac{1}{2} = 0.5 = 50\%$ (*b*) $\frac{5}{8} = 0.625 = 62.5\%$

(c) Converting Percent to Decimals. To convert percent to decimals, move the decimal point two places to the left and remove the percent sign.

Example 1.42

(*a*) $33\% = 0.33$ (*b*) $1.5\% = 0.015$ (*c*) $125\% = 1.25$

(d) Converting Percent to a Common Fraction.   To convert percent to a fraction, write a fraction with the percent in the numerator and 100 in the denominator, and remove the percent sign. Reduce the fraction to lowest terms.

Example 1.43

(a) $25\% = \dfrac{25}{100} = \dfrac{1}{4}$ (b) $87.5\% = \dfrac{87.5}{100} = \dfrac{875}{1000} = \dfrac{7}{8}$

(e) Finding a Number Which Is a Given Percent of Another Number (Base).   Express the percent rate P as a decimal or fraction.   The amount A will be the base B multiplied by P.

$$\boxed{\begin{array}{c} \text{Amount} = \text{base} \times \text{rate} \\ A = BP \end{array}} \quad \mathbf{1}$$

Example 1.44:   What is 28 percent of 40?

Solution: The base $B = 40$.   The percent expressed as a decimal is $P = 0.28$.   By Eq. 1,

$$A = 40(0.28) = 11.2$$

Example 1.45:   Find 2.5% of 84.

Solution

$$A = 2.5\% \text{ of } 84 = 0.025(84) = 2.1$$

(f) Finding the Base When a Percent of It Is Known.   We see from Eq. 1 that the base equals the amount divided by the rate (expressed as a decimal or fraction).

Example 1.46:   20% of what number is 5.6?

Solution: $A = 5.6$ and $P = 0.20$.   By Eq. 1,

$$B = \frac{A}{P} = \frac{5.6}{0.20} = 28$$

Example 1.47:   2.52 is 45% of what number?

Solution: From Eq. 1,

$$B = \frac{A}{P} = \frac{2.52}{0.45} = 5.6$$

(g) Given Two Numbers, Finding the Percent One Number Is of the Other Number.   From Eq. 1, $P = A/B$.

Example 1.48:   34.2 is what percent of 90?

Solution: By Eq. 1, with $A = 34.2$ and $B = 90$,

$$34.2 = 90P$$

$$P = \frac{34.2}{90} = 0.38 = 38\%$$

Example 1.49:   What percent of 75 is 5.4?

Solution: From Eq. 1,

$$5.4 = 75P$$

$$P = \frac{5.4}{75} = 0.072 = 7.2\%$$

(h) Percent Greater than 100. When, in Eq. 1, A is greater than B, the rate will be greater than 100. Treat such problems the same as when the rate is less than 100.

Example 1.50:   What is 250% of 24?

Solution: By Eq. 1,

$$A = 2.5(24) = 60$$

(i) Percent Difference. The percent difference between two numbers is

$$\boxed{\text{Percent difference} = \frac{\text{difference of the two numbers}}{\text{base}} \times 100} \quad \textbf{2}$$

In some cases, the base is taken as the *average* of the two numbers, and in others it is one of the two numbers.

Example 1.51:   What is the percent difference between 58 and 64?

Solution: There is no reason for preferring one of the numbers over the other for the base, so we use the average, $(58 + 64)/2 = 61$.   By Eq. 2,

$$\text{Percent difference} = \frac{64 - 58}{61} \times 100 = 9.8\%$$

When the two numbers being compared involve a *change* from one to the other, the *original* value is usually taken as the base.

$$\boxed{\text{Percent change} = \frac{\text{new value} - \text{original value}}{\text{original value}} \times 100} \quad \textbf{3}$$

Example 1.52:   A temperature rose from 58.0°C to 64.0°C.   Find the percent change in temperature.

Solution: We use the original value, 58.0, as the base.   From Eq. 3,

$$\text{Percent change} = \frac{64.0 - 58.0}{58.0} \times 100 = 10.3\% \text{ increase}$$

(j) Percent Error. When a measurement is made of some known quantity, the difference between the measured value and the known value is the *absolute error*. The *relative error* is obtained by dividing the absolute error by the known value.   The relative error expressed as a percent is called the *percent error*.

$$\boxed{\text{Percent error} = \frac{\text{measured value} - \text{known value}}{\text{known value}} \times 100} \quad \textbf{4}$$

Example 1.53: A machinist measures a 2-in.-long gage block, guaranteed to be accurate to a ten-thousandth of an inch. His reading is 1.9970 in. What are the absolute, relative, and percent errors?

Solution

$$\text{Absolute error} = \text{measured value} - \text{known value}$$

$$= 1.9970 - 2.0000$$

$$= -0.0030 \text{ in.}$$

$$\text{Relative error} = \frac{\text{absolute error}}{\text{known value}}$$

$$= \frac{-0.0030}{2.0000} = -0.0015$$

From Eq. 4,

$$\text{Percent error} = \text{relative error} \times 100$$

$$= -0.0015(100)$$

$$= -0.15\%$$

(k) Percent Concentration. In a mixture of two or more ingredients,

$$\boxed{\text{Percent concentration of Ingredient A} = \frac{\text{amount of A}}{\text{amount of mixture}} \times 100} \quad \mathbf{5}$$

Example 1.54: 250 kilograms (kg) of a certain alloy contains 18 kg of copper. What percentage of the alloy is copper?

Solution: By Eq. 5,

$$\text{Percent copper} = \frac{18}{250} \times 100 = 7.2\%$$

(l) Percent Efficiency. The power output of any machine or device is always *less* than the power input, because of inevitable power losses within the device. The *efficiency* of the device is a measure of those losses.

$$\boxed{\text{Percent efficiency} = \frac{\text{output}}{\text{input}} \times 100} \quad \mathbf{6}$$

Example 1.55: The power input to a certain speed reducer is 25 horsepower (hp) and, because of friction in the gearing, the power output is only 21 hp. Find the efficiency of the reducer.

Solution: By Eq. 6,

$$\text{Percent efficiency} = \frac{21}{25} \times 100 = 84\%$$

1.5 CONVERSION OF UNITS

To convert to a different system of units, (*a*) find a *conversion factor* linking the units to be converted with the units desired. A table of conversion factors is given in Appendix A. (*b*) Express the conversion factor as a fraction, with the *units to be converted in the denominator*. (c) Multiply the original number by the conversion factor.

Example 1.56:   Convert 18.4 centimeters (cm) to inches.

Solution: From Appendix A we find the conversion factor,

$$2.54 \text{ cm} = 1 \text{ in.}$$

Writing the conversion factor with the unit to be converted (cm) in the denominator,

$$\frac{1 \text{ in.}}{2.54 \text{ cm}} = 1$$

Since the value of this fraction is 1, we may multiply the number whose units we want to convert by this fraction without changing the value of that number.

$$18.4 \text{ cm} \times \frac{1 \text{ in.}}{2.54 \text{ cm}} = \frac{18.4}{2.54} \text{ in.} = 7.24 \text{ in.}$$

Common Error	Be sure to write the units of the given number, as well as the units of the conversion factor, so that they cancel as intended. Do not work with the numerical values only.

Example 1.57:   Convert 30 miles per hour (mi/h) to feet per second (ft/s).

Solution

$$\frac{30 \text{ mi}}{\text{h}} \times \frac{5280 \text{ ft}}{1 \text{ mi}} \times \frac{1 \text{ h}}{3600 \text{ s}} = \frac{30(5280)\text{ft}}{3600 \text{ s}} = 44 \text{ ft/s} \qquad \text{(a useful conversion factor to remember)}$$

Solved Problems

ARITHMETIC OPERATIONS WITH SIGNED NUMBERS

1.1   Evaluate the expression $-|5 - 8| + |-5 + 2|$.

   Solution

$$-|5 - 8| + |-5 + 2| = -|-3| + |-3| = -3 + 3 = 0$$

1.2   Evaluate the expression $|-8 + 6 - 2| + |5| - |-7|$.

   Solution

$$|-8 + 6 - 2| + |5| - |-7| = |-4| + 5 - 7 = 4 + 5 - 7 = 2$$

1.3   Perform the additions and subtractions as indicated.

   (a) $46 - 27 + 45$ (c) $-38 + (-6) - (-7)$

   (b) $16 - (-5)$ (d) $-(-36) + (-5) - (-53)$

Solution

(a) $46 - 27 + 45 = 64$ (c) $-38 + (-6) - 8 - (-7) = -38 - 6 - 8 + 7 = -45$

(b) $16 - (-5) = 16 + 5 = 21$ (d) $-(-36) + (-5) - (-53) = 36 - 5 + 53 = 84$

1.4 Perform the following multiplications:

(a) $5(41)$ (c) $-18(-7)$ (e) $3(-5)(-4)(-2)$

(b) $62(-15)$ (d) $-42(3)$ (f) $-6(5)(-7)$

Solution

(a) $5(41) = 205$ (c) $-18(-7) = 126$ (e) $3(-5)(-4)(-2) = -120$

(b) $62(-15) = -930$ (d) $-42(3) = -126$ (f) $-6(5)(-7) = 210$

1.5 Perform the following divisions:

(a) $27 \div 9$ (b) $72 \div 12$ (c) $45 \div (-5)$ (d) $-63 \div (-9)$ (e) $-48 \div 4$

Solution

(a) $27 \div 9 = \dfrac{27}{9} = 3$ (c) $45 \div (-5) = \dfrac{45}{-5} = -9$ (e) $-48 \div 4 = \dfrac{-48}{4} = -12$

(b) $72 \div 12 = \dfrac{72}{12} = 6$ (d) $-63 \div (-9) = \dfrac{-63}{-9} = 7$

1.6 Find the reciprocals of the following numbers:

(a) 8 (b) $\dfrac{1}{9}$ (c) -50

Solution

(a) Reciprocal of $8 = \dfrac{1}{8}$ (b) Reciprocal of $\dfrac{1}{9} = 9$ (c) Reciprocal of $-50 = -\dfrac{1}{50}$

1.7 Find the powers of the following numbers:

(a) 5^2 (c) $(-4)^2$ (e) 12^3 (g) 1^{47} (i) $(-1)^{98}$

(b) 12^2 (d) $(-2)^3$ (f) 2^5 (h) $(-1)^{47}$

Solution

    By Eq. 7,

(a) $5^2 = 5 \times 5 = 25$ (c) $(-4)^2 = 16$ (e) $12^3 = 1728$ (g) $1^{47} = 1$ (i) $(-1)^{98} = 1$

(b) $12^2 = 144$ (d) $(-2)^3 = -8$ (f) $2^5 = 32$ (h) $(-1)^{47} = -1$

1.8 Find the roots of the following numbers using a calculator, slide rule, or tables:

(a) $\sqrt{64}$ (b) $\sqrt{27}$ (c) $\sqrt{59}$ (d) $\sqrt{105}$ (e) $\sqrt{-9}$

Solution

(a) $\sqrt{64} = 8$ (c) $\sqrt{59} \cong 7.68$ (e) $\sqrt{-9}$ is imaginary.

(b) $\sqrt{27} \cong 5.20$ (d) $\sqrt{105} \cong 10.2$

Solution

Finding a common denominator and subtracting,

$$\frac{3}{8} - \frac{1}{5} = \frac{15}{40} - \frac{8}{40} = \frac{7}{40} \text{ of a tank}$$

1.17   Two resistors, 4 ohms and 5 ohms, are wired in parallel. What is the resistance of the parallel combination? (The abbreviation for ohms is Ω.)

Solution

By Eq. 202, the resistance R of the combination is given by

$$\frac{1}{R} = \frac{1}{4} + \frac{1}{5}$$

The LCD for the two fractions is 20. Adding,

$$\frac{1}{R} = \frac{5}{20} + \frac{4}{20} = \frac{9}{20}$$

Taking the reciprocal,

$$R = \frac{20}{9} = 2.2 \ \Omega$$

1.18   Resistors of $10\,\Omega$, $15\,\Omega$, and $20\,\Omega$ are wired in parallel. Find the equivalent resistance.

Solution

From Eq. 202,

$$\frac{1}{R} = \frac{1}{10} + \frac{1}{15} + \frac{1}{20}$$

$$= \frac{6}{60} + \frac{4}{60} + \frac{3}{60} = \frac{13}{60}$$

$$R = \frac{60}{13} = 4.6 \ \Omega$$

1.19   One assembly crew can put together 4/5 of a certain machine tool in a day. Another crew can assemble 3/4 of a machine per day. How much can both crews together assemble in one day?

Solution

We add the amounts done by each crew:

$$\frac{4}{5} + \frac{3}{4}$$

Finding a common denominator,

$$\frac{16}{20} + \frac{15}{20} = \frac{31}{20} = 1.55 \text{ machines per day}$$

1.20   A planer makes a 2-ft cutting stroke at 55 ft/min and returns at 275 ft/min. How long does it take for the cutting stroke and return?

Solution

From Eq. 177,

$$\text{Time} = \frac{\text{distance}}{\text{rate}}$$

So, cutting time = 2/55 min and return time = 2/275 min. Finding a common denominator and adding,

$$\frac{2}{55} + \frac{2}{275} = \frac{10}{275} + \frac{2}{275} = \frac{12}{275} \cong 0.0436 \text{ min}$$

PERCENTAGE

1.21 Convert the following decimals to percent:

(a) 0.736   (b) 0.014   (c) 2.45   (d) 0.000 48

Solution

(a) 0.736 = 73.6%   (b) 0.014 = 1.4%   (c) 2.45 = 245%   (d) 0.000 48 = 0.048%

1.22 Convert the following fractions to percent:

(a) $\dfrac{7}{8}$ (b) $\dfrac{3}{4}$ (c) $\dfrac{5}{16}$ (d) $\dfrac{5}{2}$

Solution

(a) $\dfrac{7}{8} = 0.875 = 87.5\%$ (c) $\dfrac{5}{16} = 0.3125 = 31.25\%$

(b) $\dfrac{3}{4} = 0.75 = 75\%$ (d) $\dfrac{5}{2} = 2.5 = 250\%$

1.23 Convert the following percents to decimals:

(a) 48%   (b) 4.83%   (c) 0.5%   (d) 183%

Solution

(a) 48% = 0.48   (b) 4.83% = 0.0483   (c) 0.5% = 0.005   (d) 183% = 1.83

1.24 Convert the following percents to fractions:

(a) 50%   (b) 2%   (c) 37.5%   (d) 43.75%

Solution

(a) $50\% = \dfrac{50}{100} = \dfrac{1}{2}$ (c) $37.5\% = \dfrac{37.5}{100} = \dfrac{375}{1000} = \dfrac{3}{8}$

(b) $2\% = \dfrac{2}{100} = \dfrac{1}{50}$ (d) $43.75\% = \dfrac{43.75}{100} = \dfrac{4375}{10\,000} = \dfrac{7}{16}$

1.25   Find 38% of 84.

Solution

By Eq. 1,

$$A = BP = 0.38(84) = 31.92$$

1.26   What is 2.8% of 260?

Solution

By Eq. 1,

$$A = 0.028(260) = 7.28$$

1.27   Find 280% of 58.

Solution

By Eq. 1,

$$A = 2.8(58) = 162.4$$

1.28   70% of what number is 42?

Solution

By Eq. 1,

$$42 = 0.70B$$

$$B = \frac{42}{0.70} = 60$$

1.29   1.2% of what number is 4.8?

Solution

By Eq. 1,

$$4.8 = 0.012B$$

$$B = \frac{4.8}{0.012} = 400$$

1.30   81 is 27% of what number?

Solution

By Eq. 1,

$$81 = 0.27B$$

$$B = \frac{81}{0.27} = 300$$

1.31   87.6 is what percent of 146?

Solution

By Eq. 1,

$$87.6 = 146P$$

$$P = \frac{87.6}{146} = 0.6 = 60\%$$

1.32   What percent of 88 is 22?

Solution

By Eq. 1,

$$22 = 88P$$

$$P = \frac{22}{88} = 0.25 = 25\%$$

1.33   What percent of 584 is 146?

Solution

By Eq. 1,

$$146 = 584P$$

$$P = \frac{146}{584} = 0.25 = 25\%$$

1.34   What is the percent difference between 368 and 476?

Solution

Taking the base as the average of the two numbers,

$$\frac{368 + 476}{2} = 422$$

By Eq. 2,

$$\text{Percent difference} = \frac{476 - 368}{422} \times 100 = 25.6\%$$

Note: This answer, as well as many that follow, has been *rounded*.   A complete explanation of rounding is given in Chapter 2.

1.35   A quantity changed from 157 to 168.   Find the percent change.

Solution

From Eq. 3, taking the original value as 157,

$$\text{Percent change} = \frac{168 - 157}{157} \times 100 = 7.01\% \text{ increase}$$

1.36   A quantity changed from 0.558 to 0.475.   Find the percent change.

Solution

From Eq. 3,

$$\text{Percent change} = \frac{0.475 - 0.558}{0.558} \times 100 = -14.9\% = 14.9\% \text{ decrease}$$

1.37   A 50 500-liter-capacity tank contains 5840 liters of water.   Express the amount of water in the tank as a percent of the total capacity.

Solution

By Eq. 1,

$$\frac{5840}{50\,500} \times 100 = 11.6\%$$

1.38   A crew paved 2548 ft of roadway on one day and 3158 ft on the following day.   Find the percent change in the amount they paved on the second day.

Solution

By Eq. 3,

$$\text{Percent change} = \frac{3158 - 2548}{2548} \times 100 = 23.9\% \text{ increase}$$

1.39   A 25-meter (m)-long steel girder shrank 0.25% when subjected to below-freezing temperatures. Find the shrinkage in centimeters.

Solution

By Eq. 3,

$$0.25\% \text{ of } 25 \text{ m} = 0.0025(25) = 0.0625 \text{ m}$$
$$= 6.25 \text{ cm}$$

1.40   A concrete mixture which is 12% cement by volume contains $2\frac{1}{2}$ cubic feet (ft³) of cement. What is the volume of the total mixture?

Solution

By Eq. 5,

$$B = \frac{A}{P} = \frac{2.5}{0.12} = 21 \text{ ft}^3$$

1.41   The resistance of a certain circuit loop, now 1250 Ω, is to be increased by 26%.   How much resistance should be added?

Solution

$$\text{Resistance added} = 26\% \text{ of } 1250 = 0.26(1250) = 325 \ \Omega$$

1.42   A rectifier has a dc output of 28 volts (V) with a ripple of 0.50 V peak to peak.   Express the ripple as a percent of the dc output voltage.

Solution

$$\text{Ripple} = \frac{0.50}{28} \times 100 = 1.8\%$$

1.43   A resistor is labeled as 3600 Ω with a tolerance of ±5%.   Between what two values is the actual resistance expected to lie?

Solution

$$5\% \text{ of } 3600 = 0.05(3600) = 180 \ \Omega$$

The actual resistance should be 3600 Ω ±180 Ω, or between 3420 Ω and 3780 Ω.

1.44   A certain electrolytic capacitor is stated to have a working voltage of 25 V dc −10%, +150%. Between what two voltages would the actual working voltage be expected to lie?

Solution

$$10\% \text{ of } 25 = 0.10(25) = 2.5 \text{ V}$$
$$150\% \text{ of } 25 = 1.5(25) = 37.5 \text{ V}$$

Range equals (25 − 2.5) to (25 + 37.5) or 22.5 V to 62.5 V.

1.45   The temperature of a room rises from 68°F to 73°F. Find the percent increase.

Solution

By Eq. 3,

$$\text{Percent increase} = \frac{73 - 68}{68} \times 100 = 7.4\%$$

1.46   A casting initially weighing 15.6 kg has 15% of its material machined off. What is its final weight?

Solution

The amount removed is, by Eq. 1,

$$\text{Amount removed} = 15.6(0.15) = 2.3 \text{ kg}$$

So the final weight is

$$15.6 - 2.3 = 13.3 \text{ kg}$$

1.47   An engineer is hired at a salary of $14,500, with the employment agency getting 6% of this amount as a fee. Find the amount of the fee.

Solution

$$\text{Fee} = \$14,500(0.06) = \$870.00$$

1.48   A certain common stock rose from a value of $35.49 per share to $37.45 per share. Find the percent change in value.

Solution

By Eq. 3,

$$\text{Percent change} = \frac{\$37.45 - \$35.49}{\$35.49} \times 100 = 5.52\% \text{ increase}$$

1.49   When borrowing $15,000, a company was required to pay 7.6% simple interest, deducted in advance. How much did they actually receive?

Solution

The amount deducted for interest is, by Eq. 1,

$$\text{Interest} = \$15,000(0.076) = \$1140.00$$

So they actually received

$$\$15,000 - \$1140 = \$13,860.00$$

1.50   The construction of a new production facility cost $25,400 for materials and $31,250 for labor. The labor cost was what percentage of the total?

Solution

The total cost is $25,400 + $31,250 = $56,650.   By Eq. 1,

$$P = \frac{31,250}{56,650} = 0.552 \quad \text{or} \quad 55.2\%$$

1.51   A solution is made by mixing 3.5 liters of alcohol with 5.7 liters of water. Find the percent concentration of alcohol.

Solution

By Eq. 5,

$$\text{Percent alcohol} = \frac{3.5}{3.5 + 5.7} \times 100 = 38\%$$

1.52   A certain device consumes 2.4 hp and delivers 1.8 hp. Find its efficiency.

Solution

By Eq. 6,

$$\text{Percent efficiency} = \frac{1.8}{2.4} \times 100 = 75\%$$

1.53   A certain quantity is measured at 58.8 units but is known to be actually 59.5 units. Find the absolute error, the relative error, and the percent error in the measurement.

Solution

$$\text{Absolute error} = 58.8 - 59.5 = -0.7$$

$$\text{Relative error} = -\frac{0.7}{59.5} = -0.012$$

$$\text{Percent error} = -0.012 \times 100 = -1.2\%$$
$$= 1.2\% \text{ low}$$

1.54   A piece of precision-ground shafting is known to have a diameter of 1.2500 in. You measure it with a vernier caliper and get a reading of 1.248 in. Assuming that 1.2500 is the correct diameter, what is the absolute error, the relative error, and the percent error of your reading?

Solution

$$\text{Absolute error} = \text{observed value} - \text{true value}$$
$$= 1.248 - 1.2500$$
$$= -0.002 \text{ in.}$$

$$\text{Relative error} = \frac{\text{absolute error}}{\text{true value}}$$

$$= -\frac{0.002}{1.2500}$$

$$= -0.0016$$

$$\text{Percent error} = \text{relative error} \times 100$$
$$= -0.16\%$$

1.55   You are checking a voltmeter by measuring the electromotive force (emf) of a standard cell, known to be 1.0186 V. Your reading is 1.0175 V. What is the absolute error, the relative error, and the percent error in your reading?

Solution

$$\text{Absolute error} = \text{observed value} - \text{true value}$$
$$= 1.0175 - 1.0186$$
$$= -0.0011 \text{ V}$$

$$\text{Relative error} = \frac{\text{absolute error}}{\text{true value}}$$
$$= \frac{-0.0011}{1.0186} = -0.0011$$

$$\text{Percent error} = \text{relative error} \times 100$$
$$= -0.11\%$$

1.56 An electric motor consumes 2550 watts (W). Find the horsepower it can deliver if it is 75 percent efficient.

Solution

Converting watts to horsepower,

$$2550 \text{ W} \left(\frac{1 \text{ hp}}{745.7 \text{ W}} \right) = 3.42 \text{ hp}$$

Since it is 75% efficient,

$$\text{Power output} = 0.75(3.42) = 2.57 \text{ hp}$$

1.57 A certain pump requires an input of 1.5 hp and delivers 32 000 pounds (lb) of water per hour to a reservoir 85 ft above the pump. Find its efficiency.

Solution

The work done in one hour is, by Eq. 164,

$$32\,000(85) = 2\,720\,00 \text{ ft} \cdot \text{lb/h}$$

and in 1 second (s)

$$\frac{2\,720\,000 \text{ ft} \cdot \text{lb}}{\text{h}} \times \frac{1 \text{ h}}{3600 \text{ s}} = 756 \text{ ft} \cdot \text{lb/s}$$

Since 1 hp = 550 ft·lb/s, the work done is

$$\frac{756 \text{ ft} \cdot \text{lb}}{\text{s}} \times \frac{1 \text{ hp}}{550 \text{ ft} \cdot \text{lb/s}} = 1.37 \text{ hp}$$

By Eq. 6,

$$\text{Percent efficiency} = \frac{1.37}{1.5} \times 100 = 91\%$$

CONVERSION OF UNITS

1.58 Convert 95.3 mi to kilometers.

Solution

Since 1 mi = 1.609 km,

$$95.3 \text{ mi} \left(\frac{1.609 \text{ km}}{\text{mi}} \right) = 153 \text{ km}$$

1.59   Convert 58.3 kg to pounds.

Solution

Since 1 lb = 0.4536 kg,

$$58.3 \, \cancel{kg} \left(\frac{lb}{0.4536 \, \cancel{kg}} \right) = 129 \, lb$$

1.60   Convert 59.2 gallons (gal) to liters.

Solution

Since 1 gal = 3.785 liters,

$$59.2 \, \cancel{gal} \left(\frac{3.785 \, liters}{\cancel{gal}} \right) = 224 \, liters$$

1.61   Convert 0.2450 lb/in³ to kg/m³.

Solution

Since 1 lb = 0.4536 kg and 1 m = 39.37 in.,

$$\frac{0.2450 \, \cancel{lb}}{\cancel{in^3}} \left(\frac{0.4536 \, kg}{\cancel{lb}} \right) \left[\frac{(39.37)^3 \, \cancel{in^3}}{m^3} \right] = 6782 \, kg/m^3$$

1.62   Convert 18.5 in./s to feet per minute.

Solution

$$\frac{18.5 \, \cancel{in.}}{\cancel{s}} \left(\frac{ft}{12 \, \cancel{in.}} \right) \left(\frac{60 \, \cancel{s}}{min} \right) = 92.5 \, ft/min$$

1.63   A part is to be machined 18.45 millimeters (mm) long.   Find this dimension in inches.

Solution

$$18.45 \, mm \left(\frac{1 \, in.}{25.4 \, mm} \right) = 0.7264 \, in.$$

1.64   A machine part moves at the rate of 355 ft/min.   Find the rate in inches per second.

Solution

$$\frac{355 \, ft}{min} \left(\frac{12 \, in.}{ft} \right) \left(\frac{1 \, min}{60 \, s} \right) = 71.0 \, in./s$$

1.65   Oil flows through a pipe at the rate of 26.5 gal/min.   Find the flow rate in liters/s.

Solution

$$\frac{26.5 \, gal}{min} \left(\frac{3.79 \, liter}{gal} \right) \left(\frac{1 \, min}{60 \, s} \right) = 1.67 \, liters/s$$

1.66   Find the cost in dollars of a British computer valued at £47,350 (47,350 pounds sterling) if, at the time of transaction, 1 pound sterling equals $1.86.

Solution

47,350 pounds (1.86 dollars/pound) = $88,071.00

Supplementary Problems

1.67 Evaluate the following:

(a) $-|-6-2|+|8-5|$ (b) $|6-8|-|-5+(-3)|$

1.68 Perform the indicated additions and subtractions:

(a) $25-6+3+12-13$ (b) $16-(-5)+(-12)-6+9$

1.69 Perform the following multiplications:

(a) $6(92)$ (b) $45(-7)$ (c) $-12(-11)(2)$ (d) $-6(10)(-3)(6)$

1.70 Perform the following divisions:

(a) $36 \div 6$ (b) $90 \div -5$ (c) $-72 \div -9$

1.71 Find the reciprocals of these numbers.

(a) 6 (b) -7 (c) $\dfrac{1}{12}$

1.72 Find the powers of the following:

(a) 10^2 (b) 11^3 (c) $(-6)^3$ (d) $(-2)^2$ (e) 1^{50} (f) $(-3)^5$

1.73 Find the roots of these numbers using a calculator, slide rule, or tables.

(a) $\sqrt{36}$ (b) $\sqrt{39}$ (c) $\sqrt{103}$

1.74 Find the roots of the following without using tables, slide rules, or calculators:

(a) $\sqrt[3]{8}$ (b) $\sqrt[4]{625}$ (c) $\sqrt[3]{-64}$

1.75 Simplify the following fractions:

(a) $\dfrac{7}{21}$ (b) $\dfrac{9}{12}$ (c) $\dfrac{64}{112}$

1.76 Multiply the following fractions and simplify:

(a) $\dfrac{4}{9} \times \dfrac{2}{5}$ (b) $\dfrac{6}{5} \times \dfrac{1}{3}$ (c) $3\dfrac{2}{3} \times 5\dfrac{1}{2}$

1.77 Divide the following fractions and simplify:

(a) $\dfrac{1}{2} \div \dfrac{3}{2}$ (b) $\dfrac{2}{9} \div 4$ (c) $2\dfrac{3}{8} \div 9\dfrac{1}{5}$ (d) $23 \div \dfrac{1}{17}$

1.78 Combine the following fractions as indicated:

(a) $\dfrac{1}{8} + \dfrac{1}{5}$ (b) $\dfrac{2}{9} - 3 + \dfrac{6}{11}$ (c) $1\dfrac{1}{6} - 2\dfrac{2}{3} + \dfrac{7}{11}$

1.79   Convert the following decimals to percents:

   (a) 0.234   (b) 0.0573   (c) 2.79

1.80   Convert the following fractions to percents:

   (a) $\dfrac{3}{5}$ (b) $\dfrac{2}{3}$ (c) $\dfrac{7}{3}$

1.81   Convert these percents to decimals:

   (a) 26%   (b) 13%   (c) 102.3%

1.82   Convert the following percents to fractions:

   (a) 25%   (b) 1%   (c) 18.75%

1.83   Find the given percentages of the following numbers:

   (a) 23% of 97.2   (b) 1.73% of 56.873

1.84   Evaluate:

   (a) 27% of what number is 12?   (c) 25 is 17% of what number?   (e) What percent of 24 is 6?

   (b) 72% of what number is 9.3?   (d) 2 is what percent of 7?   (f) What percent of 125 is 13?

   (g) What is the percent difference between 129 and 321?

   (h) A quantity changed from 29.3 to 57.6; what is the percent change?

   (i) A quantity changed from 107 to 23.75; what is the percent change?

1.85   Conversions:

   (a) Convert 60 mi to kilometers.   (c) Convert 27.3 gal to liters.

   (b) Convert 225 lb to kilograms.   (d) Convert 25 in./s to miles per hour.

Answers to Supplementary Problems

1.67   (a) −5   (b) −6

1.68   (a) 21   (b) 12

1.69   (a) 552   (b) −315   (c) 264   (d) 1080

1.70   (a) 6   (b) −18   (c) 8

1.71   (a) $\dfrac{1}{6}$ (b) $-\dfrac{1}{7}$ (c) 12

1.72 (*a*) 100   (*b*) 1331   (*c*) −216   (*d*) 4   (*e*) 1   (*f*) −243

1.73 (*a*) 6   (*b*) 6.24   (*c*) 10.15

1.74 (*a*) 2   (*b*) 5   (*c*) −4

1.75 (*a*) $\dfrac{1}{3}$ (*b*) $\dfrac{3}{4}$ (*c*) $\dfrac{4}{7}$

1.76 (*a*) $\dfrac{8}{45}$ (*b*) $\dfrac{2}{5}$ (*c*) $\dfrac{121}{6} = 20\dfrac{1}{6}$

1.77 (*a*) $\dfrac{1}{3}$ (*b*) $\dfrac{1}{18}$ (*c*) $\dfrac{95}{368}$ (*d*) 391

1.78 (*a*) $\dfrac{13}{40}$ (*b*) $-\dfrac{221}{99}$ (*c*) $-\dfrac{19}{22}$

1.79 (*a*) 23.4%   (*b*) 5.73%   (*c*) 279%

1.80 (*a*) 60%   (*b*) 66.7%   (*c*) 233.3%

1.81 (*a*) 0.26   (*b*) 0.13   (*c*) 1.023

1.82 (*a*) $\dfrac{1}{4}$ (*b*) $\dfrac{1}{100}$ (*c*) $\dfrac{3}{16}$

1.83 (*a*) 22.36   (*b*) 0.9839

1.84 (*a*) 44.44   (*b*) 12.917   (*c*) 147.06   (*d*) 28.6%   (*e*) 25%   (*f*) 10.4%
        (*g*) 85.33%   (*h*) 96.6% increase   (*i*) 77.8% decrease

1.85 (*a*) 96.54 km   (*b*) 102.06 kg   (*c*) 103.33 liters   (*d*) 1.42 mi/h

Chapter 2

Approximate Numbers

2.1 TERMINOLOGY

(a) Approximate Numbers. *Approximate numbers* are those which may be *expected to vary* from their written value.

Example 2.1: All measured values are approximate. A yardstick is *approximately* 36 in. long.

Example 2.2: Some fractions can be expressed only approximately in decimal form. Thus, 0.3333 is approximately equal to 1/3 because of the repeated digits that were discarded.

Example 2.3: Irrational numbers can be written only approximately in decimal form because of discarded digits. Thus, $\sqrt{2} \cong 1.4142$.

(b) Uncertainty. The *uncertainty* of a number is the amount by which it may be expected to vary. If the uncertainty is not stated directly after the number, take it to be plus or minus half the smallest unit in the number.

Example 2.4

Number	Uncertainty
25	± 0.5
18.3	± 0.05
1.3357	± 0.00005
7.34 ± 0.008	± 0.008

(c) Exact Numbers. *Exact numbers have no uncertainty.*

Example 2.5: A certain automobile has exactly four wheels and exactly one engine having exactly six cylinders.

Example 2.6: There are exactly 60 min in an hour.

(d) Significant Digits. The *significant digits* in a number are those that are known to be reasonably trustworthy. They would include the last digit (which has some uncertainty) as well as all the previous digits. *Zeros* are considered significant except when their *only* function is to locate the decimal point.

Example 2.7: The numbers 385, 284 000, 0.0836, 4.00, and 509 all have three significant digits.

(e) Accuracy. The *accuracy* of a decimal number is indicated by the number of significant digits.

Example 2.8: The numbers 18.5 and 0.004 72 are both *accurate* to three significant figures.

(f) Precision. The *precision* of a decimal number is indicated by the number of decimal places given.

Example 2.9:   The number 3.85 is said *to be precise to the nearest hundredth.* It it also said to be *precise to two decimal places.*

Example 2.10:   The accuracy, precision, and uncertainty of several numbers are given in the following table.

Number	Accurate to	Precise to nearest	Uncertainty
4.75	3 digits	Hundredth	± 0.005
475	3 digits	Unit	± 0.5
6	1 digit	Unit	± 0.5
40 000	1 digit	Ten thousand	± 5000
0.003	1 digit	Thousandth	$\pm 0.000\,5$
0.0050	2 digits	Ten thousandth	$\pm 0.000\,05$

2.2   READING SIGNIFICANT FIGURES

1. All *nonzero digits* in a number *are* significant.

   Example 2.11:   Each of the numbers

   $$3846 \qquad 4.985 \qquad 92.74 \qquad 0.8274$$
   has *four* significant digits.

2. Zeros between nonzero digits are significant.

   Example 2.12:   Each of the numbers

   $$1047 \qquad 6008 \qquad 39.06 \qquad 5.007$$
   has *four* significant figures.

3. Zeros used only as placeholders to locate the decimal point are *not* significant.

   Example 2.13:   Each of the numbers

   $$0.003\,64 \qquad 0.592 \qquad 0.0869 \qquad 0.000\,063\,7$$
   has *three* significant figures.

4. Trailing zeros after the decimal point are significant.

   Example 2.14:   Each of the numbers

   $$18.50 \qquad 2.000 \qquad 0.9400 \qquad 468.0$$
   has *four* significant digits.

5. Trailing zeros before the decimal point are not significant.

   Example 2.15:   Each of the numbers

   $$34\,800 \qquad 1590 \qquad 148\,000 \qquad 5220$$
   has *three* significant figures.

6. An overscore (‾) or tilde (˜) is sometimes placed over the last trailing zero that is significant.

   Example 2.16:   Each of the numbers

   $$468\bar{0} \qquad 8\,35\bar{0}\,000 \qquad 22\,6\tilde{0}0 \qquad 835\,\tilde{0}00$$
   has *four* significant figures.

2.3 ROUNDING OFF NUMBERS

Nonsignificant digits should be discarded and the last significant digit *rounded* as follows:

(a) Rounding Up. If the discarded number is *more* than half a unit of the last digit retained, *increase* that digit by one.

Example 2.17:   14.7456, rounded to two decimal places, is 14.75.

(b) Rounding Down. If the discarded number is *less* than half a unit, *do not change* the last retained digit.

Example 2.18:   14.7446, rounded to two decimal places, is 14.74.

(c) Rounding to Nearest Even Digit. If the discarded number is *exactly* half a unit, round to the nearest *even digit*.

Example 2.19:   14.7450 rounded to two decimal places is 14.74.

Example 2.20:   14.7350 rounded to two decimal places is 14.74.

This procedure is important only when *many* numbers are to be rounded in the same computation. It causes us to round up about as many times as we round down, and thus keeps rounding errors from building up.

2.4 ARITHMETIC OPERATIONS WITH DECIMALS

The following rules for computing with approximate numbers are for the purpose of (a) preventing further loss of accuracy or precision, and (b) ensuring that the number you finally write as an answer gives a true indication of its reliability.

(a) Addition and Subtraction. Round your answer to the number of decimal places contained in the *least precise* number.

Example 2.21:   Add 18.254 to 10.3.

Solution

$$18.254 + 10.3 = 28.554$$

which we round to 28.6.

To save work and reduce the chance for error, all numbers to be added may be rounded to *one more* decimal place than contained in the least precise number *before* adding.

Example 2.22:   Evaluate

$$68.3 - 1.0035 + 14.37 + 0.064\,556$$

Solution: Rounding to two decimal places before adding,

$$68.3 - 1.00 + 14.37 + 0.06 = 81.73$$

which we then round to 81.7.

(b) Multiplication and Division. Round your result to the number of significant figures in your *least accurate* number.

Example 2.23: $473 \times 1.5584 = 737.1232$,   which we round to 737, since 473 has only three significant figures.

Example 2.24: $25/3.583 = 6.977$,   which we round to 7.0.

Before multiplying or dividing, numbers may be shortened to *one more* significant digit than contained in the least accurate number.

Example 2.25:   Multiply $1.5768 \times 92.5 \times 0.8445$.

Solution: Rounding the first number to four significant figures and multiplying,

$$1.577 \times 92.5 \times 0.8445 = 123.2$$

which we then round to three significant figures, getting 123.

Example 2.26:   Divide 354 by 1.6336.

Solution: Rounding the divisor to four significant figures and dividing,

$$354 \div 1.634 = 216.6$$

which is then rounded to 217.

(c) Exact Numbers and Defined Numbers.   When using *exact numbers* in a computation, treat them as if they had *more* significant figures and *more* decimal places than your most accurate and precise numbers.

Example 2.27:   If a certain steel wire can support a load of 135 lb, how much weight can seven such wires support?

Solution: Multiplying,

$$135(7) = 945 \text{ lb}$$

Since the 7 is an exact number, we retain as many significant figures as contained in 135.

The same procedure would apply to numbers that are *exact by definition*.   Thus, a minute of arc is defined as *exactly* one sixtieth of a degree, and 1 in. equals *exactly* 2.54 cm.

Example 2.28:   Convert 1573.4 in. to centimeters.

Solution: Multiplying,

$$1573.4(2.54) = 3996.436$$

Since 2.54 is exact by definition, we retain as many significant figures as in 1573.4, or five.   Rounding, we get 3996.4 cm.

(d) Powers, Roots, and Reciprocals.   For each of these operations, retain as many *significant figures* in the answer as contained in the original number.

Example 2.29

(a) $(5.28)^2 = 27.9$ (c) $\sqrt{4.87} = 2.21$

(b) $(0.0655)^3 = 0.000\,281$ (d) $\sqrt[3]{86.4} = 4.42$

(e)   The reciprocal of 2.00 is $1/2.00 = 0.500$.

(f)   The reciprocal of 5.864 is $1/5.864 = 0.1705$.

2.5 POWERS OF TEN

Powers of ten can be written as ten raised to some exponent, as shown in Table 2-1.

Table 2-1   Powers of Ten

$1\,000\,000 = 10^6$	$0.1 \quad\; = 10^{-1}$
$100\,000 = 10^5$	$0.01 \quad\, = 10^{-2}$
$10\,000 = 10^4$	$0.001 \quad = 10^{-3}$
$1\,000 = 10^3$	$0.000\,1 \quad = 10^{-4}$
$100 = 10^2$	$0.000\,01 \; = 10^{-5}$
$10 = 10^1$	$0.000\,001 = 10^{-6}$
$1 = 10^0$	

Note that the exponent is always equal to the number of zeros in the decimal number.

Example 2.30

$$100\,000\,000\,000 = 10^{11}$$

For numbers smaller than one, the exponents become negative (see Table 2-1). Note that the number in the exponent is always equal to one more than the number of zeros to the right of the decimal point in the decimal number.

Example 2.31

$$\overbrace{0.000\,000\,000\,000\,01}^{13} = 10^{-14}$$

2.6 SCIENTIFIC NOTATION

(a) Definition. A number is in *scientific notation* when it is written as a number between 1 and 10 multiplied by a power of 10.

Example 2.32

$$3.75 \times 10^2$$

$$7.3385 \times 10^7$$

$$1.285 \times 10^{-3}$$

are numbers written in scientific notation.

(b) Converting Numbers to Scientific Notation. Rewrite the given number with just *one digit* to the left of the decimal point, and discard any nonsignificant zeros. The *exponent* to which ten is raised is equal to the number of places the decimal point was moved from its original location; positive if moved to the left, and negative if moved to the right.

Example 2.33:   Convert 1975 to scientific notation. We place the decimal point between the 1 and the 9, a move of three places to the left from its original position. The exponent will thus be $+3$, and we write

$$1975 = 1.975 \times 10^3$$

Example 2.34:   Convert 0.000 005 70 to scientific notation.

Solution: The decimal point must be moved six places to the right, making our exponent -6.   We write

$$0.000\,005\,70 = 5.70 \times 10^{-6}$$

The nonsignificant zeros to the left of the 5 are discarded, and the significant zero to the right of the 7 is retained.

(c)   Adding and Subtracting Numbers in Scientific Notation.   If the numbers to be added or subtracted have the *same* power of ten, add or subtract the numbers as indicated, keeping the same power of ten.

Example 2.35

$$(3.57 \times 10^5) + (7.33 \times 10^5) - (4.86 \times 10^5) = (3.57 + 7.33 - 4.86) \times 10^5$$
$$= 6.04 \times 10^5$$

If the powers of ten are *different*, they must be made equal before combining the numbers.   Each shift of the decimal point one place to the left will raise the exponent one unit, and each shift to the right will decrease the exponent one unit.

Example 2.36

$$(1.275 \times 10^6) - (3.927 \times 10^5) + (1.339 \times 10^3) = (12.75 \times 10^5) - (3.927 \times 10^5) + (0.01339 \times 10^5)$$
$$= (12.75 - 3.927 + 0.01339) \times 10^5$$
$$= 8.84 \times 10^5$$

(d)   Multiplying and Dividing Numbers in Scientific Notation.   Powers of ten are multiplied by *adding their exponents.*

Example 2.37

$$10^2 \times 10^3 = 10^{2+3} = 10^5$$
$$10^3 \times 10^{-2} \times 10^{-6} \times 10^4 = 10^{3-2-6+4} = 10^{-1}$$

Powers of ten are divided by *subtracting exponents.*

Example 2.38

$$\frac{10^3}{10^2} = 10^{3-2} = 10^1 = 10$$

Example 2.39

$$\frac{10^4}{10^7} = 10^{4-7} = 10^{-3}$$

Example 2.40

$$\frac{10^5 \times 10^3}{10^2 \times 10^{-4}} = 10^{5+3-2-(-4)} = 10^{10}$$

Numbers in scientific notation are multiplied by multiplying their decimal parts and their powers of ten separately.   Then the product is expressed in scientific notation.

Example 2.41

$$(2.58 \times 10^5) \times (7.48 \times 10^2) = (2.58 \times 7.48) \times (10^5 \times 10^2)$$
$$= 19.3 \times 10^{5+2}$$
$$= 19.3 \times 10^7$$
$$= 1.93 \times 10^8$$

Numbers in scientific notation are divided by dividing their decimal parts and their powers of ten separately.   The quotient is then expressed in scientific notation.

Example 2.42

$$\frac{48.7 \times 10^6}{3.55 \times 10^8} = \frac{48.7}{3.55} \times \frac{10^6}{10^8} = 13.7 \times 10^{6-8}$$
$$= 13.7 \times 10^{-2}$$
$$= 1.37 \times 10^{-1}$$

Solved Problems

APPROXIMATE NUMBERS

2.1   Compare the following pairs of numbers for accuracy and precision: (a) 4.863 and 7.28, (b) 4000 and 0.005, (c) 24.85 and 735.27.

Solution

(a)   4.863 and 7.28   4.863 is more accurate (four significant figures) and more precise (to nearest thousandth).

(b)   4000 and 0.005   Both have same accuracy (one significant digit).   0.005 is more precise (nearest thousandth).

(c)   24.85 and 735.27   735.27 is more accurate (five significant digits).   Both have same precision (nearest hundredth).

2.2   State the accuracy, precision, and uncertainty of the numbers: (a) 5.83, (b) 260 000, (c) 0.003 66, and (d) 4.5000.

Solution

(a)   5.83 is accurate to three digits, precise to the nearest hundredth, and has an uncertainty of ± 0.005.

(b)   260 000 is accurate to two digits, precise to the nearest ten thousand, and has an uncertainty of ± 5000.

(c)   0.003 66 is accurate to three figures, precise to the nearest hundred-thousandth, and has an uncertainty of $\pm 0.000 005$.

(d)   4.5000 is accurate to five digits, precise to the nearest ten-thousandth, and has an uncertainty of $\pm 0.000 05$.

READING SIGNIFICANT FIGURES

2.3   Determine the number of significant digits in each of the following numbers:

(a) 58.5	(c) 7.90	(e) 4800	(g) 90 000	(i) 90 $\overline{0}$00	(k) 3.000
(b) 7839	(d) 3.07	(f) 60 070	(h) 90 000.0	(j) 0.7483	(l) 0.004 003 0

Solution

	(a)	(b)	(c)	(d)	(e)	(f)	(g)	(h)	(i)	(j)	(k)	(l)
Number	58.5	7839	7.90	3.07	4800	60070	90000	90000.0	90$\bar{0}$00	0.7483	3.000	0.0040030
Significant Digits	3	4	3	3	2	4	1	6	3	4	4	5

ROUNDING OFF NUMBERS

2.4   Round the following numbers to three significant digits:

(a) 7.473   (b) 1566   (c) 0.03925   (d) 0.03935   (e) 38.555   (f) 38.455

Solution

	(a)	(b)	(c)	(d)	(e)	(f)
Number	7.473	1566	0.03925	0.03935	38.555	38.455
Rounded	7.47	1570	0.0392	0.0394	38.6	38.5

2.5   Round the following numbers to two decimal places:

(a) 18.5732   (b) 18.5752   (c) 128.435   (d) 0.0151

Solution

	(a)	(b)	(c)	(d)
Number	18.5732	18.5752	128.435	0.0151
Rounded	18.57	18.58	128.44	0.02

ARITHMETIC OPERATIONS WITH DECIMALS

2.6   Add $185 + 4.573 + 36.45 + 1.06$.

Solution

Rounding the numbers to one decimal place (one more than in 185) and adding,

$$
\begin{array}{r}
185 \\
4.6 \\
36.4 \\
1.1 \\
\hline
227.1
\end{array}
$$

which is then rounded to 227, so that it has no more decimal places than the least precise number (185).

2.7   Combine $68.35 - 2.736 + 15.277 - 97.5$.

Solution

Rounding to two decimal places (one more than in 97.5),

$$
\begin{array}{cccc}
68.35 & -2.74 & & 83.63 \\
+15.28 & -97.5 & \text{and} & -100.24 \\
\hline
83.63 & -100.24 & & -16.61
\end{array}
$$

which is then rounded to -16.6.

2.8   Multiply $25.9 \times 0.4736 \times 3.5$.

Solution

Rounding to three significant figures (one more than in 3.5) and multiplying,

$$25.9 \times 0.474 \times 3.5 = 43.0$$

which is then rounded to 43 (the number of significant figures in 3.5).

2.9   Multiply $18.5 \times 11.473 \times 0.22749$.

Solution

Rounding to four significant figures and multiplying,

$$18.5 \times 11.47 \times 0.2275 = 48.27$$

which is then rounded to 48.3.

2.10   Divide 185.23 by 15.

Solution

Rounding to three significant figures and dividing,

$$\frac{185}{15} = 12.3$$

which is then rounded to 12.

2.11   Divide 0.003 22 by 3.4977.

Solution

Rounding to four significant figures and dividing,

$$\frac{0.003\,22}{3.498} = 0.000\,920\,5$$

which is then rounded to 0.000 921.

2.12   If one bolt weighs 5.84 grams (g), how much will 8 such bolts weigh?

Solution

The "8" is exact, so our answer should contain as many significant figures as in 5.84.

$$8 \times 5.84 = 46.7 \text{ g}$$

2.13   If a field 47.3 acres in area is divided into four equal parts, how many acres should be in each part?

Solution

As the "4" is exact in this problem, our answer should contain as many significant figures as in 47.3.

$$\frac{47.3}{4} = 11.825$$

which rounds to 11.8 acres.

2.14   Subtract 593 from 596.

Solution

$596 - 593 = 3$.   Note that the number of significant figures has shrunk from 3 to 1.

2.15   Raise the following numbers to the powers indicated, retaining the proper number of significant figures in your answer.

(a) $(4.73)^2$ (b) $(0.099)^2$ (c) $(7.233)^3$ (d) $(1.5)^5$ (e) $(2.5000)^3$

Solution

(a) $(4.73)^2 = 22.4$ (c) $(7.233)^3 = 378.4$ (e) $(2.5000)^3 = 15.625$

(b) $(0.099)^2 = 0.0098$ (d) $(1.5)^5 = 7.6$

2.16   Extract the following roots by calculator, tables, or slide rule, retaining the proper number of significant figures in your answer.

(a) $\sqrt{6.47}$ (b) $\sqrt{35.880}$ (c) $\sqrt[3]{274.6}$ (d) $\sqrt[5]{45}$ (e) $\sqrt[3]{8.0000}$

Solution

(a) $\sqrt{6.47} = 2.54$ (c) $\sqrt[3]{274.6} = 6.500$ (e) $\sqrt[3]{8.0000} = 2.0000$

(b) $\sqrt{35.880} = 5.9900$ (d) $\sqrt[5]{45} = 2.1$

2.17   Find the reciprocals of (a) 3.67 (b) 0.81, (c) 2.5583, and (d) 275, retaining the proper number of significant figures in your answer.

Solution

(a) $\frac{1}{3.67} = 0.272$ (b) $\frac{1}{0.81} = 1.2$ (c) $\frac{1}{2.5583} = 0.390\,88$ (d) $\frac{1}{275} = 0.003\,64$

2.18   A batch of concrete is made by mixing 180 kg of aggregate, 175 kg of sand, 85.5 kg of cement, and 1.55 kg of dry coloring material.   What is the total weight of the mixture?

Solution

Before adding, we round the more precise numbers to one more decimal place than contained in the least precise number.

$$\text{Total weight} = 180 + 175 + 85.5 + 1.6 = 442.1 \text{ kg}$$

which is then rounded to 442 kg.

2.19   A rectangular driveway is to be 28 ft long and 18.54 ft wide.   Find its area.

Solution

Multiplying length and width,

$$\text{Area} = 28(18.54) = 519.12 \text{ ft}^2$$

which is then rounded to two digits, becoming 520 ft².

2.20   A batch of Portland cement weighs 157.6 lb and has a density of 94 lb/ft³.   Find the volume of the cement.

Solution

By Eq. 184,

$$\text{Volume} = \frac{\text{weight}}{\text{density}} = \frac{157.6}{94} = 1.6766$$

which is then rounded to two significant figures, giving

$$\text{Volume} = 1.7 \text{ ft}^3$$

2.21   Three resistors, having values of 3483 Ω, 57.45 Ω, and 2.573 Ω, are wired in series.   What is the total resistance?

Solution

The total resistance will be the sum of the individual resistances, as in Eq. 201.   Before adding, we round the more precise numbers to one more decimal place than the least precise number.

$$
\begin{array}{r}
3483 \\
57.4 \\
2.6 \\
\hline
3543.0
\end{array}
$$

and round our answer to the number of decimal places in the least precise number.   Our answer is then 3543 Ω.

2.22   The voltage drop across a resistor is measured as 114.6 V, and the current passing through it is found to be 0.56 ampere (A).   What is the wattage dissipated by the resistor?

Solution

The wattage is the product of the voltage and the current.   Multiplying,

$$\text{Wattage} = 114.6(0.56) = 64.176 \text{ W}$$

which is then rounded to two digits, giving

$$\text{Wattage} = 64 \text{ W}$$

2.23   A voltage drop of 145.3 V is measured across a 55-Ω resistor.   What is the current through the resistor?

Solution

By Ohm's law, Eq. 200, the current will be equal to the voltage drop divided by the resistance.   Dividing,

$$I = \frac{V}{R} = \frac{145.3}{55} = 2.642 \text{ A}$$

$$= 2.6 \text{ A} \qquad \text{(rounded to two digits)}$$

2.24 A pipe has an inside diameter (ID) of 3.2 cm and a wall thickness of 0.247 cm. It is surrounded by insulation having a thickness of 1.1 cm. What is the outside diameter (OD) of the insulation?

Solution

We must add 3.2, 0.494, and 2.2. We first round all figures to one more decimal place than contained in the least precise number,

$$OD = 3.2 + 0.49 + 2.2 = 5.89 \text{ cm}$$

and finally round the answer to the number of decimal places contained in the least precise number, getting 5.9 cm.

2.25 The flywheel in a certain press rotates at the speed of 1735 revolutions per minute (r/min). How many revolutions will it make in 2.4 min?

Solution

The number of revolutions equals the product of the rate and the time. Multiplying,

$$N = 1735(2.4) = 4164 \text{ revolutions}$$

which is then rounded to two significant figures, giving,

$$N = 4200 \text{ revolutions}$$

2.26 How many cubic feet of steel would be needed to make a counterweight of 146.5 lb, if the density of steel is 450 lb/ft^3?

Solution

Since the volume equals the weight divided by the density, we divide,

$$V = \frac{W}{D} = \frac{146.5}{450} = 0.3256 \text{ ft}^3$$

which is then rounded to two significant figures:

$$V = 0.33 \text{ ft}^3$$

SCIENTIFIC NOTATION

2.27 Write the following numbers as powers of ten.

(a) 1000   (b) 0.001   (c) 1 000 000   (d) 0.000 000 000 1

Solution

(a) $1000 = 10^3$ (b) $0.001 = 10^{-3}$ (c) $1\,000\,000 = 10^6$ (d) $0.000\,000\,000\,1 = 10^{-10}$

2.28 Write the following numbers in scientific notation.

(a) 3587   (b) 52 000   (c) 52 00$\bar{0}$ (d) 0.008 40   (e) 0.000 006 3

Solution

(a) $3587 = 3.587 \times 10^3$ (c) $52\,00\bar{0} = 5.2000 \times 10^4$ (e) $0.000\,006\,3 = 6.3 \times 10^{-6}$

(b) $52\,000 = 5.2 \times 10^4$ (d) $0.008\,40 = 8.40 \times 10^{-3}$

2.29   Combine the following numbers as indicated.

(a) $3.57 \times 10^3 + 9.24 \times 10^3$ (c) $3.85 \times 10^7 - 1.66 \times 10^7$

(b) $2.5 \times 10^{-2} + 4.78 \times 10^{-2}$ (d) $8.4 \times 10^4 - 6.3 \times 10^4 + 1.4 \times 10^4$

Solution

(a) $3.57 \times 10^3 + 9.24 \times 10^3 = 12.81 \times 10^3$ (c) $3.85 \times 10^7 - 1.66 \times 10^7 = 2.19 \times 10^7$

(b) $2.5 \times 10^{-2} + 4.78 \times 10^{-2} = 7.3 \times 10^{-2}$ (d) $8.4 \times 10^4 - 6.3 \times 10^4 + 1.4 \times 10^4 = 3.5 \times 10^4$

2.30   Combine the following numbers as indicated.

(a) $4.72 \times 10^3 + 6.28 \times 10^2$ (c) $5.235 \times 10^4 - 2.849 \times 10^2$

(b) $3.85 \times 10^{-3} + 7.93 \times 10^{-5}$ (d) $9.28 \times 10^2 - 3.75 \times 10^{-3}$

Solution

(a) $4.72 \times 10^3 + 6.28 \times 10^2 = 4.72 \times 10^3 + 0.628 \times 10^3$

$$= 5.35 \times 10^3$$

(b) $3.85 \times 10^{-3} + 7.93 \times 10^{-5} = 3.85 \times 10^{-3} + 0.0793 \times 10^{-3}$

$$= 3.93 \times 10^{-3}$$

(c) $5.235 \times 10^4 - 2.849 \times 10^2 = 5.235 \times 10^4 - 0.028\,49 \times 10^4$

$$= 5.207 \times 10^4$$

(d) $9.28 \times 10^2 - 3.75 \times 10^{-3} = 9.28 \times 10^2 - 0.000\,037 \times 10^2$

$$= 9.28 \times 10^2$$

2.31   Multiply the following numbers.

(a) $(7.26 \times 10^4)(2.63 \times 10^2)$ (c) $(3.75 \times 10^4)(9.248 \times 10^2)(3.2286 \times 10^{-3})$

(b) $(8.26 \times 10^3)(2.35 \times 10^{-5})$

Solution

(a) $(7.26 \times 10^4)(2.63 \times 10^2) = 19.1 \times 10^6$

$$= 1.91 \times 10^7$$

(b) $(8.26 \times 10^3)(2.35 \times 10^{-5}) = 19.4 \times 10^{-2}$

$$= 1.94 \times 10^{-1}$$

(c) $(3.75 \times 10^4)(9.248 \times 10^2)(3.2286 \times 10^{-3}) = 112 \times 10^3$

$$= 1.12 \times 10^5$$

2.32   Divide the following numbers.

(a) $(8.24 \times 10^4) \div (4.28 \times 10^2)$ (c) $(7.2 \times 10^{-7}) \div (3.84 \times 10^{-3})$

(b) $(3.72 \times 10^2) \div (7.294 \times 10^5)$ (d) $(5.285 \times 10^3) \div (7.36 \times 10^{-3})$

Solution

(a) $(8.24 \times 10^4) \div (4.28 \times 10^2) = 1.93 \times 10^2$ (c) $(7.2 \times 10^{-7}) \div (3.84 \times 10^{-3}) = 1.9 \times 10^{-4}$

(b) $(3.72 \times 10^2) \div (7.294 \times 10^5) = 0.510 \times 10^{-3}$ (d) $(5.285 \times 10^3) \div (7.36 \times 10^{-3}) = 0.718 \times 10^6$

$$= 5.10 \times 10^{-4} \qquad\qquad\qquad\qquad\qquad = 7.18 \times 10^5$$

2.33   Express 25 million ohms in scientific notation.

Solution

$$25 \text{ million} = 25\,000\,000 = 2.5 \times 10^7 \ \Omega$$

2.34   The modulus of elasticity of iron is $30\,100\,000$ lb/in². Write this number in scientific notation.

Solution

$$30\,100\,000 = 3.01 \times 10^7 \ \text{lb/in}^2$$

2.35   The coefficient of thermal expansion of iron is 6.28×10^{-6} per degree F. Write this number in decimal form.

Solution

$$6.28 \times 10^{-6} = 0.000\,006\,28 \ \text{per degree F}$$

2.36   A certain piece of road machinery weighs 17.4 tons. Express this weight in pounds, with the proper number of significant figures.

Solution

Converting to pounds,

$$(17.4 \text{ tons})\left(\frac{2000 \text{ lb}}{\text{ton}}\right) = 34\,800 \text{ lb}$$

As only three significant figures are justified, we write the answer in scientific notation as 3.48×10^4 lb.

Supplementary Problems

2.37   State the accuracy, precision, and uncertainty of the following numbers.

(a) 2.936   (b) 0.002 936   (c) 29 000   (d) 2

2.38   State the number of significant digits in the following numbers.

(a) 23.2   (c) 0.0017   (e) 25 000   (g) 0.000 000 3

(b) 17.00   (d) 19.013   (f) 2500.00

2.39   Round the following numbers to three significant figures.

(a) 27.000   (b) 125.007   (c) 0.005 716

2.40   Round the following numbers to two significant figures.

(a) 37.2763   (b) 0.007 364   (c) 5.001 73

2.41   Combine as indicated, retaining the proper number of decimal places in the answer.

   (a) $227 + 3.73 - 6.4243 + 2.7$ (b) $23.3 - 7.7204 + 6.421 - 0.0073$

2.42   Multiply the following, retaining the proper number of significant digits in the answer.

   (a) $17.0 \times 2.635 \times 4.12$ (b) $2.73649 \times 1.0007 \times 29.23456$

2.43   Divide the following, retaining the proper number of significant digits in the answer.

   (a)   265.37 by 17.21   (b)   0.0076 by 1.1927357

2.44   Raise the following numbers to the powers indicated, retaining the proper number of significant digits.

   (a) $(2.12)^2$ (b) $(0.002)^3$ (c) $(1.2000)^3$

2.45   Extract the following roots by calculator, tables, or slide rule, retaining the proper number of significant digits.

   (a) $\sqrt{1.97}$ (b) $\sqrt[3]{327.2}$ (c) $\sqrt[5]{27}$ (d) $\sqrt[4]{127}$

2.46   Find the reciprocal of the following, retaining the proper number of significant digits.

   (a)   7.23   (b)   1.0132   (c)   0.00123

2.47   Write the following numbers as powers of ten.

   (a)   1.0   (b)   1 000 000 000   (c)   0.000 000 000 000 000 000 1

2.48   Write the following numbers in scientific notation.

   (a)   2637   (b)   7427   (c)   0.003 25   (d)   700 365.635

2.49   Combine the following numbers as indicated.

   (a) $1.24 \times 10^4 + 3.76 \times 10^4 - 6.42 \times 10^4$ (c) $7.0 \times 10^2 + 3.0 \times 10^2 + 6.0 \times 10^2$

   (b) $7.9315 \times 10^{-3} + 27.76 \times 10^{-3} + 38.972 \times 10^{-3}$

2.50   Combine the following numbers as indicated.

   (a) $1.73 \times 10^3 - 2.76 \times 10^2$ (c) $0.0013 \times 10^4 - 2.91 \times 10^2 + 36.90 \times 10^5$

   (b) $1.39 \times 10^{-2} + 7.36 \times 10^4$

2.51   Multiply the following numbers.

   (a) $(1.23 \times 10^2)(6.40 \times 10^3)$ (b) $(0.013 \times 10^2)(3.915 \times 10^3)(0.1976 \times 10^{-3})$

2.52   Divide the following numbers.

   (a) $(23.643 \times 10^2) \div (7.13 \times 10^3)$ (b) $(9.133 \times 10^{-2}) \div (0.00039 \times 10^4)$

Answers to Supplementary Problems

2.37 (a) 4, thousandth, ± 0.0005 (b) 4, millionth, $\pm 0.000\,000\,5$ (c) 2, thousand, ± 500
(d) 1, unit, ± 0.5

2.38 (a) 3   (b) 4   (c) 2   (d) 5   (e) 2   (f) 6   (g) 1

2.39 (a) 27.0   (b) 125   (c) 0.005 72

2.40 (a) 37   (b) 0.0074   (c) 5.0

2.41 (a) 227   (b) 22.0

2.42 (a) 185   (b) 80.056

2.43 (a) 15.42   (b) 0.0064

2.44 (a) 4.49   (b) 0   (c) 1.7280

2.45 (a) 1.40   (b) 6.891   (c) 1.9   (d) 3.36

2.46 (a) 0.138   (b) 0.986 97   (c) 813

2.47 (a) 10^0 (b) 10^9 (c) 10^{-19}

2.48 (a) 2.637×10^3 (b) 7.427×10^3 (c) 3.25×10^{-3} (d) $7.003\,656\,35 \times 10^5$

2.49 (a) -1.42×10^4 (b) 7.466×10^{-2} (c) 1.6×10^3

2.50 (a) 1.45×10^3 (b) 7.36×10^4 (c) 3.69×10^6

2.51 (a) 7.87×10^5 (b) 1

2.52 (a) 3.32×10^{-1} (b) 2.342×10^{-2}

Chapter 3

Introduction to Algebra

3.1 DEFINITIONS AND TERMINOLOGY

(a) Constants. *Constants* are quantities that *do not change* value in a particular problem, such as 5, π, and $\sqrt{7}$. When a letter is used to represent a constant, it is usually chosen from the beginning of the alphabet (a, b, c, etc.).

(b) Variables. *Variables* are quantities that *may change* during a problem. They are usually represented by letters from the end of the alphabet.

Example 3.1: In the equation $y = ax^2 + bx + c$, the letters x and y represent variables, while a, b, and c represent constants.

(c) Symbols of Grouping. Parentheses ( ), brackets [ ], and braces { } are used to group parts of an expression together, as needed.

Example 3.2: $\{3 - 5[x + 3x(1 - x) - 2]\} - x$ shows the use of symbols of grouping.

(d) Algebraic Expressions. An *algebraic expression* is a grouping of variables, constants, signs of operation, and symbols of grouping.

Example 3.3: $4x + 3y$, $5xy/z$, and $(3 + x)^2 - \sqrt{x}$ are algebraic expressions.

(e) Terms. Plus and minus signs divide an algebraic expression into *terms* (except when the $+$ or $-$ sign is inside a symbol of grouping).

Example 3.4: The expression $x^2 - (2x + 3)^3$ has two terms.

(f) Factors. The *factors* of an algebraic expression are those quantities which, when multiplied together, result in the expression.

Example 3.5: The factors of 14 are 2 and 7. (Although 1 is a factor of any quantity it is not usually stated.)

Example 3.6: The factors of $6xy$ are 2, 3, x, and y.

(g) Coefficients. The numerical coefficient (or just *coefficient*) is the constant part of a term.

Example 3.7

(*a*) In the term $3x$, 3 is the coefficient of x.

(*b*) In the term $-3ax$, $-3a$ is the coefficient of x.

(h) Like Terms. *Like terms* are those that differ only in their coefficients.

Example 3.8: $3xy^2$ and $-5xy^2$ are like terms.

43

(i)　Monomials.　A *monomial* is an algebraic expression with only one term.

Example 3.9:　　$6xyz$ is a monomial.

(j)　Binomials.　A *binomial* is an algebraic expression with two terms.

Example 3.10:　　$2x + 5$ is a binomial.

(k)　Trinomial.　A *trinomial* is an algebraic expression having three terms.

Example 3.11:　　$x^2 + 2x - 3$ is a trinomial.

(l)　Multinomials.　A *multinomial* is any algebraic expression having more than one term.

Example 3.12:　　$x^2 - 3 + \sqrt{x}$ is a multinomial (and also a trinomial).

(m)　Polynomials.　A *polynomial* is a multinomial in which all the powers of the variables are positive integers.

Example 3.13

$2x^3 - x^2 + 4$ is a polynomial.
$2x^3 - x^{-2} + 4$ is a multinomial but not a polynomial.

3.2　INTEGRAL EXPONENTS

(a)　Definition

$$a^n \text{ means } \underbrace{a \cdot a \cdot a \cdots a}_{n \text{ factors}} \qquad \boxed{7}$$

a is called the *base*, and n is the *exponent*.

Example 3.14

$$2^3 = (2)(2)(2) = 8$$

Common Error	An exponent applies *only* to the symbol directly before it.　Thus
	$2x^2 \neq 2^2 x^2$
but	$(2x)^2 = 2^2 x^2 = 4x^2$

(b)　　| Products | $x^a \cdot x^b = x^{a+b}$ | **8** |

Example 3.15

$$10^2(10^3) = 10^{2+3} = 10^5$$

Example 3.16

$$x^5 \cdot x^2 = x^{5+2} = x^7$$

Example 3.17

$$N^2 \cdot N^x = N^{2+x}$$

(c)

| Quotients | $\dfrac{x^a}{x^b} = x^{a-b}$ | **9** |

Example 3.18

$$\frac{x^6}{x^4} = x^{6-4} = x^2$$

Example 3.19

$$\frac{7^3}{7^5} = 7^{3-5} = 7^{-2}$$

Example 3.20

$$\frac{w^a}{w^3} = w^{a-3}$$

(d)

| Powers | $(x^a)^b = x^{ab} = (x^b)^a$ | **10** |

Example 3.21

$$(2^3)^2 = 2^{3 \times 2} = 2^6$$

Example 3.22

$$(x^{-2})^5 = x^{-2(5)} = x^{-10}$$

Example 3.23

$$(z^n)^4 = z^{4n}$$

(e)

| Product Raised to a Power | $(xy)^a = x^a \cdot y^a$ | **11** |

Example 3.24

$$(MN)^2 = M^2 N^2$$

Example 3.25

$$(2x)^3 = 2^3 x^3 = 8x^3$$

| Common Error | There is *no* similar rule for the *sum* of two quantities raised to a power.
 $$(x + y)^n \neq x^n + y^n$$ |

(f)

| Quotient Raised to a Power | $\left(\dfrac{x}{y}\right)^a = \dfrac{x^a}{y^a}$ | **12** |

Example 3.26

$$\left(\frac{y}{3}\right)^3 = \frac{y^3}{3^3} = \frac{y^3}{27}$$

Example 3.27

$$\left(\frac{2x}{5y}\right)^2 = \frac{2^2 x^2}{5^2 y^2} = \frac{4x^2}{25y^2}$$

(g)

Zero Exponent	$x^0 = 1$	**13**

Example 3.28

$$157^0 = 1$$

Example 3.29

$$(3x^2 - 5xy^2 + w^3)^0 = 1$$

(h)

Negative Exponent	$x^{-a} = \dfrac{1}{x^a}$	**14**

Example 3.30

$$10^{-2} = \frac{1}{10^2}$$

Example 3.31

$$(8x^2 + 5x)^{-3} = \frac{1}{(8x^2 + 5x)^3}$$

Fractional exponents are explained in Chapter 10.

3.3 REMOVING SYMBOLS OF GROUPING

1. Multiply every term within the grouping by the factor preceding the grouping, and remove the symbol of grouping.

Example 3.32

$$2(x - 3) = 2x - 6$$

Example 3.33

$$4 + 2(x - 3) = 4 + 2x - 6 = 2x - 2$$

Example 3.34

$$x - 3(1 - y) = x - 3 + 3y$$

Common Error	Don't forget to multiply *every* term within the grouping by the preceding factor. $$-(2x + 5) \neq -2x + 5$$

2. If there are groupings within groupings, start simplifying with the *innermost* groupings.

Example 3.35

$$3[2 + 4(1 - x)] = 3[2 + 4 - 4x] = 3[6 - 4x] = 18 - 12x$$

Example 3.36

$$w + 2[2 - (x + 3)] = w + 2[2 - x - 3] = w + 2[-x - 1]$$
$$= w - 2x - 2$$

Example 3.37

$$2\{[(x - 2) - (y + 4)] + 3\} - 5 = 2\{[x - 2 - y - 4] + 3\} - 5$$
$$= 2\{[x - y - 6] + 3\} - 5 = 2\{x - y - 3\} - 5$$
$$= 2x - 2y - 6 - 5 = 2x - 2y - 11$$

3.4 ADDITION AND SUBTRACTION OF ALGEBRAIC EXPRESSIONS

Algebraic expressions are added and subtracted by *combining like terms*. Like terms are added by adding their coefficients.

Example 3.38

$$2x + 3x = 5x$$

Example 3.39

$$2x + 3y - 4z + 3x - 2y + 4z = 2x + 3x + 3y - 2y - 4z + 4z = 5x + y$$

When combining longer expressions, they can be arranged in rows with like terms one above the other.

Example 3.40:   Combine the expressions

$$(6x + 3xy - 2w + yz) - (2x + 3w - 2yz - 3y) + (y - xy + w)$$

Solution: Arranging the expressions in rows,

$$
\begin{array}{l}
6x + 3xy - 2w + yz \\
-2x - 3w + 2yz + 3y \\
 - xy + w + y \\
\hline
4x + 2xy - 4w + 3yz + 4y
\end{array}
$$

3.5 MULTIPLICATION OF ALGEBRAIC EXPRESSIONS

(a) Symbols for Multiplication

$$M \cdot N \qquad MN \qquad M(N) \qquad (M)N \qquad (M)(N) \qquad M \times N$$

The last of these should be avoided as the multiplication sign is too easily mistaken for the letter x.

(b) Multiplying Two Monomials.   Multiply the coefficients and combine the unknowns using Eq. 8.

Example 3.41

$$2x(3x) = 6x^2$$

Example 3.42

$$4x^2(-2x^3) = -8x^5$$

Example 3.43

$$3x^2y^3z(4wxy^2z^3) = 12wx^3y^5z^4$$

(c) Multiplying a Multinomial by a Monomial. Multiply each term in the polynomial by the monomial, as you did when removing symbols of grouping.

Example 3.44

$$3x(x^2 - 2) = 3x^3 - 6x$$

Example 3.45

$$-3x(x^3 - 2x + 5) = -3x(x^3) - 3x(-2x) - 3x(5)$$
$$= -3x^4 + 6x^2 - 15x$$

(d) Multiplying a Multinomial by a Multinomial. Multiply every term in one multinomial by every term in the other, and combine like terms.

Example 3.46

$$(x + 3)(x - 2) = x(x) + x(-2) + 3(x) + 3(-2)$$
$$= x^2 - 2x + 3x - 6$$
$$= x^2 + x - 6$$

Example 3.47

$$(x - 2)(x^2 - 3x + 4) = x(x^2) + x(-3x) + x(4) - 2(x^2) - 2(-3x) - 2(4)$$
$$= x^3 - 3x^2 + 4x - 2x^2 + 6x - 8$$
$$= x^3 - 5x^2 + 10x - 8$$

3.6 POWERS OF MULTINOMIALS

To raise a multinomial to some integer n, apply Eq. 7.

Example 3.48

$$(x + 2)^2 = (x + 2)(x + 2)$$
$$= x^2 + 2x + 2x + 4$$
$$= x^2 + 4x + 4$$

3.7 DIVISION OF ALGEBRAIC EXPRESSIONS

(a) Symbols for Division

$$x \div y \qquad \frac{x}{y} \qquad x/y$$

The names of the parts are

$$\text{Quotient} = \frac{\text{dividend}}{\text{divisor}} = \frac{\text{numerator}}{\text{denominator}} \xleftarrow{} \begin{array}{l}\text{Fraction}\\ \text{line}\end{array}$$

The quantity x/y is also called a *fraction* and is also spoken of as the *ratio* of x to y.

(b) Division by zero. *Division by zero* is not a permissible operation.

Example 3.49:   In the fraction

$$\frac{x + 2}{x - 2}$$

x cannot equal 2 or the illegal operation of division by zero will result.

(c) Three Signs of a Fraction. The numerator and denominator each have an algebraic sign, and a third sign applies to the entire fraction. Any two of these three signs may be changed without changing the value of a fraction.

Example 3.50:   What are the equivalent ways of writing the fraction $-\dfrac{x}{y}$?

Solution: Changing signs two at a time,

Signs of a Fraction	$-\dfrac{x}{y} = \dfrac{-x}{y} = \dfrac{x}{-y} = -\dfrac{-x}{-y}$

40

(d) Dividing a Monomial by a Monomial. Divide the coefficients. Use Eq. 9 to find the quotients of the variables.

Example 3.51

$$6x^2 \div 3x = \frac{6\,x^2}{3\,x} = \frac{6}{3} \cdot \frac{x^2}{x} = 2x$$

Example 3.52

$$5x^3y^2z \div 15xy^2z^3 = \frac{5x^3y^2z}{15xy^2z^3} = \frac{5}{15} \cdot \frac{x^3}{x} \cdot \frac{y^2}{y^2} \cdot \frac{z}{z^3} = \frac{x^2}{3z^2}$$

(e) Dividing a Multinomial by a Monomial. Divide each term of the multinomial by the monomial, and combine.

Example 3.53

$$4x^2 - 6x + 2 \div 2x = \frac{4x^2 - 6x + 2}{2x} = \frac{4x^2}{2x} - \frac{6x}{2x} + \frac{2}{2x}$$

$$= 2x - 3 + \frac{1}{x}$$

Common Error	Don't forget to divide *every* term of the multinomial by the monomial. $$\frac{x^2 + 2x + 3}{x} \neq x + 2 + 3$$

Example 3.54

$$x^2y^2 + 3x - 2y \div -6xy = \frac{x^2y^2 + 3x - 2y}{-6xy} = \frac{x^2y^2}{-6xy} + \frac{3x}{-6xy} - \frac{2y}{-6xy}$$

$$= -\frac{xy}{6} - \frac{1}{2y} + \frac{1}{3x}$$

Common Error	There is no similar rule for dividing a monomial by a multinomial.
	$$\frac{a}{b+c} \neq \frac{a}{b} + \frac{a}{c}$$

(f) Dividing a Polynomial by a Polynomial. Write the divisor and the dividend in the order of descending powers of the variable. Supply any missing terms, using a coefficient of zero. Set up as a long division, as in the following example.

Example 3.55:   Divide $x^2 + 2x^4 - 3$ by $x + 2$.

Solution

1. Write the dividend in descending order of the powers.

$$2x^4 + x^2 - 3$$

2. Supply the missing terms with coefficients of zero.

$$2x^4 + 0x^3 + x^2 + 0x - 3$$

3. Set up in long division format.

$$x + 2 \overline{)2x^4 + 0x^3 + x^2 + 0x - 3}$$

4. Divide the first term in the dividend $(2x^4)$ by the first term in the divisor (x). The result $(2x^3)$ is written above the dividend, in line with the term having the same power. It is the first term of the quotient.

$$
\begin{array}{r}
2x^3 \\
x + 2 \overline{)2x^4 + 0x^3 + x^2 + 0x - 3}
\end{array}
$$

5. Multiply the divisor by the first term of the quotient. Write the result below the dividend. Subtract it from the dividend.

$$
\begin{array}{r}
2x^3 \\
x + 2 \overline{)2x^4 + 0x^3 + x^2 + 0x - 3} \\
\underline{2x^4 + 4x^3} \\
-4x^3 + x^2 + 0x - 3
\end{array}
$$

6. Repeat steps 4 and 5, each time using the new dividend obtained, until the degree of the remainder is less than the degree of the divisor.

$$
\begin{array}{r}
2x^3 - 4x^2 + 9x - 18 \\
x + 2 \overline{)2x^4 + 0x^3 + x^2 + 0x - 3} \\
\underline{2x^4 + 4x^3} \\
-4x^3 + x^2 + 0x - 3 \\
\underline{-4x^3 - 8x^2} \\
9x^2 + 0x - 3 \\
\underline{9x^2 + 18x} \\
-18x - 3 \\
\underline{-18x - 36} \\
33
\end{array}
$$

Remainder

The result is written

$$\frac{2x^4 + x^2 - 3}{x + 2} = 2x^3 - 4x^2 + 9x - 18 + \frac{33}{x + 2}$$

Note that this method is used only for *polynomials* (multinomials in which all the powers of the variables are positive integers).

3.8 FACTORING

Factoring is the process of finding the factors of an expression. An expression is called *prime* if it has no factors other than 1 and itself.

(a) Common Factors. If each term in an expression contains the same quantity, that quantity may be *factored out*.

Common Factor	$ab + ac + ad = a(b + c + d)$	**30**

Example 3.56: Factor $x^3 - 2x + x^2$.

Solution: Since each of the three terms contains an x,

$$x^3 - 2x + x^2 = x(x^2 - 2 + x)$$

Example 3.57: Factor $3xy^2 - 9x^3y + 6x^2y^2$.

Solution: Each term contains a $3xy$, so

$$3xy^2 - 9x^3y + 6x^2y^2 = 3xy(y - 3x^2 + 2xy)$$

(b) Checking. *To check* if factoring has been done properly, merely see if the product of the factors equals the original expression.

Example 3.58: Are $x + 5$ and $x + 1$ the factors of $x^2 + 6x + 5$?

Solution: Multiplying the factors,

$$(x + 5)(x + 1) = x^2 + x + 5x + 5$$
$$= x^2 + 6x + 5 \quad \text{checks}$$

(c) Difference of Two Squares. The product of $(a - b)$ and $(a + b)$ is

$$(a - b)(a + b) = a^2 + ab - ab - b^2 \quad \text{or} \quad a^2 - b^2$$

Difference of Two Squares	$a^2 - b^2 = (a - b)(a + b)$	**31**

Example 3.59

$$x^2 - 16 = (x + 4)(x - 4)$$

Example 3.60

$$4b^2 - 9a^2 = (2b + 3a)(2b - 3a)$$

(d) Trinomials—Test for Factorability. The general quadratic trinomial is of the form

$$Ax^2 + Bx + C$$

where A, B, and C are not zero. Not all trinomials can be factored into rational factors. A trinomial is factorable if

Test for Factorability	$B^2 - 4AC$ = perfect square	**34**

Example 3.61:   Is the trinomial $5x^2 + 13x + 6$ factorable?

Solution: From Eq. 34, with $A = 5$, $B = 13$, and $C = 6$,

$$(13)^2 - 4(5)(6) = 169 - 120$$
$$= 49 = 7^2$$

So this trinomial is factorable. Its factors are

$$(5x + 3)(x + 2)$$

(e) Trinomials with a Leading Coefficient of 1. If we multiply $(x + a)$ and $(x + b)$,

Trinomials: Leading Coefficient = 1	$(x + a)(x + b) = x^2 + (a + b)x + ab$	**35**

we get a trinomial with *leading coefficient* (the coefficient of x^2) equal to 1; with the coefficient of the middle term equal to the sum of a and b; and with the last term equal to the product of a and b.

Example 3.62:   Factor $x^2 + 5x + 6$.

Solution: From Eq. 35, this trinomial, if factorable, will factor as

$$(x + a)(x + b)$$

where a and b have a sum of 5 and a product of 6.   The integers 2 and 3 meet this condition, and so

$$x^2 + 5x + 6 = (x + 2)(x + 3)$$

The *algebraic signs* of the middle and last terms of the trinomial give clues as to the factors.

1. If the sign of the last term is positive, a and b must have the same sign.
2. If the sign of the last term is negative, the sign of the middle term will tell which of the two numbers a and b is the larger, the positive one or the negative.

Example 3.63:   Factor $x^2 + x - 12$.

Solution: Since the last term is negative, a and b must have opposite signs; so the factors will have the form

$$(x + a)(x - b)$$

Since the coefficient of the middle term is $+1$, a must be greater than b by one unit. Two numbers having a product of 12 and a difference of 1 are 3 and 4, and so

$$x^2 + x - 12 = (x - 3)(x + 4)$$

(f) Trinomials with Leading Coefficient other than One. If we multiply $ax + b$ and $cx + d$,

General Quadratic Trinomial	$(ax + b)(cx + d) = acx^2 + (ad + bc)x + bd$	**36**

we get a trinomial with a leading coefficient of ac, a constant term of bd, and a middle coefficient of $(ad + bc)$.

One method of factoring such a trinomial is by *trial and error*, seeking four numbers, a, b, c, and d, that meet the requirements.

Example 3.64:   Factor $3x^2 + 10x + 3$.

Solution: The leading coefficient $ac = 3$ and the constant term $bd = 3$.   Try $a = 1, c = 3$,   and $b = 1$, $d = 3$.   Then $ad + bc = 1(3) + 1(3) = 6$, instead of the required 10.

Try $a = 1, b = 3, c = 3$, and $d = 1$.   Then $ad + bc = 1(1) + 3(3) = 10$,   so

$$3x^2 + 10x + 3 = (x + 3)(3x + 1)$$

A second method, which eliminates trial and error, is the *grouping method*.

Example 3.65:   Factor $3x^2 - 16x - 12$.

Solution

1. Multiply the leading coefficient and the constant term.

$$3(-12) = -36$$

2. Factor this number into all possible combinations of two factors.

$-1, 36$	$-4, 9$	$2, -18$
$-2, 18$	$-6, 6$	$3, -12$
$-3, 12$	$1, -36$	$4, -9$

3. Find the pair of factors whose algebraic sum equals the middle coefficient.

$$-16x = 2x - 18x$$

4. Rewrite the trinomial, splitting the middle term according to the selected factors.

$$3x^2 + 2x - 18x - 12$$

Group the first two terms together, and the last two terms together.

$$(3x^2 + 2x) + (-18x - 12)$$

5. Remove common factors from each grouping.

$$x(3x + 2) - 6(3x + 2)$$

6. Remove the common factor $(3x + 2)$ from the entire expression.

$$(3x + 2)(x - 6)$$

Example 3.66:   Factor $2x^2 - 5x - 3$.

Solution

Step 1.

$$2(-3) = -6$$

Step 2.

$-1, 6$	$1, -6$
$-2, 3$	$2, -3$

Step 3.

$$-5x = x - 6x$$

Step 4.

$$2x^2 + x - 6x - 3$$
$$(2x^2 + x) + (-6x - 3)$$

Step 5.

$$x(2x + 1) - 3(2x + 1)$$

Step 6.

$$(2x + 1)(x - 3)$$

(g) Perfect Square Trinomials. When we square a binomial,

Perfect Square Trinomials	$(a + b)^2 = a^2 + 2ab + b^2$	**37**
	$(a - b)^2 = a^2 - 2ab + b^2$	**38**

we get a trinomial where

1. The first and last terms are perfect squares.
2. The middle term equals twice the product of the square roots of the outer terms.
3. The sign of the middle term is the same as the sign of b.

Once a trinomial is recognized as a perfect square, it is easily factored by inspection.

Example 3.67:   Factor $x^2 - 4x + 4$.

Solution: The outer terms are both perfect squares, and the middle term, $4x$, is twice the product of the square roots of the first and last terms, x and 2.   Thus we have a perfect square trinomial with $a = x$ and $b = 2$, so that

$$x^2 - 4x + 4 = (x - 2)^2$$

Example 3.68

$$9x^2 + 12xy + 4y^2 = (3x + 2y)^2$$

(h) Sum and Difference of Two Cubes. The sum of two cubes is obtained when we multiply $(a + b)$ and $(a^2 - ab + b^2)$.

Sum of Two Cubes	$(a + b)(a^2 - ab + b^2) = a^3 + b^3$	**32**

Similarly

Difference of Two Cubes	$(a - b)(a^2 + ab + b^2) = a^3 - b^3$	**33**

Once recognized, they are factored by inspection.

Example 3.69:   Factor $x^3 + 8$.

Solution: This is the sum of two cubes, where

$$a^3 = x^3 \qquad \text{and} \qquad b^3 = 2^3$$
$$a = x \qquad\qquad\qquad b = 2$$

so that, by Eq. 32,

$$x^3 + 8 = (x + 2)(x^2 - 2x + 4)$$

Example 3.70

$$8y^3 - 27z^3 = (2y - 3z)(4y^2 + 6yz + 9z^2)$$

Common Error	The middle term of the trinomials in Eqs. 32 and 33 is often mistaken as $2ab$. $$a^3 + b^3 \neq (a + b)(a^2 - 2ab + b^2)$$ $\llcorner$no!

(i) Factoring by Grouping

Example 3.71: Factor $xy + 4x + 3y + 12$.

Solution: Group the two terms containing the factor x and the two containing the factor 3. Remove the common factor from each pair of terms.

$$x(y + 4) + 3(y + 4)$$

Both terms now have the common factor $y + 4$ which we factor out.

$$(y + 4)(x + 3)$$

Example 3.72

$$ax - 3 + x - 3a = (ax + x) + (-3a - 3)$$
$$= x(a + 1) - 3(a + 1)$$
$$= (a + 1)(x - 3)$$

(j) Factoring Completely. After factoring an expression, see if any of the factors can be factored *again*.

Example 3.73: Factor $x^2y - y$.

Solution: Remove the common factor

$$y(x^2 - 1)$$

then factor the difference of two squares

$$y(x + 1)(x - 1)$$

Example 3.74

$$a^4 - 8a^2 + 16 = (a^2 - 4)^2 = (a^2 - 4)(a^2 - 4)$$
$$= (a - 2)(a + 2)(a - 2)(a + 2)$$

When factoring completely, we do *not* usually factor expressions like x^2 into $x \cdot x$, or factor integers such as 6 into $2 \cdot 3$.

3.9 FURTHER OPERATIONS WITH FRACTIONS

(a) Simplifying Fractions by Changing Signs

Example 3.75: Simplify $(x - y)/(y - x)$.

Solution: Changing both the sign of the entire fraction and the sign of the denominator,

$$\frac{x - y}{y - x} = -\frac{x - y}{-(y - x)}$$
$$= -\frac{x - y}{x - y} = -1$$

(b) Simplifying Fractions (Reducing to Lowest Terms). Factor the numerator and denominator. Divide both the numerator and denominator by any factor that is *contained in both*.

Simplifying Fractions	$\dfrac{ad}{bd} = \dfrac{a}{b}$	**41**

Example 3.76

$$\frac{x^2 - 2x}{x + 3x^3} = \frac{x(x - 2)}{x(1 + 3x^2)} = \frac{x - 2}{1 + 3x^2}$$

Common Error	Note that we cannot *add* the same quantity to numerator and denominator (or *subtract* the same quantity) without changing the value of the fraction. Thus
	$\dfrac{1}{3} \neq \dfrac{1 + 1}{3 + 1} \neq \dfrac{2}{4} \neq \dfrac{1}{2}$
and	$\dfrac{3}{5} \neq \dfrac{3 - 1}{5 - 1} \neq \dfrac{2}{4} \neq \dfrac{1}{2}$

(c) Canceling. *Canceling* is the process of dividing both numerator and denominator by any factors *common to both*, and hence eliminating them.

Example 3.77:   Simplify

$$\frac{3xy - 2wx}{x^2y + wx}$$

Solution: The factor x is present in both terms of the numerator as well as both terms of the denominator, so

$$\frac{3xy - 2wx}{x^2y + wx} = \frac{x(3y - 2w)}{x(xy + w)} = \frac{3y - 2w}{xy + w}$$

Common Error	If a factor is missing from *even one term* in the numerator or denominator, it *cannot* be canceled.
	$\dfrac{xy - z}{wx} \neq \dfrac{y - z}{w}$

(d) Multiplying Fractions. The product of two fractions is a fraction whose numerator is the product of the numerators of the original fractions, and whose denominator is the product of the denominators of the original fractions.

Multiplying Fractions	$\dfrac{a}{b} \cdot \dfrac{c}{d} = \dfrac{ac}{bd}$	**42**

Example 3.78

$$\frac{3x}{2y} \cdot \frac{w}{z} = \frac{3xw}{2yz}$$

Example 3.79

$$\frac{2x}{x^2 - 1} \cdot \frac{3x - 3}{4x^2} = \frac{2x(3)(x - 1)}{(x - 1)(x + 1)4x^2} = \frac{3}{2x(x + 1)}$$

(e) Division of Fractions.   Invert the divisor and multiply.

| Division of Fractions | $\dfrac{a}{b} \div \dfrac{c}{d} = \dfrac{a}{b} \cdot \dfrac{d}{c} = \dfrac{ad}{bc}$ | **43** |

Example 3.80

$$\frac{x^2}{y} \div \frac{2w}{z} = \frac{x^2}{y} \cdot \frac{z}{2w} = \frac{x^2 z}{2yw}$$

Example 3.81

$$\frac{x^2 - 9}{2x} \div \frac{x + 3}{4x^2} = \frac{(x^2 - 9)(4x^2)}{2x(x + 3)} = \frac{(x - 3)(x + 3)(4x^2)}{2x(x + 3)} = 2x(x - 3)$$

(f) Addition and Subtraction of Fractions

1. Fractions having a *common denominator*: The algebraic sum will be a fraction whose numerator is the algebraic sum of the numerators and whose denominator is the common denominator.

| Addition and Subtraction of Fractions | $\dfrac{a}{b} \pm \dfrac{c}{b} = \dfrac{a \pm c}{b}$ | **44** |

Example 3.82

$$\frac{2}{3x} + \frac{4y}{3x} - \frac{z}{3x} = \frac{2 + 4y - z}{3x}$$

2. Fractions with different denominators: Determine the *lowest common denominator* (LCD). Multiply numerator and denominator of each fraction by the quantity that will make its denominator equal to the LCD.   Then combine as shown above.

| Adding Fractions with Different Denominators | $\dfrac{a}{b} \pm \dfrac{c}{d} = \dfrac{ad}{bd} \pm \dfrac{bc}{bd} = \dfrac{ad \pm bc}{bd}$ | **45** |

Example 3.83

$$\frac{x}{3} + \frac{2}{x} = \frac{x}{3} \cdot \frac{x}{x} + \frac{2}{x} \cdot \frac{3}{3}$$

$$= \frac{x^2}{3x} + \frac{6}{3x} = \frac{x^2 + 6}{3x}$$

Example 3.84

$$\frac{5}{x - 3} - \frac{2}{x} + \frac{x}{3(x - 3)} = \frac{5}{x - 3} \cdot \frac{3x}{3x} - \frac{2}{x} \cdot \frac{3(x - 3)}{3(x - 3)} + \frac{x}{3(x - 3)} \cdot \frac{x}{x}$$

$$= \frac{15x}{3x(x - 3)} - \frac{6(x - 3)}{3x(x - 3)} + \frac{x^2}{3x(x - 3)}$$

$$= \frac{15x - 6x + 18 + x^2}{3x(x - 3)} = \frac{x^2 + 9x + 18}{3x(x - 3)}$$

(g) Complex Fractions. Fractions are called *simple* when they have one fraction line, and *complex* (or *compound*) when they have more than one.

Example 3.85

$$\frac{\frac{x}{2}+\frac{x}{3}}{2x}$$

is a complex fraction.

Simplify complex fractions by reducing the numerator and denominator to simple fractions, and then dividing numerator by denominator as in Eq. 9.

Example 3.86

$$\frac{\frac{x}{2}+\frac{x}{3}}{2x}=\frac{\frac{3x+2x}{6}}{2x}=\frac{\frac{5x}{6}}{2x}=\frac{5x}{12x}=\frac{5}{12}$$

Solved Problems

EXPONENTS

3.1   Simplify the following expressions.   Do not use negative exponents.

(a) $B \cdot B^6$ (f) $10^a \cdot 10^b$ (k) 3^3 (p) $\left(-\frac{2}{3}\right)^3$ (u) $(3x^2y)^3$

(b) $a^2 \cdot a^3$ (g) $\dfrac{x^3}{x}$ (l) $(-2)^4$ (q) $(w^2)^3$ (v) $(5^x)^y$

(c) $x^2 \cdot x^9$ (h) $\dfrac{y^{10}}{y^5}$ (m) $(|-4|)^3$ (r) $(x^4)^5$

(d) $10^3 \cdot 10^4$ (i) $\dfrac{10^4}{10^2}$ (n) $-\left(\dfrac{2}{3}\right)^2$ (s) $(3^a)^2$

(e) $b^{2+x} \cdot b^x$ (j) $\dfrac{4^3}{4^2}$ (o) $(0.1)^3$ (t) $(4^p)^q$

Solution

(a) $B \cdot B^6 = B^7$ (e) $b^{2+x} \cdot b^x = b^{2+x+x} = b^{2+2x}$ (i) $\dfrac{10^4}{10^2} = 10^2$

(b) $a^2 \cdot a^3 = a^5$ (f) $10^a \cdot 10^b = 10^{a+b}$ (j) $\dfrac{4^3}{4^2} = 4$

(c) $x^2 \cdot x^9 = x^{11}$ (g) $\dfrac{x^3}{x} = x^2$ (k) $3^3 = 27$

(d) $10^3 \cdot 10^4 = 10^7$ (h) $\dfrac{y^{10}}{y^5} = y^5$ (l) $(-2)^4 = 16$

(m) $(|-4|)^3 = 4^3 = 64$ (q) $(w^2)^3 = w^6$ (t) $(4^p)^q = 4^{pq}$

(n) $-\left(\dfrac{2}{3}\right)^2 = -\dfrac{4}{9}$ (r) $(x^4)^5 = x^{20}$ (u) $(3x^2y)^3 = 27x^6y^3$

(o) $(0.1)^3 = 0.001$ (s) $(3^a)^2 = 3^{2a}$ (v) $(5^x)^y = 5^{xy}$

(p) $\left(-\dfrac{2}{3}\right)^3 = -\dfrac{8}{27}$

3.2	Write the following expressions with positive exponents only and simplify.

(a) x^{-3} (c) $(-b)^{-2}$ (e) $\left(\dfrac{2}{x}\right)^{-2}$ (g) $x^{-3}yz^{-2}$ (i) $\left(\dfrac{3x}{2y^2}\right)^{-2}\left(\dfrac{y}{3}\right)^{-3}$

(b) $(3x)^{-2}$ (d) $x^{-2}y^3$ (f) $(3-y)^{-3}$ (h) $3a^{-2} + 3y^{-2}$ (j) $\dfrac{x+y^{-1}}{y}$

Solution

(a) $x^{-3} = \dfrac{1}{x^3}$ (e) $\left(\dfrac{2}{x}\right)^{-2} = \dfrac{1}{\left(\dfrac{2}{x}\right)^2} = \dfrac{1}{\dfrac{4}{x^2}} = \dfrac{x^2}{4}$

(b) $(3x)^{-2} = \dfrac{1}{(3x)^2} = \dfrac{1}{9x^2}$ (f) $(3-y)^{-3} = \dfrac{1}{(3-y)^3}$

(c) $(-b)^{-2} = \dfrac{1}{b^2}$ (g) $x^{-3}yz^{-2} = \dfrac{y}{x^3z^2}$

(d) $x^{-2}y^3 = \dfrac{y^3}{x^2}$ (h) $3a^{-2} + 3y^{-2} = \dfrac{3}{a^2} + \dfrac{3}{y^2}$

(i) $\left(\dfrac{3x}{2y^2}\right)^{-2}\left(\dfrac{y}{3}\right)^{-3} = \dfrac{(3x)^{-2}}{(2y^2)^{-2}} \cdot \dfrac{y^{-3}}{3^{-3}}$

$= \dfrac{(2y^2)^2}{(3x)^2} \cdot \dfrac{3^3}{y^3} = \dfrac{4y^4}{9x^2} \cdot \dfrac{27}{y^3}$

$= \dfrac{12y}{x^2}$

(j) $\dfrac{x+y^{-1}}{y} = \dfrac{x+\dfrac{1}{y}}{y}$

$= \dfrac{x}{y} + \dfrac{1}{y^2}$

3.3	Express the following without fractions, using negative exponents where needed.

(a) $\dfrac{1}{y^2}$ (b) $\dfrac{x^{-1}}{y}$ (c) $\dfrac{a^2}{b^2}$ (d) $\dfrac{1}{w^3}$ (e) $\dfrac{a^2b^{-2}}{c^{-3}}$ (f) $\dfrac{w^5x^{-2}}{z^{-3}}$

Solution

(a) $\dfrac{1}{y^2} = y^{-2}$ (c) $\dfrac{a^2}{b^2} = a^2b^{-2}$ (e) $\dfrac{a^2b^{-2}}{c^{-3}} = a^2b^{-2}c^3$

(b) $\dfrac{x^{-1}}{y} = x^{-1}y^{-1}$ (d) $\dfrac{1}{w^3} = w^{-3}$ (f) $\dfrac{w^5x^{-2}}{z^{-3}} = w^5x^{-2}z^3$

3.4   Evaluate:

(a) $\left(\dfrac{25}{3}\right)^0$ (b) $\dfrac{10^0}{10^3}$ (c) $\dfrac{x^0}{5}$

Solution

(a) $\left(\dfrac{25}{3}\right)^0 = 1$ (b) $\dfrac{10^0}{10^3} = \dfrac{1}{1000} = 0.001$ (c) $\dfrac{x^0}{5} = \dfrac{1}{5}$

REMOVING SYMBOLS OF GROUPING

3.5   Remove the symbols of grouping and simplify.

(a) $5 + (x - y)$ (f) $-2[x - 3(2w - z)] + 6$

(b) $3 - (a - b)$ (g) $-7[-2(3x - y) - w] + x$

(c) $-2(M - N) + 6$ (h) $(x - y) - \{[x - (2 - 3y)] - x\}$

(d) $-(x - y) - (x + y)$ (i) $-\{-[-(w - 3) - x]\} + w$

(e) $6 - 3[x - 2(x - y)]$ (j) $3x - \{4x - [2x - (3x - 9) - 6] + 2\}$

Solution

(a) $5 + (x - y) = 5 + x - y$ (c) $-2(M - N) + 6 = -2M + 2N + 6$

(b) $3 - (a - b) = 3 - a + b$ (d) $-(x - y) - (x + y) = -x + y - x - y = -2x$

(e) $\begin{aligned}6 - 3[x - 2(x - y)] &= 6 - 3[x - 2x + 2y]\\ &= 6 - 3[-x + 2y]\\ &= 6 + 3x - 6y\end{aligned}$

(f) $\begin{aligned}-2[x - 3(2w - z)] + 6 &= -2[x - 6w + 3z] + 6\\ &= -2x + 12w - 6z + 6\end{aligned}$

(g) $\begin{aligned}-7[-2(3x - y) - w] + x &= -7[-6x + 2y - w] + x\\ &= 42x - 14y + 7w + x\\ &= 43x - 14y + 7w\end{aligned}$

(h) $\begin{aligned}(x - y) - \{[x - (2 - 3y)] - x\} &= x - y - \{x - 2 + 3y - x\}\\ &= x - y + 2 - 3y\\ &= x - 4y + 2\end{aligned}$

(i) $\begin{aligned}-\{-[-(w - 3) - x]\} + w &= -\{-[-w + 3 - x]\} + w\\ &= -\{w - 3 + x\} + w\\ &= -w + 3 - x + w\\ &= 3 - x\end{aligned}$

(j) $\begin{aligned}3x - \{4x - [2x - (3x - 9) - 6] + 2\} &= 3x - \{4x - [2x - 3x + 9 - 6] + 2\}\\ &= 3x - \{4x + x - 3 + 2\}\\ &= 3x - 5x + 1\\ &= 1 - 2x\end{aligned}$

ADDITION AND SUBTRACTION OF ALGEBRAIC EXPRESSIONS

3.6   Combine the following algebraic expressions as indicated and simplify.

(a) $5x + 3x$ (c) $2xy - 4xy + 7xy$ (e) $4x - 5x + 2y$ (g) $2.4x - 0.7x + 1.1x$

(b) $-7w + 3w$ (d) $3z - z - \dfrac{z}{2}$ (f) $\dfrac{3}{8}x + \dfrac{1}{4}x - \dfrac{9}{16}x$

Solution

(a) $5x + 3x = 8x$

(b) $-7w + 3w = -4w$

(c) $2xy - 4xy + 7xy = 5xy$

(d) $3z - z - \dfrac{z}{2} = \dfrac{6z}{2} - \dfrac{2z}{2} - \dfrac{z}{2} = \dfrac{3z}{2}$

(e) $4x - 5x + 2y = 2y - x$

(f) $\dfrac{3}{8}x + \dfrac{1}{4}x - \dfrac{9}{16}x = \dfrac{6x}{16} + \dfrac{4x}{16} - \dfrac{9x}{16} = \dfrac{x}{16}$

(g) $2.4x - 0.7x + 1.1x = 2.8x$

3.7 Combine the following expressions as indicated and simplify.

(a) $(3x + 5) + (2x - 3)$

(b) $(6 - x) - (7 + x)$

(c) $(2x^4 + 3x^3 - 4) + (3x^4 - 2x^3 - 8)$

(d) $(3xy - 2y + 3z) + (-3y + 8z) - (-3xy - z)$

(e) $(2x - 3y) - (5x - 3y - 2z)$

(f) $(-2x^2 - x + 6) - (7x^2 - 2x + 4) + (x^2 - 3)$

(g) $(w + 2z) - (-3w - 5x) + (x + z) - (z - w)$

Solution

(a) $(3x + 5) + (2x - 3) = 5x + 2$

(b) $(6 - x) - (7 + x) = -1 - 2x$

(c) $(2x^4 + 3x^3 - 4) + (3x^4 - 2x^3 - 8) = 5x^4 + x^3 - 12$

(d) $(3xy - 2y + 3z) + (-3y + 8z) - (-3xy - z) = 6xy - 5y + 12z$

(e) $(2x - 3y) - (5x - 3y - 2z) = -3x + 2z$

(f) $(-2x^2 - x + 6) - (7x^2 - 2x + 4) + (x^2 - 3) = -8x^2 + x - 1$

(g) $(w + 2z) - (-3w - 5x) + (x + z) - (z - w) = 5w + 2z + 6x$

MULTIPLICATION OF ALGEBRAIC EXPRESSIONS

3.8 Multiply the following monomials and simplify.

(a) $x^4 \cdot x \cdot x^2$

(b) $(a^4 b^3)(b^2)$

(c) $5x(-2x)$

(d) $(-2x)(-3x^2)$

(e) $(3x^2 y)(2xy^2)(4x^2 y^2)$

(f) $(-x^2)(-y^2)(-z^2)$

(g) $(-ab)(ac^2)(-2b^2 c)$

Solution

(a) $x^4 \cdot x \cdot x^2 = x^7$

(b) $(a^4 b^3)(b^2) = a^4 b^5$

(c) $5x(-2x) = -10x^2$

(d) $(-2x)(-3x^2) = 6x^3$

(e) $(3x^2 y)(2xy^2)(4x^2 y^2) = 24x^5 y^5$

(f) $(-x^2)(-y^2)(-z^2) = -x^2 y^2 z^2$

(g) $(-ab)(ac^2)(-2b^2 c) = 2a^2 b^3 c^3$

3.9   Multiply the following multinomials by the given monomial.

    (a) $2x(3x - 2)$ (c) $(6ab)(2a - 3b - ab)$ (e) $-w(4w + 2a - w^2)$

    (b) $(-3x)(x^2 - 2x + 2)$ (d) $-abc(2a - 3b + 6)$ (f) $(xy - 2x^2y - 3xy^2)(xy)$

Solution

    (a) $2x(3x - 2) = 6x^2 - 4x$

    (b) $(-3x)(x^2 - 2x + 2) = -3x^3 + 6x^2 - 6x$

    (c) $(6ab)(2a - 3b - ab) = 12a^2b - 18ab^2 - 6a^2b^2$

    (d) $-abc(2a - 3b + 6) = -2a^2bc + 3ab^2c - 6abc$

    (e) $-w(4w + 2a - w^2) = -4w^2 - 2aw + w^3$

    (f) $(xy - 2x^2y - 3xy^2)(xy) = x^2y^2 - 2x^3y^2 - 3x^2y^3$

3.10   Find the following products and simplify.

    (a) $(2x + 2)(5x + 3)$ (f) $(-a + 2d)(-7a^2 + 4ad - 2d^2)$

    (b) $(3a - 2x)(2a + 5x)$ (g) $(x^2 - 2x - 3)(2x^2 + x + 4)$

    (c) $(a - 5)(a + 5)$ (h) $(a - b - c)(a + b + c)$

    (d) $(x - y)(2x + 3y - 5)$ (i) $(x - 2)(x^3 - 2x^2 + 3x + 4)$

    (e) $(2x^2 - 3x - 5)(2x + 2)$

Solution

    (a) $(2x + 2)(5x + 3) = 10x^2 + 6x + 10x + 6 = 10x^2 + 16x + 6$

    (b) $(3a - 2x)(2a + 5x) = 6a^2 + 15ax - 4ax - 10x^2 = 6a^2 + 11ax - 10x^2$

    (c) $(a - 5)(a + 5) = a^2 + 5a - 5a - 25 = a^2 - 25$

    (d) $(x - y)(2x + 3y - 5) = 2x^2 + 3xy - 5x - 2xy - 3y^2 + 5y = 2x^2 + xy - 5x - 3y^2 + 5y$

    (e) $(2x^2 - 3x - 5)(2x + 2) = 4x^3 - 6x^2 - 10x + 4x^2 - 6x - 10 = 4x^3 - 2x^2 - 16x - 10$

    (f) $(-a + 2d)(-7a^2 + 4ad - 2d^2) = 7a^3 - 4a^2d + 2ad^2 - 14a^2d + 8ad^2 - 4d^3$
$$= 7a^3 - 18a^2d + 10ad^2 - 4d^3$$

    (g) $(x^2 - 2x - 3)(2x^2 + x + 4) = 2x^4 +\ \ x^3 + 4x^2$
$$ - 4x^3 - 2x^2 -\ \ 8x$$
$$ - 6x^2 -\ \ 3x - 12$$
$$= 2x^4 - 3x^3 - 4x^2 - 11x - 12$$

    (h) $(a - b - c)(a + b + c) = a^2 + ab + ac$
$$ - ab - b^2 -\ \ bc$$
$$ - ac -\ \ bc - c^2$$
$$= a^2 - b^2 - 2bc - c^2$$

    (i) $(x - 2)(x^3 - 2x^2 + 3x + 4) = x^4 - 2x^3 + 3x^2 + 4x$
$$ - 2x^3 + 4x^2 - 6x - 8$$
$$= x^4 - 4x^3 + 7x^2 - 2x - 8$$

POWERS OF MULTINOMIALS

3.11   Raise the following expressions to the power indicated and simplify.

    (a) $(x - 2)^2$ (b) $(-3a + b)^2$ (c) $(2x - y)^3$ (d) $(x^2 - 2x + 3)^2$

Solution

(a) $(x - 2)^2 = (x - 2)(x - 2) = x^2 - 2x - 2x + 4 = x^2 - 4x + 4$

(b) $(-3a + b)^2 = (-3a + b)(-3a + b) = 9a^2 - 3ab - 3ab + b^2 = 9a^2 - 6ab + b^2$

(c) $(2x - y)^3 = (2x - y)(2x - y)(2x - y) = (2x - y)(4x^2 - 4xy + y^2)$
$$= 8x^3 - 8x^2y + 2xy^2 - 4x^2y + 4xy^2 - y^3 = 8x^3 - 12x^2y + 6xy^2 - y^3$$

(d) $(x^2 - 2x + 3)^2 = (x^2 - 2x + 3)(x^2 - 2x + 3)$
$$\begin{aligned}
&= x^4 - 2x^3 + 3x^2 \\
&\quad\ \ - 2x^3 + 4x^2 - 6x \\
&\quad\qquad\quad + 3x^2 - 6x + 9 \\
\hline
&= x^4 - 4x^3 + 10x^2 - 12x + 9
\end{aligned}$$

DIVISION OF ALGEBRAIC EXPRESSIONS

3.12 Perform the divisions and simplify.

(a) $x^5 \div x^3$ (c) $w^3 \div w^3$ (e) $\dfrac{2x^2y^3}{6xy^4}$

(b) $\dfrac{x^3}{x^6}$ (d) $(-y^6) \div (-y)^6$ (f) $\dfrac{24x^3y}{4x^3y^2}$

Solution

(a) $x^5 \div x^3 = \dfrac{x^5}{x^3} = x^2$ (c) $w^3 \div w^3 = 1$ (e) $\dfrac{2x^2y^3}{6xy^4} = \dfrac{x}{3y}$

(b) $\dfrac{x^3}{x^6} = \dfrac{1}{x^3} = x^{-3}$ (d) $(-y^6) \div (-y)^6 = \dfrac{-y^6}{y^6} = -1$ (f) $\dfrac{24x^3y}{4x^3y^2} = \dfrac{6}{y}$

3.13 Perform the divisions and simplify.

(a) $(12x^2 - 18x^4) \div 6x^2$ (c) $\dfrac{10a^2b^2 - 5ab}{10ab}$ (e) $\dfrac{7x^3 + 9x^2 + 14}{-14x^2}$

(b) $\dfrac{6x + 9}{3}$ (d) $(-18a^3b - 33a^2b + 27ab^3) \div 3a^2$ (f) $\dfrac{15x^2y^2 - 30xy}{-15xy}$

Solution

(a) $(12x^2 - 18x^4) \div 6x^2 = \dfrac{12x^2}{6x^2} - \dfrac{18x^4}{6x^2} = 2 - 3x^2$

(b) $\dfrac{6x + 9}{3} = 2x + 3$

(c) $\dfrac{10a^2b^2 - 5ab}{10ab} = ab - \dfrac{1}{2}$

(d) $(-18a^3b - 33a^2b + 27ab^3) \div 3a^2 = -6ab - 11b + \dfrac{9b^3}{a}$

(e) $\dfrac{7x^3 + 9x^2 + 14}{-14x^2} = -\dfrac{x}{2} - \dfrac{9}{14} - \dfrac{1}{x^2}$

(f) $\dfrac{15x^2y^2 - 30xy}{-15xy} = -xy + 2$

3.14   Perform the division $(6x^2 + 11x + 3) \div (3x + 1)$.

Solution

$$
\begin{array}{r}
2x + 3 \\
3x + 1 \overline{)\, 6x^2 + 11x + 3} \\
\underline{6x^2 + 2x} \\
9x + 3 \\
\underline{9x + 3}
\end{array}
$$

So

$$\frac{6x^2 + 11x + 3}{3x + 1} = 2x + 3$$

3.15   Perform the division $(x^2 - 8) \div (x - 2)$.

Solution

$$
\begin{array}{r}
x + 2 \\
x - 2 \overline{)\, x^2 - 8} \\
\underline{x^2 - 2x} \\
2x - 8 \\
\underline{2x - 4} \\
-4 = \text{remainder}
\end{array}
$$

So

$$\frac{x^2 - 8}{x - 2} = x + 2 - \frac{4}{x - 2}$$

3.16   Perform the division $(20a^2 - 13a + 15)/(5a + 3)$.

Solution

$$
\begin{array}{r}
4a - 5 \\
5a + 3 \overline{)\, 20a^2 - 13a + 15} \\
\underline{20a^2 + 12a} \\
-25a + 15 \\
\underline{-25a - 15} \\
30 = \text{remainder}
\end{array}
$$

So

$$\frac{20a^2 - 13a + 15}{5a + 3} = 4a - 5 + \frac{30}{5a + 3}$$

3.17   Divide $(30x^2 - 16x^3 - 45 + 21x) \div (2x - 3)$.

Solution

$$
\begin{array}{r}
-\,8x^2 + 3x + 15 \\
2x - 3 \overline{)\, -16x^3 + 30x^2 + 21x - 45} \\
\underline{-16x^3 + 24x^2} \\
6x^2 + 21x - 45 \\
\underline{6x^2 - 9x} \\
30x - 45 \\
\underline{30x - 45}
\end{array}
$$

So

$$\frac{30x^2 - 16x^3 - 45 + 21x}{2x - 3} = -8x^2 + 3x + 15$$

3.18   Divide $(a^4 - 3a^3 + 2a^2 + 3a - 1)/(a^2 + a - 1)$.

Solution

$$
\require{enclose}
\begin{array}{r}
a^2 - 4a + 7 \\
a^2 + a - 1\enclose{longdiv}{a^4 - 3a^3 + 2a^2 + 3a - 1} \\
\underline{a^4 +\; a^3 -\; a^2} \\
-4a^3 + 3a^2 + 3a - 1 \\
\underline{-4a^3 - 4a^2 + 4a} \\
7a^2 -\; a - 1 \\
\underline{7a^2 + 7a - 7} \\
-8a + 6 = \text{remainder}
\end{array}
$$

So $$\frac{a^4 - 3a^3 + 2a^2 + 3a - 1}{a^2 + a - 1} = a^2 - 4a + 7 + \frac{6 - 8a}{a^2 + a - 1}$$

FACTORING

3.19   Factor completely.

(a) $10y - 5x$ (f) $y^3 - 6y^2$

(b) $21xy - 7x$ (g) $x^6 - 3x^4 - 5x^2$

(c) $2b^3 + b^2 - 2b$ (h) $4x^2y - 32x^2y^2$

(d) $2a^3b + 6a^2b^2 - 8ab^3$ (i) $6(a + b) - 18(a + b)^2$

(e) $a^2x^2 - 4bx^2 + 3cx^2 + 5x^2$ (j) $2(x + y)(w - z)^2 + 8(x + y)^2(w - z)$

Solution

(a) $10y - 5x = 5(2y - x)$

(b) $21xy - 7x = 7x(3y - 1)$

(c) $2b^3 + b^2 - 2b = b(2b^2 + b - 2)$

(d) $2a^3b + 6a^2b^2 - 8ab^3 = 2ab(a^2 + 3ab - 4b^2) = 2ab(a - b)(a + 4b)$

(e) $a^2x^2 - 4bx^2 + 3cx^2 + 5x^2 = x^2(a^2 - 4b + 3c + 5)$

(f) $y^3 - 6y^2 = y^2(y - 6)$

(g) $x^6 - 3x^4 - 5x^2 = x^2(x^4 - 3x^2 - 5)$

(h) $4x^2y - 32x^2y^2 = 4x^2y(1 - 8y)$

(i) $6(a + b) - 18(a + b)^2 = (a + b)[6 - 18(a + b)] = (a + b)(6 - 18a - 18b)$

(j) $2(x + y)(w - z)^2 + 8(x + y)^2(w - z) = 2(x + y)(w - z)[(w - z) + 4(x + y)]$
$$= 2(x + y)(w - z)(w - z + 4x + 4y)$$

3.20   Factor completely.

(a) $x^2 - y^2$ (d) $1 - x^6$ (g) $\dfrac{18x^2}{25} - \dfrac{32y^2}{49}$ (j) $16x^4y^2 - w^2z^4$

(b) $16x^2 - 1$ (e) $5x^4 - 20y^2$ (h) $\dfrac{2a^2}{5} - \dfrac{8b^2}{5}$ (k) $36 - (w - 3)^2$

(c) $x^4 - b^4$ (f) $(a - 6)^2 - 25$ (i) $a^2b^4 - 81a^4b^2$

Solution

(a) $x^2 - y^2 = (x + y)(x - y)$

(b) $16x^2 - 1 = (4x + 1)(4x - 1)$

(c) $x^4 - b^4 = (x^2 + b^2)(x^2 - b^2) = (x^2 + b^2)(x + b)(x - b)$

(d) $1 - x^6 = (1 + x^3)(1 - x^3) = (1 + x)(1 - x + x^2)(1 - x)(1 + x + x^2)$

(e) $5x^4 - 20y^2 = 5(x^4 - 4y^2) = 5(x^2 + 2y)(x^2 - 2y)$

(f) $(a - 6)^2 - 25 = [(a - 6) + 5][(a - 6) - 5] = (a - 1)(a - 11)$

(g) $\dfrac{18x^2}{25} - \dfrac{32y^2}{49} = 2\left(\dfrac{9x^2}{25} - \dfrac{16y^2}{49}\right) = 2\left(\dfrac{3x}{5} + \dfrac{4y}{7}\right)\left(\dfrac{3x}{5} - \dfrac{4y}{7}\right)$

(h) $\dfrac{2a^2}{5} - \dfrac{8b^2}{5} = \dfrac{2}{5}(a^2 - 4b^2) = \dfrac{2}{5}(a + 2b)(a - 2b)$

(i) $a^2b^4 - 81a^4b^2 = a^2b^2(b^2 - 81a^2) = a^2b^2(b + 9a)(b - 9a)$

(j) $16x^4y^2 - w^2z^4 = (4x^2y + wz^2)(4x^2y - wz^2)$

(k) $36 - (w - 3)^2 = [6 + (w - 3)][6 - (w - 3)] = (3 + w)(9 - w)$

3.21   Test the following trinomials for factorability.

(a) $x^2 - 30x - 64$ (c) $2x^2 - 12x + 18$ (e) $3x^2 + 7x + 2$

(b) $y^2 + 2y - 7$ (d) $3a^2 - 5a - 9$ (f) $5w^2 - 2w - 1$

Solution

(a) $x^2 - 30x - 64$

$$b^2 - 4ac = (-30)^2 - 4(1)(-64) = 1156 = (34)^2$$
$$\text{Factorable}$$

(b) $y^2 + 2y - 7$

$$b^2 - 4ac = 2^2 - 4(1)(-7) = 32$$
$$\text{Not factorable}$$

(c) $2x^2 - 12x + 18$

$$b^2 - 4ac = (-12)^2 - 4(2)(18) = 0$$
$$\text{Factorable}$$

(d) $3a^2 - 5a - 9$

$$b^2 - 4ac = 25 - 4(3)(-9) = 133$$
$$\text{Not factorable}$$

(e) $3x^2 + 7x + 2$

$$b^2 - 4ac = 49 - 4(3)(2) = 25 = 5^2$$
$$\text{Factorable}$$

(f) $5w^2 - 2w - 1$

$$b^2 - 4ac = 4 - 4(5)(-1) = 24$$
$$\text{Not factorable}$$

3.22   Factor completely.

(a) $a^2 + 5a + 6$ (e) $x^2 - 15x + 14$ (i) $x^2 + 22x - 48$

(b) $x^2 - 7x + 12$ (f) $w^2 - 7w - 8$ (j) $w^2 - 30w - 64$

(c) $x^4 - 6x^2 + 8$ (g) $y^2 + 13y + 12$

(d) $a^2 + 7a + 10$ (h) $z^2 - 7z - 18$

Solution

(a) $a^2 + 5a + 6 = (a + 3)(a + 2)$

(b) $x^2 - 7x + 12 = (x - 4)(x - 3)$

(c) $x^4 - 6x^2 + 8 = (x^2 - 2)(x^2 - 4) = (x^2 - 2)(x + 2)(x - 2)$

(d) $a^2 + 7a + 10 = (a + 5)(a + 2)$

(e) $x^2 - 15x + 14 = (x - 14)(x - 1)$

(f) $w^2 - 7w - 8 = (w + 1)(w - 8)$

(g) $y^2 + 13y + 12 = (y + 1)(y + 12)$

(h) $z^2 - 7z - 18 = (z + 2)(z - 9)$

(i) $x^2 + 22x - 48 = (x + 24)(x - 2)$

(j) $w^2 - 30w - 64 = (w - 32)(w + 2)$

3.23   Factor completely.

(a) $2x^2 + x - 6$ (e) $12b^2 - b - 6$ (i) $6x^3 + 3x^2 - 18x$

(b) $6a^2 - 7a + 2$ (f) $10x^2 + xy - 2y^2$ (j) $-18 + 42a^2 - 24a$

(c) $3w^2 - 3w - 6$ (g) $3b^2d + 5bcd + 2c^2d$

(d) $2x^4 + x^2 - 3$ (h) $\dfrac{x^2}{6} - \dfrac{5x}{6} - 6$

Solution

(a) $2x^2 + x - 6 = (x + 2)(2x - 3)$

(b) $6a^2 - 7a + 2 = (2a - 1)(3a - 2)$

(c) $3w^2 - 3w - 6 = (w - 2)(3w + 3) = 3(w - 2)(w + 1)$

(d) $2x^4 + x^2 - 3 = (x^2 - 1)(2x^2 + 3) = (x + 1)(x - 1)(2x^2 + 3)$

(e) $12b^2 - b - 6 = (3b + 2)(4b - 3)$

(f) $10x^2 + xy - 2y^2 = (2x + y)(5x - 2y)$

(g) $3b^2d + 5bcd + 2c^2d = d(3b^2 + 5bc + 2c^2) = d(3b + 2c)(b + c)$

(h) $\dfrac{x^2}{6} - \dfrac{5x}{6} - 6 = \dfrac{1}{6}(x^2 - 5x - 36) = \dfrac{1}{6}(x - 9)(x + 4)$

(i) $6x^3 + 3x^2 - 18x = 3x(2x^2 + x - 6) = 3x(x + 2)(2x - 3)$

(j) $-18 + 42a^2 - 24a = 6(-3 + 7a^2 - 4a) = 6(7a^2 - 4a - 3) = 6(7a + 3)(a - 1)$

3.24   Factor completely.

(a) $x^2 - 2xy + y^2$ (e) $a^2x^2 + 2abx + b^2$ (i) $a^2b^2 - 8ab^3 + 16b^4$

(b) $a^2 + 4a + 4$ (f) $x^4 - 2x^2y^2 + y^4$ (j) $9x^2y^2 - 6xy + 1$

(c) $x^2 + 2xy + y^2$ (g) $x^2 - 14xy + 49y^2$

(d) $2a^2 - 12a + 18$ (h) $9 - 12x + 4x^2$

Solution

(a) $x^2 - 2xy + y^2 = (x - y)^2$

(b) $a^2 + 4a + 4 = (a + 2)^2$

(c) $x^2 + 2xy + y^2 = (x + y)^2$

(d) $2a^2 - 12a + 18 = 2(a^2 - 6a + 9) = 2(a - 3)^2$

(e) $a^2x^2 + 2abx + b^2 = (ax + b)^2$

(f) $x^4 - 2x^2y^2 + y^4 = (x^2 - y^2)^2 = (x^2 - y^2)(x^2 - y^2)$
$$= (x - y)(x + y)(x - y)(x + y)$$

(g) $x^2 - 14xy + 49y^2 = (x - 7y)^2$

(h) $9 - 12x + 4x^2 = (3 - 2x)^2$

(i) $a^2b^2 - 8ab^3 + 16b^4 = b^2(a^2 - 8ab + 16b^2) = b^2(a - 4b)^2$

(j) $9x^2y^2 - 6xy + 1 = (3xy - 1)^2$

3.25   Factor completely.

(a) $x^3 - 27$ (c) $1 - 64x^3$ (e) $8x^3y^3 - 27$ (g) $(a - 1)^3 + 1$

(b) $64 + y^3$ (d) $x^4 + xy^3$ (f) $2x^3 - 16$

Solution

(a) $x^3 - 27 = (x - 3)(x^2 + 3x + 9)$

(b) $64 + y^3 = (4 + y)(16 - 4y + y^2)$

(c) $1 - 64x^3 = (1 - 4x)(1 + 4x + 16x^2)$

(d) $x^4 + xy^3 = x(x^3 + y^3) = x(x + y)(x^2 - xy + y^2)$

(e) $8x^3y^3 - 27 = (2xy - 3)(4x^2y^2 + 6xy + 9)$

(f) $2x^3 - 16 = 2(x^3 - 8) = 2(x - 2)(x^2 + 2x + 4)$

(g) $(a - 1)^3 + 1 = [(a - 1) + 1][(a - 1)^2 - (a - 1) + 1]$
$$= a(a^2 - 2a + 1 - a + 1 + 1) = a(a^2 - 3a + 3)$$

3.26   Factor by grouping.

(a) $ax + bx + 3a + 3b$ (e) $3x - 2y - 6 + xy$

(b) $2xy + wy - wz - 2xz$ (f) $x^2y^2 - 3x^2 - 4y^2 + 12$

(c) $a^3 + 3a^2 + 4a + 12$ (g) $3a^2 - ab - 2b^2 + 3a + 2b$

(d) $ab + a - b - 1$

Solution

(a) $ax + bx + 3a + 3b = x(a + b) + 3(a + b) = (x + 3)(a + b)$

(b) $2xy + wy - wz - 2xz = 2x(y - z) + w(y - z) = (2x + w)(y - z)$

(c) $a^3 + 3a^2 + 4a + 12 = a^2(a + 3) + 4(a + 3) = (a^2 + 4)(a + 3)$

(d) $ab + a - b - 1 = a(b + 1) - (b + 1) = (a - 1)(b + 1)$

(e) $3x - 2y - 6 + xy = x(3 + y) - 2(3 + y) = (x - 2)(y + 3)$

(f) $x^2y^2 - 3x^2 - 4y^2 + 12 = x^2(y^2 - 3) - 4(y^2 - 3) = (x^2 - 4)(y^2 - 3)$

(g) $3a^2 - ab - 2b^2 + 3a + 2b = (3a + 2b)(a - b) + 3a + 2b = (3a + 2b)(a - b + 1)$

FRACTIONS

3.27   In the following fractions, what values of x are not permitted?

(a) $\dfrac{9}{x}$ (b) $\dfrac{25}{x - 3}$ (c) $\dfrac{2x}{x^2 - 16}$ (d) $\dfrac{3x}{5}$ (e) $\dfrac{2x + 2}{x^2 + 5x + 6}$

Solution

(a) $\dfrac{9}{x}$ $x = 0$ is not allowed.

(b) $\dfrac{25}{x - 3}$ A value of $x = 3$ would make the denominator zero and is thus not permitted.

(c) $\dfrac{2x}{x^2 - 16}$ Values of $x = 4$ or $x = -4$ would make the denominator zero and so are not permitted.

(d) $\dfrac{3x}{5}$ All values of x are allowed.

(e) $\dfrac{2x + 2}{x^2 + 5x + 6}$ Factoring the denominator

$$\frac{2x + 2}{x^2 + 5x + 6} = \frac{2x + 2}{(x + 3)(x + 2)}$$

Values of $x = -3$ or $x = -2$ would make the denominator zero and are not permitted.

3.28   Make the following expressions equal by inserting the proper algebraic sign in the parentheses.

(a) $\dfrac{a}{a - b} = (\ )\dfrac{a}{b - a}$ (c) $\dfrac{x - y}{w - z} = (\ )\dfrac{y - x}{z - w}$

(b) $\dfrac{-x}{2y - 5} = (\ )\dfrac{x}{5 - 2y}$ (d) $\dfrac{x}{y - x} = -\dfrac{(\ )x}{x - y}$

Solution

(a) $\dfrac{a}{a - b} = (\ )\dfrac{a}{b - a}$

$$\frac{a}{a - b} = -\frac{a}{-(a - b)} = (-)\frac{a}{b - a}$$

(b) $\dfrac{-x}{2y-5} = (\ )\dfrac{x}{5-2y}$

$$\dfrac{-x}{2y-5} = \dfrac{x}{-(2y-5)} = (+)\dfrac{x}{5-2y}$$

(c) $\dfrac{x-y}{w-z} = (\ )\dfrac{y-x}{z-w}$

$$\dfrac{x-y}{w-z} = \dfrac{-(x-y)}{-(w-z)} = (+)\dfrac{y-x}{z-w}$$

(d) $\dfrac{x}{y-x} = -\dfrac{(\ )x}{x-y}$

$$\dfrac{x}{y-x} = -\dfrac{x}{-(y-x)} = -\dfrac{(+)x}{x-y}$$

3.29   Insert the missing quantity.

(a) $\dfrac{3x}{5} = \dfrac{?}{25}$ (c) $\dfrac{4x}{3y} = \dfrac{?}{-3y}$ (e) $\dfrac{4-w}{-6} = \dfrac{?}{6}$

(b) $\dfrac{?}{16x} = \dfrac{6}{32x}$ (d) $\dfrac{5-x}{y-1} = \dfrac{x-5}{?}$

Solution

(a) $\dfrac{3x}{5} = \dfrac{?}{25}$

$$\dfrac{3x}{5} \cdot \dfrac{5}{5} = \dfrac{15x}{25}$$

(b) $\dfrac{?}{16x} = \dfrac{6}{32x}$

$$\dfrac{6}{32x} \cdot \dfrac{\frac{1}{2}}{\frac{1}{2}} = \dfrac{3}{16x}$$

(c) $\dfrac{4x}{3y} = \dfrac{?}{-3y}$

$$\dfrac{4x}{3y} \cdot \dfrac{-1}{-1} = \dfrac{-4x}{-3y}$$

(d) $\dfrac{5-x}{y-1} = \dfrac{x-5}{?}$

$$\dfrac{5-x}{y-1} \cdot \dfrac{-1}{-1} = \dfrac{x-5}{1-y}$$

(e) $\dfrac{4-w}{-6} = \dfrac{?}{6}$

$$\dfrac{4-w}{-6} \cdot \dfrac{-1}{-1} = \dfrac{w-4}{6}$$

3.30   Reduce to lowest terms.

(a) $\dfrac{5x^2yz^3}{15xy^2z^3}$ (e) $\dfrac{5a^2 - 10a}{10a^2}$ (i) $\dfrac{(2a + b)(a - b)}{(b - 3a)(b - a)}$

(b) $\dfrac{3y - 9}{2y - 6}$ (f) $\dfrac{b^2 + 6b + 9}{b^2 - 9}$ (j) $\dfrac{x^3 - 1}{1 - x^2}$

(c) $-\dfrac{7x - 9}{9 - 7x}$ (g) $\dfrac{ab}{a^2b^2 - 4ab}$

(d) $\dfrac{3y - 12}{y - 4}$ (h) $\dfrac{3x^2 + 6x}{3x^2 + 12x + 12}$

Solution

(a) $\dfrac{5x^2yz^3}{15xy^2z^3} = \dfrac{x^2yz^3}{3xy^2z^3} = \dfrac{x^2y}{3xy^2} = \dfrac{x}{3y}$

(b) $\dfrac{3y - 9}{2y - 6} = \dfrac{3(y - 3)}{2(y - 3)} = \dfrac{3}{2}$

(c) $-\dfrac{7x - 9}{9 - 7x} = \dfrac{-(7x - 9)}{9 - 7x} = \dfrac{-7x + 9}{9 - 7x} = 1$

(d) $\dfrac{3y - 12}{y - 4} = \dfrac{3(y - 4)}{y - 4} = 3$

(e) $\dfrac{5a^2 - 10a}{10a^2} = \dfrac{5a(a - 2)}{10a^2} = \dfrac{a - 2}{2a}$

(f) $\dfrac{b^2 + 6b + 9}{b^2 - 9} = \dfrac{(b + 3)^2}{(b + 3)(b - 3)} = \dfrac{b + 3}{b - 3}$

(g) $\dfrac{ab}{a^2b^2 - 4ab} = \dfrac{ab}{ab(ab - 4)} = \dfrac{1}{ab - 4}$

(h) $\dfrac{3x^2 + 6x}{3x^2 + 12x + 12} = \dfrac{3x(x + 2)}{3(x^2 + 4x + 4)} = \dfrac{x(x + 2)}{(x + 2)^2} = \dfrac{x}{x + 2}$

(i) $\dfrac{(2a + b)(a - b)}{(b - 3a)(b - a)} = \dfrac{(2a + b)(-1)(-a + b)}{(b - 3a)(b - a)} = -\dfrac{(2a + b)(b - a)}{(b - 3a)(b - a)} = -\dfrac{2a + b}{b - 3a} = \dfrac{2a + b}{3a - b}$

(j) $\dfrac{x^3 - 1}{1 - x^2} = \dfrac{(x - 1)(x^2 + x + 1)}{(1 - x)(1 + x)} = \dfrac{-(x - 1)(x^2 + x + 1)}{(1 - x)(1 + x)} = -\dfrac{x^2 + x + 1}{1 + x}$

3.31   Multiply and reduce to lowest terms.

(a) $\dfrac{2x^2y}{z^3} \cdot \dfrac{3xz}{4y}$ (e) $\dfrac{5x - 15}{9x^2 + 6x - 24} \cdot \dfrac{12x - 16}{8x - 12}$

(b) $\dfrac{3x}{5} \cdot \dfrac{2x}{6}$ (f) $\dfrac{6 - 3x}{2x^2 - 8} \cdot \dfrac{x^2 + 2x + 1}{2x + 2}$

(c) $\dfrac{10}{x^2} \cdot \dfrac{5}{2xy}$ (g) $\dfrac{a + b}{b - a} \cdot \dfrac{a - b}{b - 2a} \cdot \dfrac{2a - b}{a + b}$

(d) $\dfrac{x^2 - y^2}{x^2} \cdot \dfrac{xy}{x + y}$ (h) $\dfrac{4x^2 - y^2}{x^2 - y^2} \cdot \dfrac{x + y}{2x + y}$

Solution

(a) $\dfrac{2x^2y}{z^3}\cdot\dfrac{3xz}{4y}=\dfrac{6x^3yz}{4yz^3}=\dfrac{3x^3}{2z^2}$

(b) $\dfrac{3x}{5}\cdot\dfrac{2x}{6}=\dfrac{6x^2}{5(6)}=\dfrac{x^2}{5}$

(c) $\dfrac{10}{x^2}\cdot\dfrac{5}{2xy}=\dfrac{50}{2x^3y}=\dfrac{25}{x^3y}$

(d) $\dfrac{x^2-y^2}{x^2}\cdot\dfrac{xy}{x+y}=\dfrac{(x+y)(x-y)xy}{x^2(x+y)}=\dfrac{(x-y)xy}{x^2}=\dfrac{(x-y)y}{x}$

(e) $\dfrac{5x-15}{9x^2+6x-24}\cdot\dfrac{12x-16}{8x-12}=\dfrac{5(x-3)(4)(3x-4)}{3(3x^2+2x-8)(4)(2x-3)}$

$=\dfrac{5(x-3)(3x-4)}{3(3x-4)(x+2)(2x-3)}=\dfrac{5(x-3)}{3(x+2)(2x-3)}$

(f) $\dfrac{6-3x}{2x^2-8}\cdot\dfrac{x^2+2x+1}{2x+2}=\dfrac{3(2-x)(x+1)^2}{2(x^2-4)(2)(x+1)}=\dfrac{3(-1)(-2+x)(x+1)}{4(x-2)(x+2)}$

$=-\dfrac{3(x-2)(x+1)}{4(x-2)(x+2)}=-\dfrac{3(x+1)}{4(x+2)}$

(g) $\dfrac{a+b}{b-a}\cdot\dfrac{a-b}{b-2a}\cdot\dfrac{2a-b}{a+b}=\dfrac{(a-b)(2a-b)}{(b-a)(b-2a)}\cdot\dfrac{-1}{-1}=\dfrac{(b-a)(2a-b)}{(b-a)(2a-b)}=1$

(h) $\dfrac{4x^2-y^2}{x^2-y^2}\cdot\dfrac{x+y}{2x+y}=\dfrac{(2x+y)(2x-y)(x+y)}{(x+y)(x-y)(2x+y)}=\dfrac{2x-y}{x-y}$

3.32 Divide and reduce to lowest terms.

(a) $\dfrac{3x^2}{5y}\div\dfrac{y^2}{2x}$ (d) $\dfrac{2a^2-14a}{5a^2}\div\dfrac{4a-28}{15a^3}$ (g) $\dfrac{w^2-z^2}{(w+z)^2}\div\dfrac{3w+3z}{w-z}$

(b) $\dfrac{2x}{3}\div\dfrac{5}{4x}$ (e) $\dfrac{a-3b}{a^2-6a+8}\div\dfrac{3a-9b}{a^2-16}$ (h) $(x-6)\div\dfrac{x^2-36}{3x+2}$

(c) $\dfrac{x+3}{x^2-9}\div\dfrac{6}{2x-6}$ (f) $\dfrac{a^2-16}{a^2-9}\div\dfrac{a-4a^2}{9a+27}$

Solution

(a) $\dfrac{3x^2}{5y}\div\dfrac{y^2}{2x}=\dfrac{3x^2}{5y}\cdot\dfrac{2x}{y^2}=\dfrac{6x^3}{5y^3}$

(b) $\dfrac{2x}{3}\div\dfrac{5}{4x}=\dfrac{2x}{3}\cdot\dfrac{4x}{5}=\dfrac{8x^2}{15}$

(c) $\dfrac{x+3}{x^2-9}\div\dfrac{6}{2x-6}=\dfrac{x+3}{x^2-9}\cdot\dfrac{2x-6}{6}=\dfrac{(x+3)(2)(x-3)}{(x+3)(x-3)(6)}=\dfrac{1}{3}$

(d) $\dfrac{2a^2-14a}{5a^2}\div\dfrac{4a-28}{15a^3}=\dfrac{2a(a-7)}{5a^2}\cdot\dfrac{15a^3}{4(a-7)}=\dfrac{30a^4(a-7)}{20a^2(a-7)}=\dfrac{3a^2}{2}$

(e) $\dfrac{a-3b}{a^2-6a+8}\div\dfrac{3a-9b}{a^2-16}=\dfrac{a-3b}{(a-4)(a-2)}\cdot\dfrac{(a+4)(a-4)}{3(a-3b)}=\dfrac{a+4}{3(a-2)}$

(f) $\dfrac{a^2-16}{a^2-9}\div\dfrac{a-4a^2}{9a+27}=\dfrac{(a+4)(a-4)}{(a+3)(a-3)}\cdot\dfrac{9(a+3)}{a(1-4a)}=\dfrac{9(a+4)(a-4)}{a(a-3)(1-4a)}$

(g) $\dfrac{w^2 - z^2}{(w + z)^2} \div \dfrac{3w + 3z}{w - z} = \dfrac{(w + z)(w - z)}{(w + z)^2} \cdot \dfrac{w - z}{3(w + z)} = \dfrac{(w - z)^2}{3(w + z)^2}$

(h) $(x - 6) \div \dfrac{x^2 - 36}{3x + 2} = (x - 6) \cdot \dfrac{3x + 2}{(x + 6)(x - 6)} = \dfrac{3x + 2}{x + 6}$

3.33   Add or subtract, as indicated, and simplify.

(a) $\dfrac{2y}{x} + \dfrac{5y}{x} - \dfrac{3y}{x}$ (f) $\dfrac{6}{x} + \dfrac{x}{x + 2}$

(b) $\dfrac{2x}{x + 1} - \dfrac{3}{x + 1}$ (g) $\dfrac{3}{a - 1} - \dfrac{2}{a + 3}$

(c) $\dfrac{5}{x - y} - \dfrac{4}{y - x}$ (h) $\dfrac{3a - 1}{a^2 + a - 12} + \dfrac{a + 2}{a^2 - 9}$

(d) $\dfrac{1}{R_1} + \dfrac{1}{R_2}$ (i) $\dfrac{2w}{w^2 - 4} - \dfrac{2}{w + 2}$

(e) $\dfrac{3}{x} + \dfrac{2}{y} - \dfrac{5}{z}$ (j) $\dfrac{y}{xy - x^2} + \dfrac{x}{xy - y^2}$

Solution

(a) $\dfrac{2y}{x} + \dfrac{5y}{x} - \dfrac{3y}{x} = \dfrac{2y + 5y - 3y}{x} = \dfrac{4y}{x}$

(b) $\dfrac{2x}{x + 1} - \dfrac{3}{x + 1} = \dfrac{2x - 3}{x + 1}$

(c) $\dfrac{5}{x - y} - \dfrac{4}{y - x} = \dfrac{5}{x - y} - \dfrac{-4}{x - y} = \dfrac{5 - (-4)}{x - y} = \dfrac{9}{x - y}$

(d) $\dfrac{1}{R_1} + \dfrac{1}{R_2} = \dfrac{R_2}{R_1 R_2} + \dfrac{R_1}{R_1 R_2} = \dfrac{R_1 + R_2}{R_1 R_2}$

(e) $\dfrac{3}{x} + \dfrac{2}{y} - \dfrac{5}{z} = \dfrac{3yz}{xyz} + \dfrac{2xz}{xyz} - \dfrac{5xy}{xyz} = \dfrac{3yz + 2xz - 5xy}{xyz}$

(f) $\dfrac{6}{x} + \dfrac{x}{x + 2} = \dfrac{6(x + 2)}{x(x + 2)} + \dfrac{x^2}{x(x + 2)} = \dfrac{x^2 + 6x + 12}{x(x + 2)}$

(g) $\dfrac{3}{a - 1} - \dfrac{2}{a + 3} = \dfrac{3(a + 3) - 2(a - 1)}{(a - 1)(a + 3)} = \dfrac{3a + 9 - 2a + 2}{(a - 1)(a + 3)} = \dfrac{a + 11}{(a - 1)(a + 3)}$

(h) $\dfrac{3a - 1}{a^2 + a - 12} + \dfrac{a + 2}{a^2 - 9} = \dfrac{3a - 1}{(a + 4)(a - 3)} + \dfrac{a + 2}{(a + 3)(a - 3)}$

$\qquad = \dfrac{(3a - 1)(a + 3) + (a + 2)(a + 4)}{(a + 4)(a - 3)(a + 3)} = \dfrac{3a^2 + 8a - 3 + a^2 + 6a + 8}{(a + 4)(a - 3)(a + 3)}$

$\qquad = \dfrac{4a^2 + 14a + 5}{(a + 4)(a - 3)(a + 3)}$

(i) $\dfrac{2w}{w^2 - 4} - \dfrac{2}{w + 2} = \dfrac{2w - 2(w - 2)}{(w + 2)(w - 2)} = \dfrac{4}{(w + 2)(w - 2)}$

(j) $\dfrac{y}{xy - x^2} + \dfrac{x}{xy - y^2} = \dfrac{y}{x(y-x)} + \dfrac{x}{y(x-y)} = \dfrac{y}{x(y-x)} + \dfrac{x}{-y(-x+y)} = \dfrac{y}{x(y-x)} - \dfrac{x}{y(y-x)}$

$$= \dfrac{y}{x(y-x)} \cdot \dfrac{y}{y} - \dfrac{x}{y(y-x)} \cdot \dfrac{x}{x} = \dfrac{y^2}{xy(y-x)} - \dfrac{x^2}{xy(y-x)}$$

$$= \dfrac{y^2 - x^2}{xy(y-x)} = \dfrac{(y+x)(y-x)}{xy(y-x)} = \dfrac{y+x}{xy}$$

3.34 Simplify.

(a) $\dfrac{x + \dfrac{1}{y}}{y - \dfrac{1}{x}}$ (b) $\dfrac{\dfrac{1}{a} + 2}{\dfrac{1}{a} + 3}$ (c) $\dfrac{\dfrac{1}{x} - x}{x}$ (d) $\dfrac{\dfrac{x+2}{a}}{\dfrac{x+2}{b}}$ (e) $\dfrac{1}{1 - \dfrac{1}{1-x}}$

Solution

(a) $\dfrac{x + \dfrac{1}{y}}{y - \dfrac{1}{x}} = \dfrac{\dfrac{xy+1}{y}}{\dfrac{xy-1}{x}} = \dfrac{xy+1}{y} \cdot \dfrac{x}{xy-1} = \dfrac{x(xy+1)}{y(xy-1)}$

(b) $\dfrac{\dfrac{1}{a} + 2}{\dfrac{1}{a} + 3} = \dfrac{\dfrac{1+2a}{a}}{\dfrac{1+3a}{a}} = \dfrac{1+2a}{a} \cdot \dfrac{a}{1+3a} = \dfrac{1+2a}{1+3a}$

(c) $\dfrac{\dfrac{1}{x} - x}{x} = \dfrac{1-x^2}{x} \cdot \dfrac{1}{x} = \dfrac{1-x^2}{x^2}$

(d) $\dfrac{\dfrac{x+2}{a}}{\dfrac{x+2}{b}} = \dfrac{x+2}{a} \cdot \dfrac{b}{x+2} = \dfrac{b}{a}$

(e) $\dfrac{1}{1 - \dfrac{1}{1-x}} = \dfrac{1}{\dfrac{1-x}{1-x} - \dfrac{1}{1-x}} = \dfrac{1}{\dfrac{-x}{1-x}} = \dfrac{1-x}{-x} = \dfrac{x-1}{x}$

Supplementary Problems

3.35 Simplify the following without using negative exponents.

(a) $c^7 \cdot c^5$ (c) 4^4 (e) $(b^x)^3$ (g) $\dfrac{5^4}{5}$ (i) $(x^2)^6$ (k) $x^{3+2x} \cdot x^4$

(b) $3^4 \cdot 3$ (d) $(0.2)^2$ (f) $\dfrac{y^4}{y^2}$ (h) $-\left(\dfrac{2}{3}\right)^3$ (j) $(7^y)^y$ (l) $\left(\dfrac{2a}{x^2}\right)^3$

3.36 Write the following with positive exponents only and simplify.

(a) y^{-4} (b) $\left(\dfrac{3}{x}\right)^{-3}$ (c) $3^{-3}yx^{-2}$ (d) y^3x^{-2} (e) $\dfrac{x + 2y^{-1}}{x}$ (f) $(x-3)^{-3}$

3.37   Express without fractions, using negative exponents.

(a) $\dfrac{2}{x^2}$ (b) $\dfrac{x^2a}{y^3}$ (c) $\dfrac{2}{z^4}$

3.38   Evaluate.

(a) $5^4 \cdot 5^2$ (b) $\dfrac{5}{x^0}$ (c) $(3^2)^{-4}$ (d) $\dfrac{5^1}{5^3}$

3.39   Remove symbols of grouping and simplify.

(a) $6 - (x + y)$ (c) $4 - [x + 3(y - 2)]$

(b) $5(x + 3y) - 2$ (d) $2y + \{3[x^2 - 2 - (y + 3) + 6]\}$

3.40   Combine the following as indicated.

(a) $2y + 7.2y$ (b) $a - \dfrac{3}{2}a + 4a$ (c) $\dfrac{3}{4} - \dfrac{x}{3} + \dfrac{2x}{6}$

3.41   Combine the following.

(a) $(2x - 5) + (x + 4)$ (c) $(a + b) - (a^2 - b^2) + (3a - 4b)$

(b) $(3y^2 - 4y) - (4x^2 + 2x)$ (d) $(2a + 4b - c) + (c - b + 4a) - (a^2 + 4c - b)$

3.42   Multiply the following monomials.

(a) $x \cdot x^3 \cdot x^5$ (b) $(x^4y^2)(x)$ (c) $5x(-2y)(3y^2)$ (d) $(-2)(-xy)(x^2y)$

3.43   Multiply the following.

(a) $4x^2(4y + 3x - 1)$ (b) $2x(6 - 2x + 6y^2)$ (c) $x^2y^6(x^2zy^3 - 3xy)$

3.44   Find the following products.

(a) $(x - y)(x + y)$ (c) $(a^4 - b^2 + c)(-a + 4b^2 - c)$

(b) $(2y^2 + x^3)(x - y + 3)$ (d) $(x - y - z)(-z - y + x)$

3.45   Raise the following to the indicated power.

(a) $(y + 4)^2$ (b) $(2x^2 - x + 1)^3$

3.46   Perform the indicated divisions.

(a) $a^2 \div a$ (b) $\dfrac{x^6}{x^3}$ (c) $(-y^3) \div (y)^6$ (d) $\dfrac{6x^3y}{2x}$

3.47   Perform the divisions.

(a) $(4x^3 - 3x) \div x^2$ (b) $\dfrac{5a - 7b^3 + a}{a^2}$ (c) $\dfrac{35x^6y^{12} - 50yx^7}{5x^4y}$

3.48   Perform the division.

    (a) $(3x^2 - 2x + 2) \div (x + 1)$ (b) $(6x^3 - 2x + 4) \div (x + 2)$

3.49   Factor completely.

    (a) $5a - 10b$ (d) $4(x - y) - 9(x - y)^2$ (g) $4x^4 - 20y^2$

    (b) $6x^2 - x^2 b$ (e) $2(a + b)(x - y) + 4x(a + b)^2(x - y)$ (h) $\dfrac{a^4}{3} + \dfrac{ab^2}{3}$

    (c) $4a^3 b^2 - x^2 b^2 + b^2 x^3$ (f) $9x^2 - 1$ (i) $49 - (x - 3)^2$

3.50   Test the following trinomials for factorability.

    (a) $y^2 - 20y + 12$ (b) $2a^2 - 12a + 18$ (c) $5x^2 - 5x - 1$ (d) $x^2 + 4x - 2$

3.51   Factor completely.

    (a) $x^2 + 5x + 6$ (c) $w^2 - 14w - 15$ (e) $9a^2 + 6a + 1$ (g) $4b^2 d - ad + a^2 bd^2$

    (b) $y^2 - 3y + 2$ (d) $b^2 - 8b - 48$ (f) $3x^2 - 8x + 5$ (h) $-18 + 42x^2 - 40x$

3.52   Factor completely.

    (a) $4a^2 + 5a + 1$ (b) $x^2 + 2xy + y^2$ (c) $9 - 12a + 4a^2$ (d) $9a^2 b^2 - 6ab + 1$

3.53   Factor completely.

    (a) $x^3 + 8$ (b) $64y^3 + 8$ (c) $8x^3 + 27$

3.54   Factor by grouping.

    (a) $xw - 4yw + 3a - ab$ (c) $2x + 4y - 4x + 8y$

    (b) $9x - 2abx + 18xy - 4abxy$ (d) $2x^2 - 5xy + 2y^2 + 4a - 6b^2$

3.55   In the following fractions, what values of x are not permitted?

    (a) $\dfrac{3}{x}$ (b) $\dfrac{50}{x + 3}$ (c) $\dfrac{4x - 1}{x^2 - 4x + 4}$

3.56   Make the following equal by inserting the proper algebraic sign in the parentheses.

    (a) $\dfrac{a - b}{x - z} = (\ )\dfrac{b - a}{z - x}$ (b) $\dfrac{x}{a - c} = -\dfrac{(\ )x}{c - a}$

3.57   Insert the missing quantity.

    (a) $\dfrac{2y}{10} = \dfrac{?}{20}$ (b) $\dfrac{4a}{28} = \dfrac{10a}{?}$ (c) $\dfrac{z}{2} = \dfrac{?}{10}$

3.58   Reduce to lowest terms.

    (a) $\dfrac{9x^3 y^2 z}{6yx^2 z^3}$ (b) $\dfrac{3x^2 - 2yx}{x^2 + 3xz}$ (c) $\dfrac{x^2 + 3x + 2}{x + 1}$ (d) $\dfrac{x^3 - 1}{x^2 - 1}$

3.59   Multiply or divide, as indicated, and reduce to lowest terms.

(a) $\dfrac{3xy}{x^2y^3} \cdot \dfrac{xz}{xy}$ (c) $\dfrac{a+3x}{x^2+5x+4} \div \dfrac{3a^2+9ax}{x+4}$

(b) $\dfrac{10}{x^5} \cdot \dfrac{12}{2y^3}$ (d) $\dfrac{y+2}{\dfrac{y^2+4y+4}{12y+24}}$

3.60   Add or subtract, as indicated, and simplify.

(a) $\dfrac{3a}{b} - \dfrac{6a^2}{b} + \dfrac{a^3}{b}$ (c) $\dfrac{x}{y} + \dfrac{x}{z} - \dfrac{y}{x}$ (e) $\dfrac{a}{xy+9} + \dfrac{b}{xy-9}$

(b) $\dfrac{9}{x+y} + \dfrac{12x}{2x+2y}$ (d) $\dfrac{4x-1}{x^2+x-6} + \dfrac{x-6}{x^2-9}$

3.61   Simplify.

(a) $\dfrac{a+\dfrac{1}{b}}{b-\dfrac{1}{a}}$ (b) $\dfrac{\dfrac{1}{x}+2}{\dfrac{1}{x}-2}$ (c) $\dfrac{1}{2-\dfrac{1}{x}}$

Answers to Supplementary Problems

3.35   (a) c^{12} (b) $3^5 = 243$ (c) 256 (d) 0.04 (e) b^{3x} (f) y^2 (g) $5^3 = 125$

(h) $-\dfrac{8}{27}$ (i) x^{12} (j) 7^{y^2} (k) x^{2x+7} (l) $\dfrac{8a^3}{x^6}$

3.36   (a) $\dfrac{1}{y^4}$ (b) $\dfrac{x^3}{27}$ (c) $\dfrac{y}{27x^2}$ (d) $\dfrac{y^3}{x^2}$ (e) $1 + \dfrac{1}{2xy}$ (f) $\dfrac{1}{(x-3)^3}$

3.37   (a) $2x^{-2}$ (b) ax^2y^{-3} (c) $2z^{-4}$

3.38   (a) $15\,625$ (b) 5 (c) $\dfrac{1}{6561}$ (d) $\dfrac{1}{25}$

3.39   (a) $6 - x - y$ (b) $5x + 15y - 2$ (c) $-x - 3y + 10$ (d) $3x^2 - y + 3$

3.40   (a) $9.2y$ (b) $\dfrac{7a}{2}$ (c) $\dfrac{3}{4}$

3.41   (a) $3x - 1$ (b) $3y^2 - 4y - 4x^2 - 2x$ (c) $-a^2 + 4a + b^2 - 3b$ (d) $-a^2 + 6a + 4b - 4c$

3.42   (a) x^9 (b) x^5y^2 (c) $-30xy^3$ (d) $2x^3y^2$

3.43   (a) $16x^2y + 12x^3 - 4x^2$ (b) $12x - 4x^2 + 12xy^2$ (c) $x^4y^9z - 3x^3y^7$

3.44   (a) $x^2 - y^2$ (b) $x^4 - x^3y + 3x^3 - 2y^3 + 2xy^2 + 6y^2$
　　　　 (c) $-a^5 + 4a^4b^2 - a^4c + ab^2 - ac + 5b^2c - 4b^4 - c^2$ (d) $x^2 - 2xy - 2xz + y^2 + 2yz + z^2$

3.45   (a) $y^2 + 8y + 16$ (b) $8x^6 - 12x^5 + 18x^4 - 13x^3 + 9x^2 - 3x + 1$

3.46   (a) a (b) x^3 (c) $-\dfrac{1}{y^3}$ (d) $3x^2y$

3.47   (a) $4x - \dfrac{3}{x}$ (b) $\dfrac{6}{a} - \dfrac{7b^3}{a^2}$ (c) $7x^2y^{11} - 10x^3$

3.48   (a) $3x - 5 + \dfrac{7}{x+1}$ (b) $6x^2 - 12x + 22 + \dfrac{-40}{x+2}$

3.49   (a) $5(a - 2b)$ (b) $x^2(6 - b)$ (c) $b^2(4a^3 - x^2 + x^3)$ (d) $(x - y)[4 - 9(x - y)]$
　　　　 (e) $(a + b)(x - y)[2 + 4x(a + b)]$ (f) $(3x + 1)(3x - 1)$ (g) $4(x^4 - 5y^2)$ (h) $\dfrac{a(a^3 + b^2)}{3}$
　　　　 (i) $(x + 4)(10 - x)$

3.50   (a) Not factorable   (b) factorable   (c) not factorable   (d) not factorable

3.51   (a) $(x + 2)(x + 3)$ (b) $(y - 1)(y - 2)$ (c) $(w + 1)(w - 15)$ (d) $(b + 4)(b - 12)$
　　　　 (e) $(3a + 1)(3a + 1)$ (f) $(3x - 5)(x - 1)$ (g) $d(4b^2 - a + a^2bd)$ (h) $(6x + 2)(7x - 9)$

3.52   (a) $(4a + 1)(a + 1)$ (b) $(x + y)(x + y)$ (c) $(2a - 3)(2a - 3)$ (d) $(3ab - 1)(3ab - 1)$

3.53   (a) $(x + 2)(x^2 - 2x + 4)$ (b) $(4y + 2)(16y^2 - 8y + 4)$ (c) $(2x + 3)(4x^2 - 6x + 9)$

3.54   (a) $w(x - 4y) + a(3 - b)$ (b) $(x + 2xy)(9 - 2ab)$ (c) $-2(x + 6y)$
　　　　 (d) $(2x - y)(x - 2y) + 2(2a - 3b^2)$

3.55   (a) 0 (b) -3 (c) 2

3.56   (a) $(+)$ (b) $(+)$

3.57   (a) $\dfrac{4y}{20}$ (b) $\dfrac{10a}{70}$ (c) $\dfrac{5z}{10}$

3.58   (a) $\dfrac{3xy}{2z^2}$ (b) $\dfrac{3x - 2y}{x + 3z}$ (c) $x + 2$ (d) $\dfrac{x^2 + x + 1}{x - 1}$

3.59   (a) $\dfrac{3z}{xy^3}$ (b) $\dfrac{60}{x^5 y^3}$ (c) $\dfrac{1}{3a(x+1)}$ (d) 12

3.60   (a) $\dfrac{a^3 - 6a^2 + 3a}{b}$ (b) $\dfrac{3(2x+3)}{x+y}$ (c) $\dfrac{x^2 z + x^2 y + y^2 z}{xyz}$ (d) $\dfrac{5x^2 - 21x + 15}{(x+3)(x-3)(x-2)}$

        (e) $\dfrac{axy - 9a + bxy + 9b}{(xy-9)(xy+9)}$

3.61   (a) $\dfrac{a(ab+1)}{b(ab-1)}$ (b) $\dfrac{1+2x}{1-2x}$ (c) $\dfrac{x}{2x-1}$

Chapter 4

Equations

4.1 DEFINITIONS AND TERMINOLOGY

(a) Equations. An *equation* results when two expressions are set equal to each other.

Example 4.1

$$x^2 - 3x = 5$$

is an equation.

(b) Parts of an Equation. An equation has three parts: the *left-hand side* (LHS) or *member*, the *right-hand side* (RHS), and an *equal sign*. The equal sign states that the LHS is equal to the RHS.

(c) Conditional Equations. A *conditional equation* is one whose sides are equal *only for certain values* of the unknowns.

Example 4.2: The sides of the equation

$$x - 2 = 0$$

are equal only when $x = 2$.

(d) Roots. The value of the unknown that makes the sides of a conditional equation (hereafter referred to as just "equation") equal is the *solution* to the equation.

Example 4.3: The solution to the equation

$$x + 5 = 9$$

is $x = 4$, because this value of x makes the two sides equal. This value, 4, is said to *satisfy* the given equation. It is also called a *root* of that equation.

(e) Multiple Roots. If there is more than one value of the unknown that satisfies the given equation, all such values are referred to as the *solution set*, or *roots*, of the equation.

Example 4.4: The equation $x^2 = 9$ has the solution set $x = 3$ and $x = -3$.

(f) Identities. An *identity* is an equation that is true for all values of the unknown.

Example 4.5

$$x^2 - 9 = (x + 3)(x - 3)$$

is an identity.

(g) Degree. The *degree* of an equation is the same as that of the highest-degree term in the equation.

80

Example 4.6:   The equation

$$x^2 - 2x + 3 = 0$$

is a second-degree equation.

(h) Number of Unknowns. Equations are also classified by the *number of unknowns* they contain.

Example 4.7:   The equation

$$3x - 2y - 5 = 0$$

is a first-degree equation in two unknowns, x and y.

(i) Type of Expressions. Equations are further classified by the type of expressions they contain.

Example 4.8:   Samples of the various types of equations treated in this Outline, and their locations, are

Equation	Type	Chapter
1. $2x - 5 = 5x + 3$	Linear (first-degree) equation, one unknown	4
2. $x^2 - 7x + 5 = 0$	Quadratic (second-degree) equation, one unknown	13
3. $3 \sin x = 2$	Trigonometric equation, one unknown	8
4. $\begin{cases} 2x - 3y = 5 \\ x + 4y = 7 \end{cases}$	Pair of linear equations, two unknowns	11
5. $\begin{cases} 4x - 3y + 2z = 5 \\ x - 2y + 3z = -3 \\ 2x - 3y + z = 2 \end{cases}$	Three linear equations, three unknowns	11
6. $\log 2x + \log x = 5$	Logarithmic equation, one unknown	14
7. $2.5^x = 4$	Exponential equation, one unknown	14
8. $\sqrt{x + 2} = 3x$	Radical equation, one unknown	10, 13
9. $x^4 - 2x^2 + 3 = 0$	Equations of quadratic type, one unknown	13

4.2 CHECKING AN APPARENT SOLUTION

A solution is checked by substituting back in the original equation.

Example 4.9:   Determine if the value $x = -5/3$ is a solution to the equation

$$\frac{x - 3}{4} = \frac{2x + 1}{2}$$

Solution: Substituting $-5/3$ for x,

$$\frac{\left(-\dfrac{5}{3}\right) - 3}{4} \overset{?}{=} \frac{2\left(-\dfrac{5}{3}\right) + 1}{2}$$

$$\frac{-\dfrac{14}{3}}{4} \overset{?}{=} \frac{-\dfrac{7}{3}}{2}$$

$$-\frac{7}{6} = -\frac{7}{6}\quad\quad \text{checks}$$

Common Error	Be sure to substitute in the *original* equation before it has been subjected to any (possibly incorrect) manipulation.

Check every solution. Apparent solutions that do not satisfy the original equation are called *extraneous*, and should be discarded. They are most frequently introduced when both sides of an equation are multiplied by an expression containing the unknown.

Example 4.10: Check the apparent solutions $x = 2$ and $x = -2$ to the radical equation

$$\sqrt{5 + 2x} = x + 1$$

Solution: When $x = 2$,

$$\sqrt{5 + 2(2)} \overset{?}{=} 2 + 1$$
$$\sqrt{9} = 3 \quad \text{checks}$$

When $x = -2$,

$$\sqrt{5 + 2(-2)} \overset{?}{=} -2 + 1$$
$$\sqrt{1} \neq -1 \quad \text{does not check}$$

4.3 SOLVING EQUATIONS

(a) Mathematical Operations. In general, equations are solved by performing valid mathematical operations on *both sides* of the equation, with the object of isolating the unknown on one side of the equals sign. One such operation is to *add* or *subtract* the same quantity from both sides of the equation.

Example 4.11: Solve for x.

$$x - 5 = 7$$

Solution: Adding 5 to both sides,

$$x - 5 + 5 = 7 + 5$$
$$x = 7 + 5$$
$$= 12$$

Check: Substituting 12 for x,

$$12 - 5 \overset{?}{=} 7$$
$$7 = 7 \quad \text{checks}$$

Other operations include *multiplying* or *dividing* both sides by the same quantity.

Example 4.12: Solve for x.

$$\frac{5x}{2} = 3$$

Solution: Multiplying both sides by 2,

$$5x = 2(3) = 6$$

Dividing both sides by 5,

$$x = \frac{6}{5}$$

Check

$$\frac{5\left(\dfrac{6}{5}\right)}{2} \stackrel{?}{=} 3$$

$$\frac{6}{2} = 3 \qquad \text{checks}$$

Other operations that may be done to both sides of an equation include taking roots, raising to any power, and taking reciprocals, logarithms, and antilogarithms, always with the exception that the operation must not result in an illegal operation such as division by zero.

Common Error	The mathematical operations you perform must be done *to both sides* of the equation in order to preserve the equality.

Common Error	The mathematical operations you perform on both sides of an equation must be done to each side *as a whole*, not term by term.

Example 4.13:   Solve for x.

$$\frac{1}{x} = \frac{1}{2} + \frac{1}{3}$$

Solution: Taking reciprocals of both sides, being sure to take the reciprocal of the RHS as a whole,

$$x = \frac{1}{\dfrac{1}{2} + \dfrac{1}{3}} = \frac{1}{\dfrac{5}{6}}$$

$$= \frac{6}{5}$$

Incorrect Solution: Taking reciprocals of both sides, but taking the reciprocal of the RHS term by term,

$$x \neq 2 + 3 \neq 5$$

(b) Transposing.   In the equation

$$x - 5 = 3$$

if we add 5 to both sides, we get

$$x = 3 + 5 = 8$$

Note that the -5 has *moved* to the other side of the equals sign, changing its algebraic sign on the way.

$$x \overbrace{\left(-5\right)} = 3 \left(+5\right)$$

Thus we may move terms across the equals sign, provided that we change the algebraic sign of the term.   This process is called *transposing*.

Example 4.14:   Solve for x.

$$5x - 2 = 3x - 1$$

Solution: Transposing the -2,

$$5x = 3x - 1 + 2 = 3x + 1$$

Transposing the $3x$,

$$5x - 3x = 1$$

Combining like terms,

$$2x = 1$$

Dividing both sides by 2,

$$x = \frac{1}{2}$$

Check

$$5\left(\frac{1}{2}\right) - 2 \overset{?}{=} 3\left(\frac{1}{2}\right) - 1$$

$$\frac{5}{2} - \frac{4}{2} \overset{?}{=} \frac{3}{2} - \frac{2}{2}$$

$$\frac{1}{2} = \frac{1}{2} \qquad \text{checks}$$

4.4 FRACTIONAL EQUATIONS

(a) Removing Fractions.   Fractions can be quickly removed by multiplying both sides of the equation by the lowest common denominator (LCD).

Example 4.15:   Solve for x:

$$\frac{3}{x} = \frac{5}{2x} + \frac{2}{3}$$

Solution: The LCD is $6x$.   Multiplying both sides by $6x$,

$$6x\left(\frac{3}{x}\right) = 6x\left(\frac{5}{2x}\right) + 6x\left(\frac{2}{3}\right)$$

$$18 = 15 + 4x$$

Transposing the 15,

$$18 - 15 = 4x$$

$$3 = 4x$$

Dividing both sides by 4,

$$x = \frac{3}{4}$$

Check

$$\frac{3}{3/4} \overset{?}{=} \frac{5}{2(3/4)} + \frac{2}{3}$$

$$4 \overset{?}{=} \frac{10}{3} + \frac{2}{3}$$

$$4 = 4 \qquad \text{checks}$$

(b) Checking.   Be sure to check any apparent solutions, because multiplying by a LCD containing the unknown will often introduce an extraneous solution.

Example 4.16:   Solve for x.

$$\frac{4}{x-4} = \frac{x}{x-4} - 3$$

Solution: Multiplying by the LCD, $x - 4$,

$$(x-4)\left(\frac{4}{x-4}\right) = (x-4)\left(\frac{x}{x-4}\right) - (x-4)(3)$$

$$4 = x - 3(x-4)$$

$$4 = x - 3x + 12.$$

Transposing and collecting terms,

$$2x = 12 - 4$$

$$2x = 8$$

$$x = 4$$

Check

$$\frac{4}{4-4} \stackrel{?}{=} \frac{4}{4-4} - 3$$

This causes division by zero, which shows that $x = 4$ is not a solution of the given equation.

(c) Cross-Multiplying.   If both sides of an equation are simple fractions, such as

$$\frac{a}{b} = \frac{c}{d}$$

multiplying both sides by the LCD, bd, gives

$$ad = bc$$

The same result could have been obtained by multiplying the numerator of each side by the denominator of the opposite side.

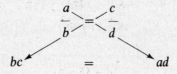

This process is called *cross-multiplying*.

Example 4.17:   Solve for x.

$$\frac{4}{x} = \frac{3}{x+2}$$

Solution: Cross-multiplying,

$$4(x+2) = 3x$$

$$4x + 8 = 3x$$

Transposing,

$$4x - 3x = -8$$

$$x = -8$$

4.5 LITERAL EQUATIONS

An equation containing letters other than the letter representing the unknown is called a *literal equation*.

Example 4.18: $ax + b = c$ is a literal equation.

Literal equations are solved by the same procedures as those used for equations containing only one letter.

Example 4.19: Solve for x.

$$ax + b = cx + d$$

Solution: Transposing b and cx,

$$ax - cx = d - b$$

Factoring,

$$x(a - c) = d - b$$

Dividing by $(a - c)$,

$$x = \frac{d - b}{a - c}$$

Note that this solution is not valid when a equals c because division by zero would result.

4.6 FORMULAS

(a) Definition. A *formula* is an equation expressing a relation between two or more mathematical or physical quantities. Some of the more common formulas used in technology are listed in the Summary of Formulas in the front of this Outline.

Example 4.20

| Ohm's Law | $I = \dfrac{V}{R}$ | **200** |

is a formula relating the current I through a resistance R having a voltage V across its terminals.

(b) Solving a Formula for a Different Quantity. Use the same methods as for literal equations to solve a formula for some quantity other than the one given.

Example 4.21: Solve Ohm's law for R.

Solution

$$I = \frac{V}{R}$$

Multiplying both sides by R,

$$RI = R\left(\frac{V}{R}\right)$$

$$RI = V$$

Dividing by I,

$$R = \frac{V}{I}$$

Example 4.22:   To convert temperatures from degrees Fahrenheit, F, to degrees Celsius, C, we use the formula

$$\boxed{C = \frac{5}{9}(F - 32)} \quad \textbf{188}$$

Solve this formula for F.

Solution: Multiplying both sides by 9/5,

$$F - 32 = \frac{9}{5}C$$

Transposing the -32,

$$\boxed{F = \frac{9}{5}C + 32} \quad \textbf{189}$$

(c)   Substituting in Formulas When Units Are Specified.   Often a formula is given in which the units of the various quantities are specified.   When substituting in such a formula, convert all quantities to the specified units *beforehand*.

Example 4.23:   The pressure head, in feet, of a liquid of density γ (pounds per cubic foot), having a pressure p [pounds per square inch(lb/in^2)], is

$$\text{Pressure head} = \frac{144p}{\gamma}$$

Find the pressure head of water ($\gamma = 62.4$ lb/ft^3) at a pressure of 15.5 kg/cm^2.

Solution: Converting the pressure to lb/in^2,

$$p = \frac{15.5 \text{ kg}}{cm^2} \cdot \frac{(2.54 \text{ cm})^2}{in^2} \cdot \frac{2.20 \text{ lb}}{kg} = 220 \text{ lb/in}^2$$

Substituting,

$$\text{Pressure head} = \frac{144(220)}{62.4}$$

$$= 508 \text{ ft}$$

(d)   Substituting in Formulas When Units Are Not Specified.   Most formulas do not specify units.   In this case, units should be carried along with the values into the formula.   Conversion factors should be used to cause units to cancel properly, leaving the answer with the units you desire.

Example 4.24:   The formula for uniform motion is

$$\text{Distance} = \text{rate} \times \text{time}$$

or

$$\boxed{D = Rt} \quad \textbf{177}$$

Find the distance in kilometers traveled by a car going 8.12 m/s for 3.45 h.

Solution: Substituting the given values, with units, into Eq. 177,

$$D = \frac{8.12 \text{ m}}{\text{s}} (3.45 \text{ h})$$

Since the units of time do not cancel, we introduce the conversion from seconds to hours.

$$D = \frac{8.12 \text{ m}}{\text{s}} (3.45 \text{ h}) \frac{3600 \text{ s}}{\text{h}}$$

$$= 101\,000 \text{ m} = 101 \text{ km}$$

(e) Empirical Formulas. When a formula is determined from *experimental data*, it is called an *empirical formula*.

Example 4.25: The formula for the flow rate Q over a sharp-crested rectangular weir of width L is

$$Q = 3.33LH^{3/2}$$

where H is the height of the water above the weir. The *form* of this particular formula had been obtained from fluid flow theory, but the coefficient 3.33 was found by *experiment and observation*. This formula is thus an empirical one.

4.7 RATIO AND PROPORTION

(a) Ratio. A *ratio* is the quotient of one quantity divided by another quantity of the same kind. If the two quantities to be compared are a and b, the ratio of a to b is written

$$\frac{a}{b} \qquad \text{or} \qquad a : b$$

The quantities a and b are called the *terms* of the ratio.

Example 4.26: A field is 50 m long and 25 m wide. What is the ratio of length to width?

Solution: The ratio is $50 : 25$. Ratios are usually reduced to lowest terms and so this ratio would be given as $2 : 1$.

Example 4.27: A rod is 2 ft long with a 2-in. diameter. What is the ratio of length to diameter?

Solution: Before forming the ratio, we must convert both quantities to the same units. It does not matter whether we choose feet, inches, or any other unit of length. Converting to inches, our ratio is $24 : 2$ or $12 : 1$.

(b) Proportion. A *proportion* is the equation obtained when one ratio is set equal to another. If the ratio $a : b$ equals the ratio $c : d$, we have the proportion

$$a : b = c : d$$

which reads, "the ratio of a to b equals the ratio of c to d," or "a is to b as c is to d."

Example 4.28: The ratio of the wing span S of a certain aircraft to the mean chord C is 16 to 3. Express this fact as a proportion.

Solution: From the definition of a proportion,

$$S : C = 16 : 3$$

(called the *Aspect Ratio*.)

Proportions are also written

$$a : b : : c : d$$

and

$$\frac{a}{b} = \frac{c}{d}$$

(c) Means and Extremes. The two inside terms of a proportion are called the *means* and the two outside terms are the *extremes*.

Example 4.29:   In the proportion

$$w : x = y : z \qquad \text{or} \qquad \frac{w}{x} = \frac{y}{z}$$

x and y are the means and w and z are the extremes.

(d) Finding a Missing Term. Manipulate a proportion as you would any other equation to isolate the unknown.

Example 4.30:   Find x if

$$2 : x = 5 : 7$$

Solution: Rewriting in fractional form,

$$\frac{2}{x} = \frac{5}{7}$$

Taking reciprocals of both sides,

$$\frac{x}{2} = \frac{7}{5}$$

Multiplying by 2,

$$x = \frac{2(7)}{5} = 2.8$$

(e) Mean Proportional. When the means of a proportion are equal, as in

$$\boxed{a : b = b : c} \qquad \mathbf{50}$$

the term b is called the *mean proportional* between a and c.   Solving for b, we get

$$\boxed{b = \pm \sqrt{ac}} \qquad \mathbf{51}$$

Example 4.31:   Find the mean proportional between 2 and 8.

Solution: From Eq. 51,

$$b = \pm \sqrt{2(8)} = \pm 4$$

So

$$2 : 4 = 4 : 8$$

and

$$2 : -4 = -4 : 8$$

The mean proportional b is also called the *geometric mean* between a and c because a, b, and c form a *geometric progression* (a series of numbers in which each term is obtained by multiplying the previous term by the same quantity).

Solved Problems

DEFINITIONS AND TERMINOLOGY

4.1   State the degree of the following equations:

(a) $x + 9x^2 = 0$ (b) $x^3 + 6x + 3x^2 = 0$ (c) $9x + 20x^2 + x^5 = 0$

Solution

(a)   Second degree   (b)   third degree   (c)   fifth degree

4.2   State the number of unknowns in the following equations:

(a) $x^2 + 3x = 0$ (b) $x^3 + 3x + y = 0$ (c) $x^2 + 3y + y^2 - z^5 = 0$

Solution

(a)   One   (b)   two   (c)   three

CHECKING

4.3   Check the following apparent solutions:

(a) $x + 5 = 7$ $(x = 2)$ (b) $3x + 9 = 27$ $(x = 4)$ (c) $\dfrac{10}{x + 1} - 8 = -3$ $(x = 1)$

Solution

(a) $2 + 5 \overset{?}{=} 7$

$7 = 7$ checks

(b) $3(4) + 9 \overset{?}{=} 27$

$21 \neq 27$ does not check

(c) $\dfrac{10}{1 + 1} - 8 \overset{?}{=} -3$

$5 - 8 = -3$ checks

SOLVING EQUATIONS

4.4   Solve for x.

(a) $x + 3 = 24$ (d) $62x - 20 = 39$ (g) $2x + 2 = 2(x - 1) + 4$

(b) $3x + 8 = 20$ (e) $3(x - 2) = 12$ (h) $2(3x + 4) - 6x = 0$

(c) $5x - 2 = 10$ (f) $2(x - 1) = 3(x + 2)$

Solution

(a) $x + 3 = 24$

Subtracting 3 from both sides, $x = 24 - 3 = 21$.

(b) $3x + 8 = 20$

Subtracting 8 from both sides, $3x = 20 - 8 = 12$.   Dividing by 3, $x = 12/3 = 4$.

(c) $5x - 2 = 10$

Adding 2 to both sides, $5x = 10 + 2 = 12$.   Dividing by 5, $x = 12/5 = 2.4$.

(d) $62x - 20 = 39$

Adding 20 to both sides, $62x = 39 + 20 = 59$.   Dividing by 62, $x = 59/62$.

(e) $3(x - 2) = 12$

Dividing both sides by 3, $x - 2 = 12/3 = 4$.   Transposing -2, $x = 4 + 2 = 6$.

(f) $2(x - 1) = 3(x + 2)$

Clearing parentheses, $2x - 2 = 3x + 6$.   Transposing $2x$ and 6,

$$-2 - 6 = 3x - 2x$$

$$x = -8$$

(g) $2x + 2 = 2(x - 1) + 4$

Removing parentheses, $2x + 2 = 2x - 2 + 4$.   Transposing,

$$2x - 2x = 2 - 2$$

$$0 = 0$$

The given equation is an identity, which is satisfied by any value of x.

(h) $2(3x + 4) - 6x = 0$

Clearing parentheses,

$$6x + 8 - 6x = 0$$

$$8 = 0$$

This indicates that the given equation has no solution.

FRACTIONAL EQUATIONS

4.5   Solve for x.

(a) $\dfrac{x}{5} = 4$

(b) $\dfrac{x}{2} + 3 = 5$

(c) $7 - \dfrac{x}{4} = 0$

(d) $-3 + \dfrac{x}{2} = 3$

(e) $\dfrac{1}{3x} + 6 = 8$

(f) $\dfrac{1}{5x} + 3 = -13$

(g) $\dfrac{1}{5x + 3} - 5 = 20$

(h) $\dfrac{x - 4}{x + 2} = 3$

(i) $\dfrac{1}{x} + \dfrac{1}{2} = 1$

(j) $\dfrac{a}{2} + \dfrac{3a + 1}{5} = \dfrac{a + 3}{10}$

(k) $\dfrac{x - 3}{x - 4} = \dfrac{x + 4}{x - 5}$

(l) $\dfrac{7}{x + 1} = \dfrac{2x + 3}{x - 2} + \dfrac{3x - 2x^2}{x^2 - x - 2}$

(m) $\dfrac{8}{x + 5} = \dfrac{7}{x} + \dfrac{3x - 2}{x^2 + 5x}$

Solution

(a) $x/5 = 4$

    Multiplying both sides by 5, $x = 5(4) = 20$.

(b) $x/2 + 3 = 5$

    Transposing the 3, $x/2 = 5 - 3 = 2$. Multiplying by 2, $x = 2(2) = 4$.

(c) $7 - \dfrac{x}{4} = 0$

    Transposing $-x/4$, $7 = x/4$. Multiplying by 4, $x = 4(7) = 28$.

(d) $-3 + \dfrac{x}{2} = 3$

    Transposing -3, $x/2 = 3 + 3 = 6$. Multiplying by 2, $x = 2(6) = 12$.

(e) $\dfrac{1}{3x} + 6 = 8$

    Transposing 6, $1/3x = 2$. Multiplying by $3x$, $1 = 6x$. Dividing by 6, $x = 1/6$.

(f) $\dfrac{1}{5x} + 3 = -13$

    Transposing 3, $1/5x = -16$. Multiplying by $5x$, $1 = -80x$. Dividing by -80, $x = -1/80$.

(g) $\dfrac{1}{5x + 3} - 5 = 20$

    Transposing 5, $1/(5x + 3) = 25$. Multiplying by $(5x + 3)$, $1 = 125x + 75$. Transposing 75, $-74 = 125x$. Dividing by 125, $x = -74/125$.

(h) $\dfrac{x - 4}{x + 2} = 3$

    Multiplying both sides by $(x + 2)$, $x - 4 = 3(x + 2) = 3x + 6$. Transposing $3x$ and -4, $x - 3x = 6 + 4$. Combining like terms, $-2x = 10$. Dividing by -2, $x = -5$.

(i) $\dfrac{1}{x} + \dfrac{1}{2} = 1$

    Transposing 1/2, $1/x = 1 - (1/2) = 1/2$. Taking the reciprocal of both sides, $x = 2$.

(j) $\dfrac{a}{2} + \dfrac{3a + 1}{5} = \dfrac{a + 3}{10}$

    Multiplying by the LCD, 10, $5a + 2(3a + 1) = a + 3$. Removing parentheses, $5a + 6a + 2 = a + 3$. Collecting terms, $11a + 2 = a + 3$. Transposing,

$$11a - a = 3 - 2$$

$$10a = 1$$

$$a = \frac{1}{10}$$

(k) $\dfrac{x - 3}{x - 4} = \dfrac{x + 4}{x - 5}$

Multiplying by $(x - 4)(x - 5)$,

$$(x - 3)(x - 5) = (x - 4)(x + 4)$$

$$x^2 - 8x + 15 = x^2 - 16$$

$$-8x = -31$$

$$x = \frac{31}{8}$$

(l) $\quad \dfrac{7}{x + 1} = \dfrac{2x + 3}{x - 2} + \dfrac{3x - 2x^2}{x^2 - x - 2}$

Multiplying by the LCD, $(x + 1)(x - 2)$,

$$7(x - 2) = (2x + 3)(x + 1) + 3x - 2x^2$$

Clearing parentheses and collecting terms,

$$7x - 14 = 2x^2 + 5x + 3 + 3x - 2x^2$$

$$7x - 14 = 8x + 3$$

Transposing,

$$7x - 8x = 3 + 14$$

$$-x = 17$$

$$x = -17$$

(m) $\quad \dfrac{8}{x + 5} = \dfrac{7}{x} + \dfrac{3x - 2}{x^2 + 5x}$

Multiplying by the LCD, $x(x + 5)$,

$$8x = 7(x + 5) + 3x - 2$$

Removing parentheses and collecting terms,

$$8x = 7x + 35 + 3x - 2$$

$$= 10x + 33$$

Transposing,

$$-2x = 33$$

$$x = -\frac{33}{2}$$

LITERAL EQUATIONS

4.6 Solve for x.

$$\frac{ab}{x} = 2c$$

Solution

Multiplying both sides by x, $2cx = ab$. Dividing by $2c$, $x = ab/2c$.

4.7 Solve for x.

$$w = \left(\frac{y}{2}\right)(x - 2w)$$

Solution

Multiplying both sides by $2/y$, $x - 2w = 2w/y$. Transposing $-2w$, $x = 2w/y + 2w$.

4.8   Solve for x.

$$\frac{(a+b)}{x} = a^2 + ab$$

Solution

Taking reciprocals of both sides,

$$\frac{x}{a+b} = \frac{1}{a^2 + ab}$$

Multiplying by $a + b$,

$$x = \frac{a+b}{a(a+b)} = \frac{1}{a}$$

PROPORTION

4.9   Find x if $x : 3 = 2 : 9$.

Solution

$$\frac{x}{3} = \frac{2}{9}$$

Multiplying by 3,

$$x = \frac{3(2)}{9} = \frac{2}{3}$$

4.10   Find x if $8 : 5 = 32 : x$.

Solution

$$\frac{8}{5} = \frac{32}{x}$$

Taking reciprocals of both sides,

$$\frac{5}{8} = \frac{x}{32}$$

Multiplying by 32,

$$x = \frac{32(5)}{8} = 20$$

4.11   Show that the product of the means equals the product of the extremes.

Solution

Given

$$\frac{a}{b} = \frac{c}{d}$$

Multiplying both sides by bd, we get

$$\boxed{ad = bc} \quad \mathbf{46}$$

4.12   Show that the means or extremes may be interchanged without affecting a proportion.

Solution

Given

$$\frac{a}{b} = \frac{c}{d}$$

or
$$ad = bc$$

Dividing by cd,

$$\boxed{\frac{a}{c} = \frac{b}{d}} \quad \mathbf{48}$$

or dividing instead by ab,

$$\boxed{\frac{d}{b} = \frac{c}{a}} \quad \mathbf{47}$$

4.13   Show that the means may be interchanged with the extremes.

Solution

Given

$$\frac{a}{b} = \frac{c}{d}$$

or
$$ad = bc$$

Dividing by ac,

$$\frac{d}{c} = \frac{b}{a}$$

or
$$\boxed{b : a = d : c} \quad \mathbf{49}$$

4.14   Find the mean proportional between 3 and 27.

Solution

From Eq. 51,

$$b = \pm\sqrt{3(27)} = \pm 9$$

4.15   Find the mean proportional between 4 and 36.

Solution

By Eq. 51,

$$b = \pm\sqrt{4(36)} = \pm 12$$

4.16   Two numbers are said to be in the *Golden Ratio*, or *Golden Section*, when the ratio of the smaller to the larger equals the ratio of the larger to their sum.   Express this statement as a proportion.

Solution

Let x = the smaller number, y = the larger number,   and $x + y$ = their sum.   Then

$$\frac{x}{y} = \frac{y}{x + y}$$

4.17   If two numbers x and y are in the Golden Ratio, the numerical value of this ratio is found to be $x/y = \sqrt{5}/2 - 1/2 \cong 0.618$ (see Chapter 13, Problem 13.9).

If a line 12.5 m long is to be divided in the Golden Ratio, find the length of each segment.

Solution

If x equals the shorter segment,

$$\frac{x}{12.5 - x} = 0.618$$

Multiplying by $12.5 - x$

$$x = 0.618(12.5) - 0.618x$$

Transposing,

$$x + 0.618x = 7.73$$

Factoring,

$$x(1 + 0.618) = 7.73$$

Dividing,

$$x = \frac{7.73}{1.618} = 4.77 \text{ m}$$

and the longer piece is $12.5 - 4.77 = 7.73$ m.

4.18   The series of numbers   2, 5, 12.5, 31.25, ...   forms a *geometric progression*, where the ratio of any term to the one preceding it is a constant.   Find this ratio.

Solution

Dividing each term by the preceding one,

$$\frac{5}{2} = 2.5 \qquad \frac{12.5}{5} = 2.5 \qquad \frac{31.25}{12.5} = 2.5$$

This quantity is called the *common ratio*.

SOLVED PROBLEMS FROM TECHNOLOGY

4.19   A rod of cross-section area a and length L will stretch by an amount e when subject to a tensile load P.   The modulus of elasticity is given by

$$\boxed{E = \frac{PL}{ae}} \quad \textbf{192}$$

Solve this equation for the elongation e.

Solution

Multiplying both sides by e,

$$Ee = \frac{PL}{a}$$

Dividing by E,

$$e = \frac{PL}{aE}$$

4.20   When a bar of length L_0 having a coefficient of linear thermal expansion α is increased in temperature by an amount Δt, it will expand to a new length L, where

$$\boxed{L = L_0(1 + \alpha \Delta t)} \quad \mathbf{195}$$

Solve this equation for α.

Solution

Dividing both sides by L_0,

$$1 + \alpha \Delta t = \frac{L}{L_0}$$

Transposing the 1,

$$\alpha \Delta t = \frac{L}{L_0} - 1$$

Dividing by Δt,

$$\alpha = \frac{L/L_0 - 1}{\Delta t}$$

4.21   The correction for the sag in a surveyor's tape L ft in length, weighing w lb/ft, and pulled with a force of P lb is

$$C = \frac{w^2 L^3}{24P^2} \text{ ft}$$

Solve this equation for P.

Solution

Multiplying both sides by P^2,

$$CP^2 = \frac{w^2 L^3}{24}$$

Dividing by C,

$$P^2 = \frac{w^2 L^3}{24C}$$

Taking the square root of both sides,

$$P = \sqrt{\frac{w^2 L^3}{24C}} = \sqrt{\frac{w^2 L^2(L)}{4(6C)}} = \frac{wL}{2}\sqrt{\frac{L}{6C}}$$

4.22   If the resistance of a conductor is R_1 at temperature t_1, the resistance will change to a value R when the temperature changes to t, where

$$\boxed{R = R_1[1 + \alpha(t - t_1)]} \quad \textbf{208}$$

and α is the temperature coefficient of resistance at temperature t_1.   Solve this equation for t.

Solution

   Removing the brackets,

$$R = R_1 + R_1 \, \alpha(t - t_1)$$

Transposing R_1,

$$R_1\alpha(t - t_1) = R - R_1$$

Dividing by $R_1\alpha$,

$$t - t_1 = \frac{R - R_1}{R_1\alpha}$$

Transposing $-t_1$,

$$t = \frac{R - R_1}{R_1\alpha} + t_1$$

4.23   The impedance of a circuit having a resistance R, an inductance L, and a capacitance C is

$$\boxed{Z = \sqrt{R^2 + \left(\omega L - \frac{1}{\omega C}\right)^2}} \quad \textbf{213}$$

where ω is the angular frequency.   Solve this equation for L.

Solution

   Squaring both sides,

$$R^2 + \left(\omega L - \frac{1}{\omega C}\right)^2 = Z^2$$

Transposing R^2,

$$\left(\omega L - \frac{1}{\omega C}\right)^2 = Z^2 - R^2$$

Taking the square root,

$$\omega L - \frac{1}{\omega C} = \sqrt{Z^2 - R^2}$$

Transposing $-1/\omega C$,

$$\omega L = \sqrt{Z^2 - R^2} + \frac{1}{\omega C}$$

Dividing by ω,

$$L = \frac{1}{\omega}\sqrt{Z^2 - R^2} + \frac{1}{\omega^2 C}$$

4.24   The formula for the equivalent resistance R for the parallel combination of two resistors R_1 and R_2 is

$$\boxed{\dfrac{1}{R} = \dfrac{1}{R_1} + \dfrac{1}{R_2}} \quad \mathbf{202}$$

Solve this formula for R_1.

Solution

$$\frac{1}{R_1} = \frac{1}{R} - \frac{1}{R_2}$$

The common denominator on the right-hand side is RR_2.

$$\frac{1}{R_1} = \frac{1}{R} \cdot \frac{R_2}{R_2} - \frac{1}{R_2} \cdot \frac{R}{R}$$

$$= \frac{R_2}{RR_2} - \frac{R}{RR_2} = \frac{R_2 - R}{RR_2}$$

Taking the reciprocal of both sides,

$$R_1 = \frac{RR_2}{R_2 - R}$$

4.25   What resistance must be placed in parallel with $10\ \Omega$ so that the resistance of the parallel combination will be $8\ \Omega$?

Solution

Using the formula obtained in Problem 4.24,

$$R_1 = \frac{8(10)}{10 - 8} = \frac{80}{2} = 40\ \Omega$$

4.26   The formula for the displacement S of a freely falling body having an initial velocity V_0 and acceleration a is

$$\boxed{S = V_0 t + \frac{1}{2} a t^2} \quad \mathbf{178}$$

Solve this equation for a.

Solution

Transposing $V_0 t$,

$$\frac{1}{2} a t^2 = S - V_0 t$$

Multiplying by $2/t^2$,

$$a = \frac{2(S - V_0 t)}{t^2}$$

4.27   The formula for the amount of heat flowing through a wall by conduction is

$$q = \frac{kA(t_1 - t_2)}{L}$$

Solve this equation for t_2.

Solution

Multiplying by L/kA,

$$t_1 - t_2 = \frac{qL}{kA}$$

Transposing t_1,

$$-t_2 = \frac{qL}{kA} - t_1$$

Multiplying by -1,

$$t_2 = t_1 - \frac{qL}{kA}$$

4.28   The formula for the amount S obtained when investing an amount P for n years at a rate i is

$$\boxed{S = P(1 + ni)} \quad \mathbf{166}$$

Solve this equation for n.

Solution

Dividing by P,

$$1 + ni = \frac{S}{P}$$

Transposing 1,

$$ni = \frac{S}{P} - 1$$

Dividing by i,

$$n = \frac{S}{Pi} - \frac{1}{i}$$

4.29   A bar 6.00 m long having a cross-sectional area of 15.8 square centimeters (cm^2) is subject to a tensile load of 17 600 newtons (N).   The modulus of elasticity of the metal is 2.11×10^6 newtons per square centimeter (N/cm^2).   Use Eq. 192 to find the elongation in millimeters.

Solution

From Eq. 192,

$$e = \frac{PL}{aE} = \frac{(17\,600 \text{ N})(6.00 \text{ m})}{(15.8 \text{ cm}^2)(2.11 \times 10^6 \text{ N/cm}^2)} = 3.17 \times 10^{-3} \text{ m}$$

$$= 3.17 \text{ mm}$$

4.30 A distance of 25.0 yards (yd) is to be laid off with a steel tape. The 100-ft-long tape weighs 26.0 oz and is pulled with a force of 14.0 lb. Find the tape correction in feet, using the equation from Problem 4.21.

Solution

Before substituting in the equation, we make certain the units of our given quantities are as specified.

$$w = \frac{26.0 \text{ oz}}{100 \text{ ft}} \cdot \frac{1 \text{ lb}}{16 \text{ oz}} = 0.0163 \text{ lb/ft}$$

$$L = 25.0 \text{ yd} \times 3 \text{ ft/yd} = 75.0 \text{ ft}$$

$$P = 14.0 \text{ lb}$$

Substituting,

$$C = -\frac{(0.0163)^2(75.0)^3}{24(14.0)^2} = -0.0238 \text{ ft}$$

4.31 The resistance of a copper coil is 125 Ω at 20°C. The temperature coefficient of resistance is 0.003 93 at 20°C. Use Eq. 208 to find the resistance at 60°C.

Solution

Substituting into Eq. 208,

$$R = 125[1 + 0.003\,93(60 - 20)]$$
$$= 125[1 + 0.157] = 145 \ \Omega$$

4.32 The torque T delivered by a motor of a given horsepower P rotating at N rev/min is

$$T = \frac{33\,000P}{2\pi N} \text{ ft} \cdot \text{lb}$$

Find the torque delivered by a 1.5-hp motor rotating at 200 radians per second.

Solution

We first convert the speed to the units required in the formula.

$$N = \frac{200 \text{ rad}}{s} \cdot \frac{1 \text{ rev}}{2\pi \text{ rad}} \cdot \frac{60 \text{ s}}{\text{min}} = 1910 \text{ rev/min}$$

Substituting,

$$T = \frac{33\,000(1.5)}{2\pi\,(1910)} = 4.12 \text{ ft} \cdot \text{lb}$$

4.33 Using Eq. 178, find the displacement after 12 s of a body thrown downward with a speed of 25 ft/s.

Solution

Substituting, using $a = 32.2 \text{ ft/s}^2$,

$$s = \frac{25 \text{ ft}}{s} (12 \text{ s}) + \left(\frac{1}{2}\right) \frac{32.2 \text{ ft}}{s^2} (12 \text{ s})^2$$

$$= 300 + 2320 = 2620 \text{ ft}$$

4.34   The formula for the pressure loss h in a pipe is

$$h = \frac{6270 f L Q^2}{D^5} \text{ ft}$$

where f is the friction factor, L is the length of pipe in feet, Q is the flow rate in cubic feet per second, and D is the pipe diameter in inches. Compute the pressure drop in a 2.55-in.-diameter, 32.5-ft-long pipe, where $f = 0.0260$ and the flow rate is 215 gal/min.

Solution

We convert the flow rate to the units required in the formula (1 ft^3 = 7.48 gal),

$$Q = \frac{215 \text{ gal}}{\text{min}} \cdot \frac{1 \text{ min}}{60 \text{ s}} \cdot \frac{1 \text{ ft}^3}{7.48 \text{ gal}}$$

$$= 0.479 \text{ ft}^3/\text{s}$$

Substituting,

$$h = \frac{6270(0.0260)(32.5)(0.479)^2}{(2.55)^5}$$

$$= 11.3 \text{ ft}$$

4.35   Using Eq. 188, convert 68.5°F to degrees Celsius.

Solution

Substituting,

$$C = \frac{5}{9}(68.5 - 32)$$

$$= \frac{5}{9}(36.5) = 20.3°C$$

4.36   Use Eq. 167 to find the amount S obtained when $1500 is allowed to accumulate for 3 years at a compound interest rate of 7%.

Solution

Substituting, with $i = 0.07$,

$$S = P(1 + i)^n$$
$$= \$1500(1 + 0.07)^3$$
$$= \$1500(1.07)^3 = \$1837.56$$

4.37   Use the equation obtained in Problem 4.28 to find the number of years it would take for $1000 to accumulate to $1500, at a simple interest rate of 7.5%.

Solution

Substituting, with $i = 0.075$,

$$n = \frac{S}{Pi} - \frac{1}{i}$$

$$= \frac{1500}{1000(0.075)} - \frac{1}{0.075}$$

$$= 20 - 13.3$$

$$= 6.7 \text{ years}$$

DIMENSIONLESS RATIOS

In technology, it is often more convenient to work with the ratio of two quantities of the same kind, because the units then cancel, leaving the ratio dimensionless. Examples of some dimensionless ratios are

Poisson's Ratio	Fuel-Air Ratio	Turn Ratio
Humidity Ratio	Compression Ratio	Damage Ratio
Endurance Ratio	Load Ratio	Gear Ratio
Radian Measure of Angles		

4.38 The light-gathering ability, or "speed," of a lens is dependent not only upon its diameter but on the *ratio* of the focal length to diameter (the *focal ratio* or *f value*). Find the diameter of an $f/3.5$ lens having a focal length of 50.0 mm.

Solution

The focal ratio is

$$1 : 3.5$$

or

$$\frac{3.5}{1} = \frac{\text{focal length}}{\text{diameter}} = \frac{50.0}{\text{diameter}}$$

$$\text{Diameter} = \frac{50.0}{3.5} = 14.3 \text{ mm}$$

SPECIFIC VALUES

The word *specific* is often used to denote a ratio. Some specific values found in technology are

Specific Heat	Specific Volume	Specific Conductivity
Specific Weight	Specific Reluctance	Specific Speed

4.39 The specific gravity of a solid or liquid is the ratio of density of the substance to the density of water, at a standard temperature.

$$\boxed{\text{Specific gravity} = \frac{\text{density of substance}}{\text{density of water}}} \quad \textbf{186}$$

Taking the density of water as 62.4 lb/ft^3, find the density of iron having a specific gravity of 7.2.

Solution

If d = the density of iron, by Eq. 186,

$$\frac{d}{62.4} = 7.2$$

$$d = 7.2(62.4) = 450 \text{ lb/ft}^3$$

Supplementary Problems

4.40 State the degree of the following equations:

(a) $2x + x^4 + x = 0$ (b) $x^5 + x^3 + x^6 = 0$ (c) $x^2 + 2x + 1 = 0$

4.41   State the number of unknowns in the following equations:

     (a) $r^2 + s^3 + t^4 = 0$ (b) $5x + zy + 10z = 0$ (c) $7x^2 + 8p + z + 10w + 17q = 0$

4.42   Check the following apparent solutions:

     (a) $2x + 10 = 0$ $(x = -5)$ (b) $-7 + 21z = 0$ $\left(z = \dfrac{1}{3}\right)$ (c) $\dfrac{12}{x+1} + 7 = 14$ $(x = 1)$

4.43   Solve the following equations:

     (a) $x + 6 = 18$ (i) $9(y - 1) = 7$ (q) $\dfrac{x-5}{x+5} = \dfrac{x+1}{x-1}$

     (b) $7x + 13 = 11$ (j) $5(x + 3) = 4(x - 2)$ (r) $\dfrac{3}{x+2} + \dfrac{4x-5}{x+1} = \dfrac{2x+4x^2}{x^2+3x+2}$

     (c) $2x - 9 = 13$ (k) $\dfrac{x-1}{x-3} = 5$ (s) $\dfrac{1}{x-7} = \dfrac{7}{x} + \dfrac{2x-3}{x^2-7x}$

     (d) $11z + 59 = 9$ (l) $\dfrac{1}{z} + \dfrac{1}{3} = 1$ (t) $x + 3(x - 4) = 4$

     (e) $\dfrac{y}{7} = 11$ (m) $\dfrac{2}{5z} + 7 = 11$ (u) $\dfrac{2x-9}{3} = \dfrac{3x+4}{2}$

     (f) $\dfrac{x}{5} + 1 = 15$ (n) $\dfrac{1}{2x} - 4 = -7$ (v) $\dfrac{2x+3}{2x-4} = \dfrac{x-1}{x+2}$

     (g) $11 - \dfrac{y}{2} = 0$ (o) $\dfrac{1}{3x-3} + 2 = 17$ (w) $\dfrac{3}{x} - \dfrac{4}{5x} = \dfrac{1}{10}$

     (h) $-2 - \dfrac{z}{5} = 2$ (p) $\dfrac{b}{3} + \dfrac{2b+1}{7} = \dfrac{b-5}{11}$ (x) $\dfrac{2x+1}{x} + \dfrac{x-4}{x+1} = 3$

4.44   Solve the following equations for x:

     (a) $5(x - 3z) = 3x + 2z$ (c) $z = \dfrac{y}{3}(x + 4z)$ (e) $s = \dfrac{xL - a}{x - L}$

     (b) $\dfrac{bc}{x} = 3a$ (d) $\dfrac{b+c}{x} = b^2 + bc$ (f) $A = \dfrac{m}{x}(p + x)$

4.45   A bar 3.40 ft long having a cross-sectional area of 1.79 in^2 is subjected to a tensile load P. The modulus of elasticity of the metal is 2.94×10^6 lb/in^2 and the elongation is 0.025 77 in. Find the tensile load P in pounds.

4.46   A distance of 36.0 yd is to be laid off with a steel tape. The 100-ft-long tape weighs 3.00 lb and is pulled with a force of 20.2 lb. Find the tape correction in inches.

4.47   What resistance must be placed in parallel with 22 Ω so that the resistance of the parallel combination will be 3 Ω?

4.48   The resistance of a conductor is 102 Ω at 11°C. If the temperature coefficient of resistance is 2.37×10^{-3} per °C at 11°C, find the temperature at which the resistance will be 163 Ω.

4.49   Find the displacement after 3 min of a body thrown downward with a speed of 11 m/s. Take the acceleration due to gravity as 9.80 meters per second squared (m/s^2).

4.50   The pressure loss in a pipe is 7.90 ft, the friction factor is 0.0102, the length of the pipe is 112.9 in., and the flow rate is 98.2 gal/min.   Find the diameter of the pipe.   Use the formula from Problem 4.34.

4.51   Find the amount P which will accumulate to $2700 in 6 years at a compound interest rate of 5%.

4.52   Find the new length of a 15.8-m beam when its temperature is increased by 70°C, if the coefficient of linear expansion is 1.14×10^{-5} per degree Celsius.

Answers to Supplementary Problems

4.40   (*a*) Fourth degree   (*b*) sixth degree   (*c*) second degree

4.41   (*a*) Three   (*b*) three   (*c*) five

4.42   (*a*) Checks   (*b*) checks   (*c*) does not check

4.43   (*a*) 12   (*b*) $-\dfrac{2}{7}$ (*c*) 11   (*d*) -4.5 (*e*) 77   (*f*) 70   (*g*) 22   (*h*) -20

(*i*) 1.78   (*j*) -23 (*k*) 3.5   (*l*) $\dfrac{3}{2}$ (*m*) $\dfrac{1}{10}$ (*n*) $-\dfrac{1}{6}$ (*o*) 1.022   (*p*) -1.13

(*q*) 0   (*r*) $\dfrac{7}{4}$ (*s*) 6.5   (*t*) 4   (*u*) -6 (*v*) $-\dfrac{2}{13}$ (*w*) 22   (*x*) $\dfrac{1}{4}$

4.44   (*a*) $\dfrac{17z}{2}$ (*b*) $\dfrac{bc}{3a}$ (*c*) $\dfrac{3z}{y} - 4z$ (*d*) $\dfrac{1}{b}$ (*e*) $\dfrac{Ls - a}{s - L}$ (*f*) $\dfrac{mp}{A - m}$

4.45   3320 lb

4.46 -1.39 in.

4.47   3.47 Ω

4.48   263°C

4.49 1.60×10^5 m

4.50   1.29 in.

4.51 $2014.78

4.52   15.9 m

Chapter 5

Word Problems

5.1 SUGGESTIONS FOR SOLVING WORD PROBLEMS

The usual steps to be followed in solving a word problem are:

1. Study the problem statement.
2. Locate and label the unknown.
3. Estimate the answer.
4. Write and solve the equation.
5. Check the answer.

These steps are explained in more detail below.

5.2 STUDY THE PROBLEM STATEMENT

1. *Visualize* the situation described in the problem. Picture it in your mind.
2. *Draw a diagram.* Place upon it as much of the given information as possible. Show the unknown on the diagram if you can.
3. Find and decode the *key words* in the problem statement. Look up the meanings of unfamiliar words in a dictionary, and find the meanings of technical words in a handbook or textbook.

> **Example 5.1:** A problem contains the statement "*A clockwise couple of 357 newton-meters* $(N \cdot m)$ *is applied to the free end of a cantilever beam.*" You cannot solve the problem containing this statement without obtaining the meanings of the expressions
>
> Clockwise couple
>
> Newton-meters
>
> Free end
>
> Cantilever beam

4. Locate the words and expressions standing for mathematical operations such as

... the sum of ...

... is equal to ...

... the product of ...

Realize that there are several ways of verbally stating most mathematical operations.

> **Example 5.2:** The expression $(a + b)$ might be written
>
> | | |
> |---|---|
> | a plus b | a and b |
> | a increased by b | the sum of a and b |
> | a added to b | |

106

5. Don't give up.　Students sometimes quit when they don't see how to solve a problem after reading it once.　In fact it is unusual to be able to see the solution at this stage, and you should proceed to the next steps anyway.

5.3 LOCATE AND LABEL THE UNKNOWN

1. Search the problem statement to determine exactly *what is to be found*.　Look for phrases such as

> What is ... ?
>
> Find the ... .
>
> How much ... ?
>
> How long ... ?
>
> At what rate ... ?

2. Assign a *symbol* to the unknown.　Specify *units* of measure.

Example 5.3:　A problem statement contains the following sentence: "*Find the amount of land in the first purchase.*"　Label the unknown.

Solution:　If we choose acres as our units, we write, let　x = number of acres in first purchase.

Common Error	The unknown must be precisely specified so that it cannot be mistaken for other quantities in the problem.　Units, where applicable, must be included.

Example 5.4:　The following statements are not good enough for labeling the unknown in the previous example.

1. Let x = first purchase.　It is not clear whether x refers to land or money.
2. Let x = land.　Too vague.　Does this refer to the *length* of the land, or its *cost*, or its *area*?
3. Let x = area of the land.　Units are missing.
4. Let x = acres of land.　It is not clear whether this represents the first purchase, the second purchase, or the final amount.

Common Error	Students often neglect to define and label the unknown at the start of a problem.　*This is a step that must never be omitted.*

3. Write all other unknown quantities in the problem in terms of your original unknown.

Example 5.5:　A problem contains the question "*How many kilograms of alloy A and of alloy B must be melted together to produce 100 kg of the new (previously specified) alloy?*"

　Although two quantities are unknown (the amounts of alloy A and alloy B needed), we need only one unknown, because if　x = kilograms of alloy A required,　then　$100 - x$ = kilograms of alloy B required.

5.4 ESTIMATE THE ANSWER

Make some simplifying assumptions to get a *rough estimate* of the answer.

Example 5.6:   A problem states "*Two cars, 100 mi apart, start traveling toward each other at the same time. One car travels at 40 mi/h and the other at 60 mi/h. Find the time required for them to meet.*"   Guess the answer.

Solution: One way to get an approximate answer is to assume that both cars are traveling at the same speed, say 50 mi/h, with each one covering half the distance.   Then by Eq. 177,

$$\text{Time} = \frac{\text{distance}}{\text{rate}} \cong \frac{50 \text{ mi}}{50 \text{ mi/h}} = 1 \text{ h}$$

It is often possible to *bracket* the answer with two numbers between which the answer must lie.

Example 5.7:   A problem states "*How much pure tin must be added to 40 kg of solder containing 30 percent tin to raise the tin content to 50 percent?*"   Bracket the answer.

Solution: The solder already contains 12 kg of tin (30 percent of 40 kg of solder).   If we added, say, another 10 kg we would have 22 kg of tin in 50 kg of solder, or $22/50 \times 100 = 44$ percent tin,   which is less than the required 50 percent.   Adding 20 kg of tin would give us $32/60 \times 100 = 53$ percent tin,   which is too high.   We would thus expect the answer to lie between 10 and 20 kg of tin.   Further, we would expect it to lie closer to 20 kg than to 10 because 53 percent is closer to the required concentration than 44.

5.5   WRITE AND SOLVE THE EQUATION

(a)   Problem Statement.   Equations are sometimes given verbally in the problem statement.

Example 5.8:   Four times a certain number, decreased by five, is fifteen.   Find the number.

Solution: If $x =$ the number,   then the equation given in the problem statement is

$$\underbrace{4x}_{\substack{\text{four times a} \\ \text{certain number}}} \quad \underbrace{-5}_{\substack{\text{decreased} \\ \text{by five}}} \quad \underbrace{= 15}_{\text{is fifteen}}$$

Solving for x,

$$4x = 15 + 5 = 20$$

$$x = 5$$

Equations may be even more subtly hidden in the problem statement.

Example 5.9:   A problem contains the sentence "*A pilot flies to a distant airfield at 120 mi/h and returns to his starting point at 155 mi/h.*"   The equation implied here is

$$\text{Outward distance} = \text{return distance}$$

(b)   Mathematical Relationships.   Often the relationships between the quantities in a problem will not be given because they are mathematical ones which you are expected to know or to be able to find. A Table of Formulas is given in the front of this Outline.

Example 5.10:   One side of a right triangle is 18 cm long and the hypotenuse (the longest side) is 5 cm longer than the remaining side of the triangle.   Find the lengths of the unknown sides.

Solution: To solve this problem, we need to know the Pythagorean Theorem, Eq. 72.   If we let $x =$ length of one side, cm   the length of the hypotenuse is then $x + 5$.   By the Pythagorean Theorem,

$$x^2 + 18^2 = (x + 5)^2$$

Solving,

$$x^2 + 324 = x^2 + 10x + 25$$

$$10x = 299$$

$$x = 29.9 \text{ cm}$$

and the hypotenuse is

$$x + 5 = 34.9 \text{ cm}$$

(c) Formulas from Technology. Physical quantities behave according to certain laws, which are summarized as formulas. A sampling of formulas taken from several branches of technology are given in the Table of Formulas in the beginning of this Outline.

Example 5.11:   What resistance must be placed in series with 100 Ω to obtain a total resistance of 500 Ω?

Solution: To solve this problem, we must know that the equivalent resistance R of two resistors R_1 and R_2 placed in series is

$$\boxed{R = R_1 + R_2} \quad \mathbf{201}$$

Solving the problem is then a simple matter.

$$500 = R_1 + 100$$
$$R_1 = 400 \ \Omega$$

5.6   CHECKING

(a)   Check Against the Problem Statement

Example 5.12:   In seeking two numbers whose sum is 15 and whose product is 56, a student got the answers 6 and 9.   Do these check?

Solution: They obviously do not check.   Their sum is 15 but their product is not 56.

Common Error	Checking an answer by substituting in your equation is not good enough. The equation itself may be wrong. Check your answer with the problem statement.

(b)   Other Checks. Check the answer against your guess.   If they are very different, rework the problem.

Check the dimensions (units of measure) of the answer.   If, for example, you are solving for a rate, and the units of the answer come out in miles, you know there is a problem.

In geometric problems, check your answer against the diagram.   If it was drawn approximately to scale, the answer should appear the proper size.

Check the answer against your own knowledge and experience.   You should not get answers that show cars traveling at 500 mi/h, alloys containing 125 percent tin, and $100 investments that earn $600 per year.

SOLVING TYPICAL WORD PROBLEMS

5.7   NUMBER PROBLEMS

Number problems give the equation right in the problem statement.

Example 5.13:   If three times a certain number is decreased by five, the result is twenty-two.   Find the number.

Solution: Let x = the number.   Then

$$3x - 5 = 22$$

Solving for x,

$$3x = 27$$

$$x = 9$$

Checking, three times nine is 27, which, decreased by 5, is 22.

5.8   UNIFORM MOTION PROBLEMS

(a)   Basic Ideas about Uniform Motion. Motion is called *uniform* when the speed is constant.   The distance traveled at constant speed is related to the rate of travel and the elapsed time by

Uniform Motion	Distance = rate × time $D = Rt$

177

Example 5.14:   A car travels for 2 h at 50 kilometers per hour (km/h).   How far does it travel?

Solution: From Eq. 177, $D = 2(50) = 100$ km.

Common Error	Equation 177 applies only for *uniform* motion.

Example 5.15:   A stone is released and falls under the action of gravity for 10 s.   How far will it travel?

Solution: This is *not* a case of uniform motion, so Eq. 177 does not apply.   The equation for the displacement of a freely falling body is Eq. 178 and is discussed under quadratic equations, Chapter 13.

We also need the formula

Average speed = $\dfrac{\text{total distance traveled}}{\text{total elapsed time}}$

181

Example 5.16:   A man walks 5 mi in 1.5 h and then runs another mile in 15 min (0.25 h).   Find his average speed for the entire trip.

Solution

$$\text{Total distance} = 5 + 1 = 6 \text{ mi}$$

$$\text{Total time} = 1.5 + 0.25 = 1.75 \text{ h}$$

From Eq. 181,

$$\text{Average speed} = \frac{6}{1.75} = 3.43 \text{ mi/h}$$

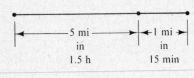

Fig. 5-1

Common Error	The average speed *is not* the average of the individual speeds.

Example 5.17: A boat travels 40 km upstream at 10 km/h and returns to the starting point at 20 km/h. Find the average speed for the round trip.

Solution: Let x = average speed, km/h.   By Eq. 177,

$$\text{Upstream time} = \frac{40}{10} = 4 \text{ h}$$

$$\text{Downstream time} = \frac{40}{20} = 2 \text{ h}$$

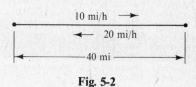

Fig. 5-2

Then

$$\text{Total time} = 4 + 2 = 6 \text{ h}$$

Also,

$$\text{Total distance} = 2(40) = 80 \text{ km}$$

By Eq. 181,

$$x = \frac{80}{6} = 13\frac{1}{3} \text{ km/h}$$

The average speed is *not* the average of the speeds,

$$x \neq \frac{10 + 20}{2} = 15 \text{ km/h}$$

(b)   A Typical Motion Problem

Example 5.18: A pilot flies to a distant airfield against a 30 mi/h headwind, and returns to his starting point with the aid of a 30 mi/h tailwind. His plane can go 120 mi/h in still air, and the total trip took 4 h flying time. Find the distance between the two airfields.

Solution: Let d = distance between airfields, mi.   The equation given in the problem statement is

$$\text{Time out} + \text{time back} = 4 \text{ h}$$

But, by Eq. 177,

$$\text{Time out} = \frac{\text{distance}}{\text{rate}} = \frac{d}{120 - 30} = \frac{d}{90}$$

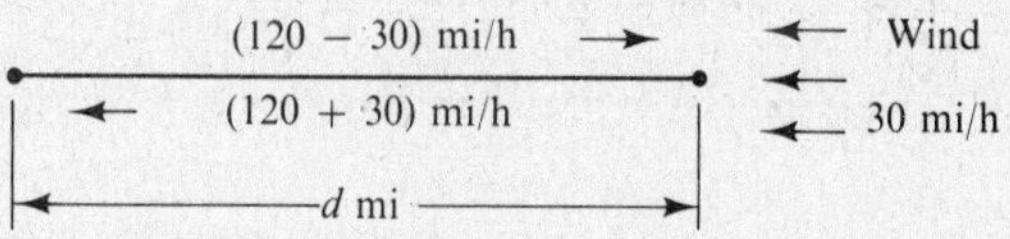

Similarly,

$$\text{Time back} = \frac{\text{distance}}{\text{rate}} = \frac{d}{120 + 30} = \frac{d}{150}$$

Fig. 5-3

So

$$\frac{d}{90} + \frac{d}{150} = 4$$

Solving for d, we multiply by the LCD, 450,

$$5d + 3d = 8d = 1800$$

$$d = 225 \text{ mi}$$

Check

$$\text{Time out} = \frac{225}{90} = 2.5 \text{ h}$$

$$\text{Time back} = \frac{225}{150} = 1.5 \text{ h}$$

$$\text{Total time} = 2.5 + 1.5 = 4 \text{ h} \qquad \text{checks}$$

5.9 MIXTURE PROBLEMS

(a) Basic Relationships. The total amount of mixture is equal to the sum of the amounts of the ingredients.

$$\boxed{\text{Total amount of mixture} = \text{amount of } A + \text{amount of } B + \cdots} \qquad \textbf{157}$$

Example 5.19:   Ten pounds of cement is mixed with 20 lb of sand and 40 lb of aggregate. What is the weight of the mixture?

Solution: By Eq. 157,

$$\text{Weight} = 10 + 20 + 40 = 70 \text{ lb}$$

For each ingredient,

$$\boxed{\text{Final amount of each ingredient} = \text{initial amount} + \text{amount added} - \text{amount removed}} \qquad \textbf{158}$$

Example 5.20:   A tank containing 10 liters of solution, half alcohol and half water, is drained of 4 liters. Then 5 liters of alcohol are added. How much alcohol is contained in the final mixture.

Solution

$$\text{Initial volume of alcohol} = 0.5(10) = 5 \text{ liters}$$

$$\text{Amount removed} = 0.5(4) = 2 \text{ liters}$$

$$\text{Amount added} = 5 \text{ liters}$$

By Eq. 158,

$$\text{Final amount of alcohol} = 5 - 2 + 5 = 8 \text{ liters}$$

The percent concentration of each ingredient is

$$\boxed{\text{Percent concentration of ingredient } A = \frac{\text{amount of ingredient } A \text{ in mixture}}{\text{total amount of mixture}} \times 100} \qquad \textbf{5}$$

Example 5.21:   An 8-gal radiator contains $1\frac{1}{2}$ gal of antifreeze. What is the percent concentration of antifreeze in the radiator?

Solution: By Eq. 5,

$$\text{Percent antifreeze} = \frac{1.5}{8} \times 100 = 18\frac{3}{4}\%$$

When *two mixtures* are combined into a third mixture, the amount of any ingredient A in the final mixture is

$$\boxed{\text{Final amount of } A = \text{amount of } A \text{ in mixture 1} + \text{amount of } A \text{ in mixture 2}} \qquad \textbf{159}$$

Example 5.22:   One hundred kilograms of steel containing 2% nickel is melted with 200 kg of steel containing 5% nickel. How much nickel is in the final alloy?

Solution: By Eq. 159,

$$\text{Final amount of nickel} = 0.02(100) + 0.05(200) = 2 + 10 = 12 \text{ kg}$$

(b) A Typical Mixture Problem

Example 5.23:   How much solder containing 30% tin and 70% lead must be combined with another solder containing 60% tin and 40% lead to make 100 lbs of 50/50 solder?

Solution: Let x = pounds of 30/70 solder needed.   Figure 5-4 shows the three alloys and the amount of tin in each.   By Eq. 157, the weight of the 60/40 solder is $100 - x$.   The weight of tin it contains is, by Eq. 5, $0.6(100 - x)$.   The weight of the tin in the 30/70 solder is $0.3x$.   By Eq. 159, the sum of these must give the weight of tin in the final mixture,

$$0.3x + 0.6(100 - x) = 0.5(100)$$

Multiplying by 10,

$$3x + 6(100 - x) = 5(100)$$

Clearing parentheses,

$$3x + 600 - 6x = 500$$

$$3x = 100$$

$$x = 33\frac{1}{3} \text{ lb of 30/70 solder}$$

$$100 - x = 66\frac{2}{3} \text{ lb of 60/40 solder}$$

Check

$$\text{Total weight} = 33\frac{1}{3} + 66\frac{2}{3} = 100 \text{ lb} \qquad \text{checks}$$

$$\text{Pounds of tin} = 0.3\left(33\frac{1}{3}\right) + 0.6\left(66\frac{2}{3}\right) = 10 + 40 = 50 \qquad \text{checks}$$

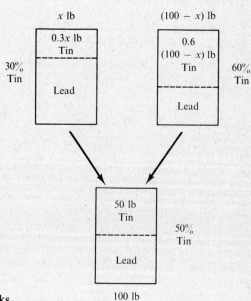

Fig. 5-4

5.10 STATICS PROBLEMS

(a) Basic Relationships.   The *moment of a force* about some point a is the product of the force F and the perpendicular distance d from the force to the point.

$$\boxed{M_a = Fd} \quad \textbf{172}$$

Example 5.24:   Find the moment of the force in Fig. 5-5 about

(*a*) point a

(*b*) point b

(*c*) point c

Solution

(*a*) $M_a = (15.3 \text{ lb})(2.43 \text{ ft}) = 37.2 \text{ ft·lb}$

(*b*) $M_b = (15.3 \text{ lb})(0) = 0$

(*c*) $M_c = (15.3 \text{ lb})(18.4) = 282 \text{ in.·lb}$

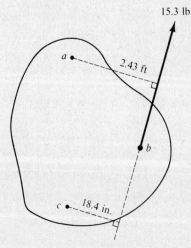

Fig. 5-5

If a body is in *equilibrium* (is not moving or moves with a constant velocity),

The sum of all horizontal forces acting on the body = 0	**173**
The sum of all vertical forces acting on the body = 0	**174**
The sum of the moments about any point on the body = 0	**175**

Example 5.25:   The symmetrical cart in Fig. 5-6 is not moving. Find the force F_g exerted by the ground on *each* of the four wheels, and the force F_s exerted by the spring on the cart.

Solution: By Eq. 173,

$$18.4 + F_s = 0$$

$$F_s = -18.4 \text{ lb}$$

The minus sign indicates that F_s acts in a direction opposite to the 18.4-lb force on the left of the cart.
     By Eq. 174,

$$51.6 + 4F_g = 0$$

$$F_g = \frac{-51.6}{4} = -12.9 \text{ lb}$$

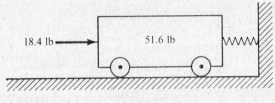

Fig. 5-6

Example 5.26:   Find the scale reading R in Fig. 5-7.

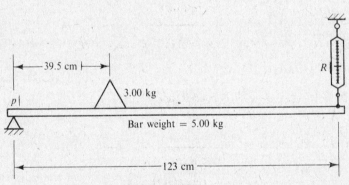

Fig. 5-7

Solution: We assume the weight of the bar to be concentrated at its center of gravity, 61.5 cm from either end (123 cm divided by 2). Taking moments about point p, we get by Eq. 175

$$123R = 3.00(39.5) + 5.00(61.5)$$

$$R = \frac{3.00(39.5) + 5.00(61.5)}{123} = 3.46 \text{ kg}$$

(b) A Typical Statics Problem

Example 5.27:   A bar of uniform cross section is 42.6 in. long and weighs 8.06 lb. A weight of 12.5 lb is suspended from one end. The bar and weight combination is to be suspended from a cable attached at the balance point. How far from the weight should the cable be attached, and what is the tension T in the cable?

Solution: We draw a diagram, Fig. 5-8, and identify the unknown.   Let x = distance from the weight to the attachment point a in inches.

We treat the weight of the bar as if it were concentrated at the midpoint (the center of gravity) of the bar. Its distance from a is $21.3 - x$.   Then by Eq. 175,

$$12.5x = 8.06(21.3 - x)$$

Solving for x,

$$12.5x = 8.06(21.3) - 8.06x$$

$$(12.5 + 8.06)x = 8.06(21.3)$$

$$x = \frac{8.06(21.3)}{12.5 + 8.06} = 8.35 \text{ in.}$$

Finding the tension in the cable, by Eq. 174,

$$T = 12.5 + 8.06$$

$$= 20.6 \text{ lb}$$

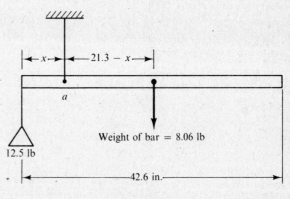

Fig. 5-8

5.11 WORK PROBLEMS

(a)   Basic Relationships

> If a job can be completed in n days, then $1/n$ of a job can be done in one day.   **161**

Example 5.28:   A pipe can fill a certain swimming pool in four days.   How much can be filled in one day?

Solution: From Eq. 161, using $n = 4$,

$$\text{Amount filled in one day} = \frac{1}{4} \text{ pool}$$

The second idea is the obvious one that

> The total amount of work done equals the sum of the fractions of the job done by each worker.   **163**

Example 5.29:   If worker A does 1/3 of a job per day and worker B does 1/4 of a job per day, find the total amount of work done by both in one day.

Solution: From Eq. 163,

$$\text{Total work} = \frac{1}{3} + \frac{1}{4} = \frac{4}{12} + \frac{3}{12} = \frac{7}{12} \text{ of a job}$$

These ideas assume that the work is done at a *uniform rate*.

(b)   A Typical Work Problem

Example 5.30:   Smith can assemble a certain machine in four days and Jones can assemble an identical one in five days.   How long will it take the two men working together to assemble three machines?

Solution: Let x = number of days to assemble three machines.   In one day, by Eq. 161,

Smith can assemble 1/4 machine

Jones can assemble 1/5 machine

By Eq. 163, together they can assemble

$$\frac{1}{4} + \frac{1}{5} = \frac{9}{20} \text{ machines per day}$$

In x days, they can assemble

$$\frac{9}{20} x \text{ machines}$$

So

$$\frac{9}{20} x = 3$$

Solving,

$$x = \frac{20(3)}{9} = 6\frac{2}{3} \text{ days}$$

Check

$$\frac{9/20 \text{ machines}}{\text{day}} \times 6\frac{2}{3} \text{ days} = \frac{9}{20} \times \frac{20}{3} = \frac{9}{3} = 3 \text{ machines}$$

5.12   FLUID FLOW

(a) **Basic Relationships.**   The total amount A which has flowed past some point in elapsed time T is the product of the elapsed time and the flow rate Q.

Amount of flow	$A = QT$	**160**

Example 5.31:   Water flows from a tank at the rate of 13.5 liters/min.   How much has flowed out after an hour?

Solution: By Eq. 160,

$$A = \frac{13.5 \text{ liters}}{\text{min}} \times 60 \text{ min} = 810 \text{ liters}$$

The amount of flow may be expressed in terms of weight, as well as in terms of volume.

Example 5.32:   Oil flows past a point in a pipeline at a rate of 82.0 lb/h.   How much oil passes in 48 min?

Solution: By Eq. 160,

$$A = \frac{82.0 \text{ lb}}{\text{h}} \times \frac{48}{60} \text{ h} = 65.6 \text{ lb}$$

For a tank, the amount remaining in the tank equals the initial amount plus the amount added minus the amount removed.

Final amount = initial amount + amount added − amount removed	**158**

Example 5.33:   A tank contains 255 liters of brine initially.   Brine is then added at the rate of 124 liters/h and at the same time drained off at the rate of 95 liters/h.   How much brine is in the tank after 2 h?

Solution: By Eq. 160,

$$\text{Amount added} = 124(2) = 248 \text{ liters}$$

$$\text{Amount removed} = 95(2) = 190 \text{ liters}$$

Then by Eq. 158,

$$\text{Final amount} = 255 + 248 - 190 = 313 \text{ liters}$$

(b)   A Typical Fluid Flow Problem

Example 5.34:   A 1000-gal tank is half full.   Water runs in at a rate of 212 gal/h.   After 2 h, a drain is opened and water runs out at a rate of 133 gal/h.   In how many hours will the tank be full?

Solution: Let x = number of hours to fill tank,   starting from when the tank is half full.   By Eq. 160,

$$\text{Amount added} = 212x$$

$$\text{Amount removed} = 133(x - 2)$$

By Eq. 158,

$$1000 = 500 + 212x - 133(x - 2)$$

$$500 = 212x - 133x + 266$$

$$79x = 234$$

$$x = 2.96 \text{ h}$$

Solved Problems

5.1   Write $(a - b)$ seven different ways.

Solution

> a decreased by b
> a minus b
> b subtracted from a
> The amount by which b is less than a
> The amount by which a is greater than b
> The difference between a and b
> a diminished by b

5.2   Write ab four different ways.

Solution

> The product of a and b
> a multiplied by b
> a times b
> a increased by a factor of b

5.3   Write a/b three different ways.

Solution

> a divided by b
> The quotient of a and b
> The ratio of a to b

5.4 Discuss what is wrong with the following statements: (*a*) The speed limit is 50 mi. (*b*) The clock is running 10 min fast. (*c*) The clock is ahead by 10 min per day. (*d*) The energy output was 550 kilowatts (kW).

Solution

(*a*) The unit given (miles) is not a unit of speed.   A correct unit would be miles per hour, for example.

(*b*) The rate of gain of the clock should be given as amount gained per unit time, such as 10 min/day.

(*c*) The word "ahead" implies an amount, not a rate, and so does not agree with the units given.   The "per day" should be omitted.

(*d*) The kilowatt is a unit of power, not energy.

5.5 Rewrite the following verbal expressions as algebraic expressions:

(*a*) Three more than four times a certain number.

(*b*) Three consecutive integers.

(*c*) Two numbers whose sum is 350.

(*d*) Two numbers whose difference is 25.

(*e*) The amounts of alcohol and of water in 200 gal of an alcohol-water solution.

(*f*) A fraction whose denominator is 2 more than 4 times its numerator.

(*g*) The distance traveled in x h by a car going 50 km/h.

(*h*) The angles in a triangle, if one angle is twice another.

(*i*) The number of gallons of antifreeze in a radiator containing x gallons of a mixture which is 40% antifreeze.

Solution

(*a*) If $x =$ the number,   the expression becomes $4x + 3$.

(*b*) If $x =$ the first integer,   the others are $x + 1$ and $x + 2$.

(*c*) If $x =$ one number, $350 - x$ is the other (not $x - 350$).

(*d*) If $x =$ the smaller number, $x + 25$ is the larger.

(*e*) If $x =$ gallons of alcohol, $200 - x =$ gallons of water.

(*f*) If $x =$ the numerator, $4x + 2 =$ the denominator.   The fraction is then $x/(4x + 2)$.

(*g*) Since   distance = rate × time,   distance = $50x$ km.

(*h*) If $x =$ one angle, $2x =$ a second angle,   then $180 - x - 2x = 180 - 3x$ is the third angle, since the three must add up to 180 degrees.

(*i*) Gallons of antifreeze will be 40 percent of the total, or $0.4x$.

NUMBER PROBLEMS

5.6 Twelve less than 5 times a certain number is 8.   Find the number.

Solution

Let $x =$ the number, then

$$5x - 12 = 8$$

$$5x = 20$$

$$x = 4$$

5.7 The sum of 8 and some number equals 3 times that number.   Find it.

Solution

Let x = the number, then

$$x + 8 = 3x$$
$$2x = 8$$
$$x = 4$$

5.8 When 2 is added to 3 times a certain number, the result is equal to subtracting 5 from 4 times that number.   Find the number.

Solution

Let x = the number, then

$$3x + 2 = 4x - 5$$
$$3x - 4x = -5 - 2$$
$$-x = -7$$
$$x = 7$$

Check

$$3(7) + 2 \overset{?}{=} 4(7) - 5$$
$$21 + 2 \overset{?}{=} 28 - 5$$
$$23 = 23 \quad \text{checks}$$

5.9 Find three consecutive integers whose sum is 27.

Solution

Let x = first integer

then $x + 1$ = second integer

$x + 2$ = third integer

and $x + (x + 1) + (x + 2) = 27$

$$3x + 3 = 27$$
$$x = 8$$

$$x + 1 = 9 \qquad \text{and} \qquad x + 2 = 10$$

So the three integers are 8, 9, and 10.

5.10 The denominator of a fraction is 5 greater than 2 times its numerator, and the reduced value of the fraction is 1/3.   Find the numerator.

Solution

Let x = the numerator.   Then

$$\frac{x}{2x + 5} = \frac{1}{3}$$

Solving,

$$3x = 2x + 5$$
$$x = 5$$

MOTION PROBLEMS

5.11 A truck can travel from a construction site to a borrow pit at the rate of 50 mi/h, but returns loaded at 35 mi/h. What is the average speed for the round trip?

Solution

Let x = average speed, mi/h. By Eq. 177, the time to travel to the pit is $d/50$, where d is the one-way distance, and the time to return is $d/35$. The total travel time is then

$$\frac{d}{50} + \frac{d}{35} = \text{total time}$$

and $2d = \text{total distance}$

By Eq. 181, the average speed is

$$V_{avg} = \frac{2d}{\dfrac{d}{50} + \dfrac{d}{35}}$$

Multiplying numerator and denominator of this fraction by $1/d$,

$$V_{avg} = \frac{2}{\dfrac{1}{50} + \dfrac{1}{35}} = 41 \text{ mi/h}$$

5.12 The pointer of a recording voltmeter can travel to the right at the rate of 8 centimeters per second (cm/s). What must be the return rate if the total time for the pointer to traverse the full 10-cm scale and return to zero must not exceed 1.5 s?

Solution

Let x = return rate, cm/s. We are given the information that the time for full-scale deflection + return time = 1.5 s. Then, by Eq. 177,

$$\frac{10}{8} + \frac{10}{x} = 1.5$$

Solving, $\dfrac{10}{x} = 1.5 - 1.25 = 0.25$

$$x = \frac{10}{0.25} = 40 \text{ cm/s}$$

5.13 A shaper is set to have a forward cutting speed of 50 ft/min, and a stroke of 18 in. It is observed to make 80 cuts (and returns) in 3 min. What is the return speed?

Solution

Let x = return speed, ft/min. The time for 1 cut and return is

$$\frac{3 \text{ min}}{80 \text{ cuts}} = 0.0375 \text{ min/cut}$$

So, for one cut and return,

Cutting time + return time = 0.0375

but by Eq. 177,

$$\text{Cutting time} = \frac{1.5}{50} \quad \text{and} \quad \text{Return time} = \frac{1.5}{x}$$

so $\dfrac{1.5}{50} + \dfrac{1.5}{x} = 0.0375$

Solving,
$$1.5x + 1.5(50) = 0.0375(50)x$$
$$1.875x - 1.5x = 75$$
$$0.375x = 75$$
$$x = 200 \text{ ft/min}$$

MIXTURE PROBLEMS

5.14 One hundred twenty liters of fuel, containing 3 percent oil, is available for a certain 2-cycle engine. This fuel is to be used for another engine requiring a 10 percent oil mixture. How many liters of oil must be added?

Solution

Let
$$x = \text{liters of oil added}$$
Then

$$0.03(120) = \text{liters of oil in original mixture}$$
$$120 + x = \text{liters of final mixture}$$
$$0.1(120 + x) = \text{liters of oil in final mixture}$$

By Eq. 158,
$$0.03(120) + x = 0.1(120 + x)$$
Solving,
$$3.6 + x = 12 + 0.1x$$
$$0.9x = 8.4$$
$$x = 9.3 \text{ liters}$$

5.15 A concrete mixture is to be made which contains 40 percent sand, by weight. A ton of mixture containing 25 percent sand is already on hand. How many pounds of sand must be added to this mixture to arrive at the required 40 percent?

Solution

Let
$$x = \text{pounds of sand to be added}$$
Then

$$0.25(2000) = \text{pounds of sand in original mixture}$$
$$2000 + x = \text{total weight of final mixture}$$
$$0.4(2000 + x) = \text{pounds of sand required in final mixture.}$$

Then by Eq. 158,
$$0.25(2000) + x = 0.4(2000 + x)$$
Solving for x,
$$500 + x = 800 + 0.4x$$
$$0.6x = 300$$
$$x = 500 \text{ lb sand}$$

5.16 A certain automatic soldering machine requires a solder containing half tin and half lead. How much pure tin must be added to 40 kg of a solder containing 70% lead and 30% tin to raise the tin content to 50 percent?

Solution

Let $x = $ kg of tin to be added. Then $40 + x = $ final weight of the alloy, kg. By Eq. 158,

Final amount of tin = initial amount + amount added $= 0.3(40) + x$

Then by Eq. 5,

$$\text{Percent tin} = \frac{\text{final amount of tin}}{\text{final amount of alloy}} \times 100$$

so

$$50 = \frac{0.3(40) + x}{40 + x}(100)$$

Solving for x,

$$12 + x = 20 + 0.5x$$

$$0.5x = 8$$

$$x = 16 \text{ kg of tin to be added}$$

5.17   A certain ore contains 22% copper, and a second ore contains 39% copper. How many tons of each must be mixed to obtain 85 tons of ore containing 34% copper?

Solution

Let

$$x = \text{tons of } 22\% \text{ ore needed}$$

so

$$85 - x = \text{tons of } 39\% \text{ ore needed}$$

By Eq. 159,

$$0.22x + 0.39(85 - x) = 0.34(85)$$

Solving for x,

$$0.22x + 33.2 - 0.39x = 28.9$$

$$0.17x = 4.3$$

$$x = 25 \text{ tons of } 22\% \text{ ore}$$

$$85 - x = 60 \text{ tons of } 39\% \text{ ore}$$

STATICS PROBLEMS

5.18   A beam of length L has a concentrated load P at a distance d from one end as in Fig. 5-9. Neglecting the weight of the beam, write an expression for the reactions at each support.

Solution

Let

$$R_1 = \text{reaction at left support}$$

$$R_2 = \text{reaction at right support}$$

Taking moments about the left end, by Eq. 175,

$$R_2 L = Pd$$

$$R_2 = \frac{Pd}{L}$$

Similarly, taking moments about the right end,

$$R_1 L = P(L - d)$$

$$R_1 = \frac{P(L - d)}{L}$$

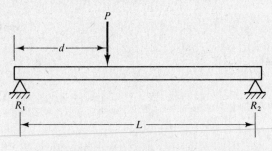

Fig. 5-9 Beam with concentrated load.

5.19   A cantilever beam has an additional support 6 m from the built-in end, as in Fig. 5-10. The beam is 15 m long and has a concentrated load of 10 000 kg at the free end. Find the vertical reactions at the built-in end and at the support.

Solution

Let R_1 and R_2 be the vertical reactions at the built-in end and the support, respectively, as in Fig. 5-10. The fulcrum can produce only an upward force on the beam, so it is necessary for R_1 to be in the downward direction to keep the beam from rotating.

Taking moments about the support, we get by Eq. 175,

$$6R_1 = 9(10\,000)$$

$$R_1 = 15\,000 \text{ kg}$$

Taking moments about the built-in end,

$$6R_2 = 15(10\,000)$$

$$R_2 = 25\,000 \text{ kg}$$

As a check, we note that the sum of the two downward forces, 10 000 kg and 15 000 kg, equals the single upward force, 25 000 kg, as Eq. 174 says it must.

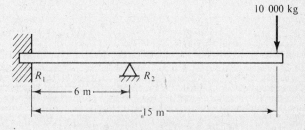

Fig. 5-10   Cantilever beam

WORK PROBLEMS

5.20   It is estimated that bulldozer A can prepare a certain building site in 5.5 days, and that the larger bulldozer B can prepare the same site in 4 days. If we assume that they do not get in each other's way, how long will it take the two machines, working together, to prepare the site?

Solution

Let t = number of days required,   working together.   By Eq. 161, if machine A can do the whole job in 5.5 days, it can do 1/5.5 of a job in 1 day,   and $t/5.5$ of a job in t days. Similarly, machine B can do $t/4$ of a job in t days.   Together they do 1 job in t days, so from Eq. 163,

$$\frac{t}{5.5} + \frac{t}{4} = 1 \text{ job}$$

Solving for t, we multiply by the LCD,

$$4t + 5.5t = 5.5(4)$$

$$t = \frac{(5.5)(4)}{9.5} = 2.3 \text{ days}$$

5.21   Technician A can wire a receiver in 10 h.   After working for 3 h he is joined by technician B who, working alone, could wire the receiver in 6 h.   If each continued working at his usual rate, how many additional hours would it take to finish the job?

Solution

Let x = additional hours needed.   By Eq. 161, technician A can do 1/10 of the job per hour, and since he works for $(3 + x)$ hours, will do $(3 + x)/10$ = fraction of job done by A.   Similarly, $x/6$ = fraction of job done by B.   Then by Eq. 163, the sum of these two fractions must equal one job.

$$\frac{3 + x}{10} + \frac{x}{6} = 1$$

Solving,

$$18 + 6x + 10x = 60$$

$$16x = 42$$

$$x = 2\frac{5}{8} \text{ h}$$

5.22 A punch press can produce a box of stampings in 15 h. A new machine is to be ordered having a speed such that both machines working together would stamp a box of parts in 5 h. How long would it take the new machine alone to stamp a box of parts?

Solution

Let x = number of hours for new machine to stamp one box of parts. By Eq. 161, the old machine can stamp 1/15 box/h, and the new machine, $1/x$ box/h. Working together, they stamp $(1/15 + 1/x)$ box/h, and so in 5 h,

$$5\left(\frac{1}{15} + \frac{1}{x}\right) = 1 \text{ box}$$

Solving,

$$\frac{1}{x} = \frac{1}{5} - \frac{1}{15} = \frac{2}{15}$$

$$x = \frac{15}{2} = 7.5 \text{ h}$$

FLUID FLOW PROBLEMS

5.23 A settling tank contains 6000 ft^3 of water. At what rate (gal/min) must water be pumped out to empty the tank in 5 h, if water is simultaneously running in at the rate of 10 gal/min?

Solution

Let Q = rate of removal, gal/min. Then, by Eq. 160,

$$\text{Amount added} = 10(5)(60) = 3000 \text{ gal}$$

$$\text{Amount removed} = Q(5)(60) = 300Q \text{ gal}$$

and

$$\text{Initial amount} = 6000 \text{ ft}^3 \ (7.48 \text{ gal/ft}^3)$$

$$= 44\,880 \text{ gal}$$

By Eq. 158,

$$\text{Final amount} = 0 = 44\,880 + 3000 - 300Q$$

$$300Q = 47\,880$$

$$Q = 160 \text{ gal/min}$$

FINANCIAL PROBLEMS

5.24 What salary should a person receive so that she would take home $15,000 after deducting 22 percent for taxes?

Solution

Let x = salary before taxes, dollars.

Then

$$x - 0.22x = \$15,000$$

Solving,

$$0.78x = \$15,000$$

$$x = \$19,230.77$$

5.25   A consultant had to pay income taxes of $4380 plus 32 percent of the amount by which his taxable income exceeded $20,000. His tax bill was $5132. What was his taxable income?

Solution

Let $x =$ taxable income, dollars. From the problem statement,

$$\text{Tax} = \$4380 + (\text{taxable income} - \$20,000)(0.32)$$

$$\$5132 = \$4380 + 0.32(x - \$20,000)$$

$$\$5132 - \$4380 = 0.32x - \$6400$$

$$0.32x = \$7152$$

$$x = \$22,350.00$$

5.26   A company invested $98,000, part at 6% and the remainder at $8\frac{1}{2}\%$, simple interest. It received $6180 in interest after a year. How much was invested at each rate?

Solution

Let $x =$ dollars invested at 6%; then, $\$98,000 - x =$ dollars invested at $8\frac{1}{2}\%$. From Eq. 1, the 6% investment will earn $0.06x$ dollars, and the $8\frac{1}{2}\%$ investment will earn $0.085(98,000 - x)$ dollars. Together

$$0.06x + 0.085(\$98,000 - x) = \$6180$$

Solving,

$$0.06x + 0.085(\$98,000) - 0.085x = \$6180$$

$$0.025x = \$2150$$

$$x = \$86,000 = \text{amount at } 6\%$$

$$\$98,000 - x = \$12,000 = \text{amount at } 8\tfrac{1}{2}\%$$

5.27   A power company changed its rates from 6.3¢ per kilowatthour (kWh) to 5.1¢ per kWh plus $1.10 per month service charge. How much power can you purchase before your monthly bill would exceed the bill under the former rate structure?

Solution

Let $x =$ kilowatthours of power purchased. By Eq. 165,

$$\text{Former cost} = 0.063x \text{ dollars}$$

$$\text{New cost} = 0.051x + 1.10 \text{ dollars}$$

To find the value of x at which the old and new costs are equal, we equate the two.

$$0.063x = 0.051x + 1.10$$

$$0.012x = 1.10$$

$$x = 92 \text{ kWh}$$

DC CIRCUIT PROBLEMS

5.28   A voltage divider is to be constructed using two resistors, as in Fig. 5-11. The ratio of the voltages across each resistor is to be 2 to 3, and the current drain from the battery is to be 1 milliampere (mA). Find the required values for the two resistors. The voltage from the battery is 6V.

Solution

Let R = one resistance, Ω. Since the voltage drop across a resistor is proportional to the resistance, the two resistors must be in the ratio of 2 to 3; so let $(2/3)R$ = other resistor, ohms. By Eq. 201, the total resistance is

$$R + \frac{2}{3}R$$

and by Ohm's law, Eq. 200, $R = V/I$.

$$R + \frac{2}{3}R = \frac{6}{0.001}$$

Solving,

$$\frac{5}{3}R = 6000$$

$$R = 3600\ \Omega = \text{first resistance}$$

$$\frac{2}{3}R = 2400\ \Omega = \text{second resistance}$$

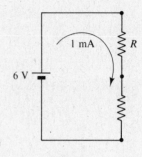

Fig. 5-11

5.29 A voltage divider is to be constructed using three resistors, as in Fig. 5-12. The voltage drop across R_1 is to be twice that across R_3, while the drop across R_2 is to be 1V greater than 3 times the drop across R_3. The current I though the divider is 5 mA (milliamperes must be converted to amperes to fit the units of the problem: 5 mA = 0.005 A). Find the values of the three resistors.

Solution

Let $R_1 = 2R_3$. Since voltage drop across a resistor = IR, $IR_2 = 3(IR_3) + 1$. Dividing by I,

$$R_2 = \frac{3IR_3}{I} + \frac{1}{I}$$

Since $I = 0.005$ A,

$$R_2 = 3R_3 + \frac{1}{0.005} = 3R_3 + 200$$

By Ohm's law, Eq. 200,

$$R_1 + R_2 + R_3 = \frac{V}{I}$$

$$2R_3 + (3R_3 + 200) + R_3 = \frac{24}{0.005} = 4800$$

Solving,

$$6R_3 = 4600$$

$$R_3 = 767\ \Omega$$

$$R_1 = 2R_3 = 1534\ \Omega$$

$$R_2 = 3R_3 + 200 = 2500\ \Omega$$

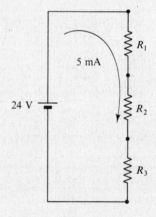

Fig. 5-12

A TEMPERATURE PROBLEM

5.30 On a certain day, the temperature difference on the Fahrenheit scale between inside and outside was 80°, and it was noted that the inside temperature was greater than freezing by 1.5 times the amount that the outside temperature was below freezing. What were the inside and outside temperatures?

Solution

Let $\qquad\qquad\qquad\qquad\qquad\qquad x =$ inside temperature, °F

Then $\qquad\qquad\qquad\qquad\qquad x - 80 =$ outside temperature, °F

The inside temperature is then $x - 32$ degrees above freezing, while the outside temperature is $32 - (x - 80)$, or $112 - x$ degrees below freezing. Thus $x - 32 = 1.5(112 - x)$. Solving,

$$x - 32 = 168 - 1.5x$$

$$2.5x = 200$$

$$x = 80°F \text{ inside}$$

$$x - 80 = 0°F \text{ outside}$$

Supplementary Problems

NUMBER PROBLEMS

5.31 Four more than six times a certain number is fifty-eight. Find the number.

5.32 Find a number such that the sum of twelve and that number is five times that number.

5.33 Find three consecutive odd integers whose sum is 33.

MOTION PROBLEMS

5.34 A motorist travels 105 km to another town at an average speed of 85 km/h. What must be his return rate if the total time for the second trip is not to exceed 2 h?

5.35 A truck departs at 8:15 A.M. traveling at an average speed of 45 mi/h, and a car leaves the same terminal 45 min later to overtake the truck. If the car averages 55 mi/h, at what time and at what distance from the terminal will it overtake the truck?

MIXTURE PROBLEMS

5.36 Two thousand kilograms of an alloy containing 4.6% tin is available. How many kilograms of pure tin must be added to raise the percentage of tin to 9.5%?

5.37 How many liters of a solution containing 24.7% alcohol, and how many liters of another solution containing 69.2% alcohol must be mixed together to make 25.5 liters of solution containing 38.8% alcohol? (All percentages are by volume.)

STATICS PROBLEMS

5.38 A horizontal uniform bar of negligible weight has a 5-lb weight hanging from one end and a 10-lb weight hanging from the other end. The bar is seen to balance 2.2 ft from the 10-lb weight. Find the length of the bar.

5.39   A horizontal uniform beam of negligible weight is 3.87 m long and is supported by columns at either end.   A concentrated load of 5.25 kg is applied to the beam.   At what distance from one end must this load be located so that the vertical reaction at that same end is 3.04 kg?

WORK PROBLEMS

5.40   If worker A can do a certain job in 6.8 days, and workers A and B can do the same job in 3.3 days working together, how long would it take worker B to do the job when working alone?

5.41   A tank can be filled by a certain pipe in 8.4 h.   Three hours after this pipe is opened, it is supplemented by a smaller pipe which, by itself, could fill the tank in 15.7 h.   Find the total time, measured from the opening of the larger pipe, to fill the tank.

FLUID FLOW PROBLEMS

5.42   A tank, initially full, is being emptied at the rate of 540 liters/h.   When 500 liters remained in the tank, an input valve was opened and liquid entered the tank at the rate of 1000 liters/h, while continuing to drain at the former rate.   The time to refill the tank, measured from the time the input valve was opened, is 5.3 h.   Find the capacity of the tank.

5.43   At what rate must liquid be drained from a tank in order to empty it in 5 h if the tank takes 3 h to fill at the rate of 185 gal/min?

FINANCIAL PROBLEMS

5.44   The labor costs for a certain project were $1100 per day for 15 technicians and helpers.   If the technicians earned $80/day and the helpers $70/day, how many technicians were employed on the project?

5.45   A company has $50,000 invested in bonds, and earns $4226 in interest annually.   Part of the money is invested at 7.5% and the remainder at 9.2% simple interest.   How much is invested at each rate?

5.46   How much must a company earn in order to have $500,000 left after paying 28 percent in taxes?

Answers to Supplementary Problems

5.31	9		**5.39**	1.63 m
5.32	3		**5.40**	6.4 days
5.33	9, 11, and 13		**5.41**	6.52 h
5.34	136 km/h		**5.42**	2938 liters
5.35	12:22 P.M., 186 mi		**5.43**	111 gal/min
5.36	108 kg		**5.44**	Five technicians
5.37	17.4 liters of 24.7% solution, 8.1 liters of 69.2% solution		**5.45**	$22,000 at 7.5% and $28,000 at 9.2%
5.38	6.6 ft		**5.46**	$694,444

Chapter 6

Functions and Graphs

6.1 DEPENDENT AND INDEPENDENT VARIABLES

Example 6.1:   In the equation

$$y = 2x + 3$$

y is called the *dependent variable* because its value depends upon the value given to x.   For example, if we let $x = 2$,   then $y = 7$;   or if $x = 3$,   then $y = 9$.   The variable x is called the *independent variable*.

Example 6.2:   If we rearrange the previous equation as follows,

$$x = \frac{y - 3}{2}$$

x is now said to be the dependent variable and y the independent variable.

Example 6.3:   If both variables are on the same side of the equal sign, as in the equation

$$2x - y + 3 = 0$$

neither x nor y is called dependent or independent.

6.2 ORDERED PAIRS OF NUMBERS

In technological problems, we must often deal with *pairs* of numbers, rather than with single quantities.

Example 6.4:   The transient current in a certain circuit is 3 A after an elapsed time of 2 s.   Thus, $(2, 3)$ would be called a number pair.

When the *order* in which a pair of numbers is written conveys information about the numbers, the pair of numbers is called an *ordered pair*.   In ordered pairs, the independent variable is usually written first, followed by the dependent variable.   Ordered pairs are also called *corresponding values*.

Example 6.5:   For the equation in Example 6.1, the values $x = 2$ and $y = 7$ could be written as an ordered pair

$$(2, 7)$$

6.3 FUNCTIONS

If two variables x and y are somehow related so that a single value of y can be found for a given value of x, then the equation, rule, or procedure used to find y is called a *function*.

Functions are usually given in the form of *equations*.

Example 6.6:   The equation $y = 3x^2 - 4x + 2$ is a function.   We may find one and only one value for y corresponding to any given value of x.   For example, y equals 2 when x is zero.

Not all equations are functions.   For an equation to be a function, it must be *single-valued* (giving *just one* value of y for any value of x).

Example 6.7:   The equation $y = \pm\sqrt{x}$ is not a function because it yields two values of y for each value of x. It is called a *relation*.

Functions are sometimes given as *verbal statements*, *tables of ordered pairs*, or *graphs*.

Example 6.8:   The statement "*the circumference of a circle is equal to pi (π) times the diameter*" is a function.

Example 6.9:   The table of ordered pairs

x	1	2	3	4
y	3	5	8	12

is a function.

Example 6.10:   The graph, Fig. 6-1, is a function.

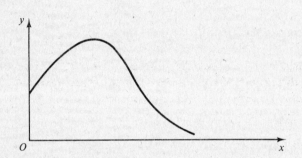

Fig. 6-1   A functional relationship in graphical form.

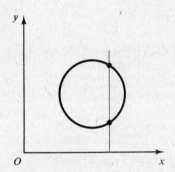

Fig. 6-2

Example 6.11:   The graph, Fig. 6-2, is *not* a function, because there are two values of y for single values of x.

It is sometimes helpful to visualize a function as a machine, as in Fig. 6-3. A value for the independent variable x is dropped into the hopper of the *function machine*, the crank is turned, and a single value of y comes out the chute. The method by which the machine chooses the value of the dependent variable y depends upon the particular function built into the machine. Of course, the machine must always give the same y for the same x.

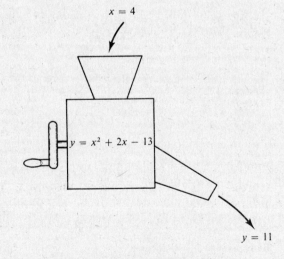

Fig. 6-3   The function machine.

6.4 IMPLICIT AND EXPLICIT FORM

When one variable is isolated on one side of the equals sign, the equation is said to be in *explicit form.*

Example 6.12:　　The equations

$$y = 2x + 6$$
$$z = 2x - 5w$$
$$x = y^2 - 8y$$

are all in explicit form.

When a variable is *not* isolated, the equation is in *implicit form.*

Example 6.13:　　The equations

$$x + y = 5$$
$$y = 2x - 3y$$
$$w + x = y + z$$

are all in implicit form.

6.5 FUNCTIONAL NOTATION

Just as we use the symbol x (or some other letter) to represent a *number*, without saying which number we are specifying, we need a symbol to represent a *function* without having to specify which particular function we are talking about.　Such a notation is

$$y = f(x)$$

This is read "y is a function of x" or "y equals f of x."　(It does *not* mean f times x.)

Of course, instead of x and y we may have any two variables, written in the form

$$(\text{Dependent variable}) = f(\text{independent variable})$$

Example 6.14:　　The equation　$y = 3x^2 + 4x$　can also be written,　$y = f(x) = 3x^2 + 4x$.

Example 6.15:　　The statement "*the amount of heat, Q, flowing through a wall depends upon the temperature difference T across the wall, the thickness t of the wall, and its cross-sectional area A,*" expressed in functional notation, is

$$Q = f(T, t, A)$$

Example 6.16:　　Given the equation

$$w = f(x, y) = \frac{x + 4}{3y}$$

rewrite the equation in the form,　$x = f(w, y)$.

Solution:　　This notation is a shorthand way of saying that the equation is to be solved for x.　Multiplying both sides by $3y$,

$$x + 4 = 3wy$$

Transposing,

$$x = f(w, y) = 3wy - 4$$

Example 6.17:   If

$$w = f(x, y) = 2x + y$$

and

$$x = f(y, z) = y - 4z$$

find

$$w = f(y, z)$$

Solution: We are asked, in shorthand, to find an equation for w in terms of y and z, but not x. So, in the equation

$$w = 2x + y$$

we substitute $y - 4z$ for x (from the second equation).

$$w = 2(y - 4z) + y$$
$$= 2y - 8z + y$$

and so

$$w = f(y, z) = 3y - 8z$$

Implicit functions can be written in the form

$$f(x, y) = 0$$

Example 6.18:   The equation $2x + 3y + 5 = 0$ can be represented by

$$f(x, y) = 0$$

If more than one function is to be put into functional notation, subscripts, or letters other than f, can be used.

Example 6.19

$$y = f_1(x)$$
$$y = g(x)$$
$$y = r_2(w)$$

6.6 EVALUATING FUNCTIONS

The substitution of numerical values for the independent variable in a function is conveniently represented in functional notation.

Example 6.20:   If $f(x) = x^2 + 2x$, find $f(3)$ and $4f(3)$.

Solution: The notation $f(3)$ means that 3 is to be substituted for x in the equation; so

$$f(3) = (3)^2 + 2(3) = 15$$

and

$$4f(3) = 4(15) = 60$$

Example 6.21:   If $f(x) = 3x - 1$, find $f(1), f(2)$, and $f(3)$.

Solution: Substituting,

$$f(1) = 3(1) - 1 = 2$$
$$f(2) = 3(2) - 1 = 5$$
$$f(3) = 3(3) - 1 = 8$$

Example 6.22:   For the function in the previous example, find

$$\frac{f(3) - 2f(1)}{3f(2)}$$

Solution: Substituting the previously computed values into the given expression,

$$\frac{f(3) - 2f(1)}{3f(2)} = \frac{8 - 2(2)}{3(5)} = \frac{4}{15}$$

Example 6.23:   If $f(z) = az^3 - a/z$, find $f(a)$.

Solution

$$f(a) = a(a)^3 - \frac{a}{a} = a^4 - 1$$

Example 6.24:   If $f(w, y) = 2wy - 3y^2$, find $f(4, 1)$.

Solution: Since $w = 4$ and $y = 1$,

$$f(4, 1) = 2(4)(1) - 3(1)^2 = 5$$

6.7 DOMAIN AND RANGE

Some functions are not defined for certain values of the variables.

Example 6.25:   The function

$$y = \sqrt{x}$$

is not defined for negative values of x, as these would give imaginary values for y.

The set of permissible values of x is called the *domain*, and the set of permissible values of y is called the *range*.   If the domain or range is not given with a function, they are usually taken to be all the real numbers except those that will result in division by zero, that will give imaginary numbers, or that will result in an illegal mathematical operation.

Example 6.26:   Find the domain and range of the function $y = \sqrt{x - 4}$.

Solution: The quantity under the radical sign, $x - 4$,   will be negative for any values of x less than 4, and thus will result in imaginary values for y.   Hence the domain of x is all real numbers equal to or greater than 4. Accordingly, the range of y includes all the positive real numbers and zero.

6.8 VARIATION

(a) Direct Variation.   If two variables are related by an equation of the form

$$\boxed{y = kx} \quad \mathbf{52}$$

where k is a constant, we say that y *varies directly* as x, or, that y is *directly proportional* to x.   The constant k is called the *constant of proportionality*.

Example 6.27:   The formula for the circumference of a circle of radius r is Eq. 94,

$$C = 2\pi r$$

Comparing this formula with Eq. 52, we see that the circumference varies directly as the radius, and that the constant of proportionality is 2π.

Example 6.28:   If y is directly proportional to x, and y is 8 when x is 2, find y when x is 5.

Solution: Since y varies directly as x, we use Eq. 52,

$$y = kx$$

To find the constant of proportionality, substitute the given values for x and y, 2 and 8,

$$8 = k(2)$$

so

$$k = 4$$

Our equation is then

$$y = 4x$$

When $x = 5$,

$$y = 4(5) = 20$$

Example 6.29:   If z varies directly as w, fill in the missing numbers in the following table of values:

w	0	1		4	
z			5	2	9

Solution: We find the constant of proportionality from the given pair of values $(4, 2)$.   Since

$$z = kw$$

substituting gives

$$2 = k(4)$$

$$k = \frac{1}{2}$$

so

$$z = \frac{w}{2}$$

With this equation we find the missing values.

When $w = 0$,

$$z = 0$$

When $w = 1$,

$$z = \frac{1}{2}$$

When $z = 5$,

$$w = 2(5) = 10$$

When $z = 9$,

$$w = 2(9) = 18$$

Direct variation may also be written using the special symbol $\propto$.

$$y \propto x$$

is read "y varies directly as x" or "y is directly proportional to x."

(b) Inverse Variation. If y varies *inversely* as x, we may write this relationship as

$$\boxed{y = \frac{k}{x}} \quad \textbf{53}$$

Example 6.30:   If q is inversely proportional to p, and q is 9 when p is 3, find q when p is 6.

Solution: p and q are related by Eq. 53,

$$q = \frac{k}{p}$$

We find the constant of proportionality by substituting the given values of p and q, 3 and 9.

$$9 = \frac{k}{3}$$

$$k = 3(9) = 27$$

So

$$q = \frac{27}{p}$$

When $p = 6$

$$q = \frac{27}{6} = 4\frac{1}{2}$$

(c) Joint Variation. When y varies *jointly* as x and w, we have

$$\boxed{y = kxw} \quad \mathbf{54}$$

Example 6.31: If y varies jointly as x and w, how will y change when x is tripled and w is halved?

Solution: Let y' be the new value of y obtained when x is replaced by $3x$ and w is replaced by $w/2$, while the constant of proportionality k, of course, does not change. Substituting in Eq. 54,

$$y' = k(3x)\left(\frac{w}{2}\right) = \frac{3}{2}kxw$$

$$= \frac{3}{2}y \quad \text{since } kxw = y$$

So the new y is $1\frac{1}{2}$ times as large as the former value.

(d) Combined Variation. Often the value of y depends on two or more variables. The variation may be direct or inverse. Further, the variables may sometimes be raised to powers, as in the following examples.

Example 6.32: "y varies directly as the square of x and inversely as the cube of w" is written

$$y = \frac{kx^2}{w^3}$$

Example 6.33: L is directly proportional to M and the square of N and inversely proportional to the square root of P. We write

$$L = \frac{kMN^2}{\sqrt{P}}$$

Example 6.34: If y varies directly as the square of x and inversely as the cube of w, by what factor will y change if x is doubled and w is halved?

Solution: First write the equation linking y to x and w, including a constant of proportionality k.

$$y = k\frac{x^2}{w^3}$$

We get a new value for y, let's call it y', when x is replaced with $2x$ and w is replaced with $w/2$.

$$y' = k\frac{(2x)^2}{(w/2)^3} = k\frac{4x^2}{w^3/8} = 32k\frac{x^2}{w^3}$$

$$= 32y$$

We see that y' is 32 times larger than the original y.

Example 6.35: If y varies directly as the square root of w and inversely as the cube of x, and y is 11.6 when w is 38.4 and x is 6.28, find y when $w = 73.9$ and $x = 15.2$.

Solution: From the problem statement,

$$y = k\frac{\sqrt{w}}{x^3}$$

We must first solve for k. Substituting the known values for x, y, and w,

$$11.6 = k \frac{\sqrt{38.4}}{(6.28)^3}$$

Solving for k,

$$k = \frac{11.6(6.28)^3}{\sqrt{38.4}} = 463.6$$

So

$$y = 463.6 \frac{\sqrt{w}}{x^3}$$

When $w = 73.9$ and $x = 15.2$,

$$y = 463.6 \frac{\sqrt{73.9}}{(15.2)^3} = 1.13$$

6.9 THE RECTANGULAR COORDINATE SYSTEM

In Chapter 1 we graphed numbers on the number line.   In order to graph *pairs* of numbers, we need a second number line at right angles to the first, both number lines intersecting at their zero points, as in Fig. 6-4.   With this, we obtain a *rectangular coordinate system*.   The two number lines are called *axes*.   The point of intersection of the two axes is called the *origin*.   The horizontal axis is the x axis, whose numbers increase to the right, and the vertical axis is the y axis, whose numbers increase in the upward direction.   The two axes divide the plane into four *quadrants*, numbered counterclockwise from the upper right, with Roman numerals.

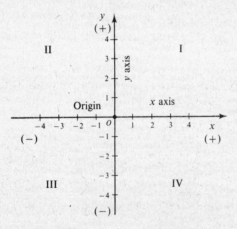

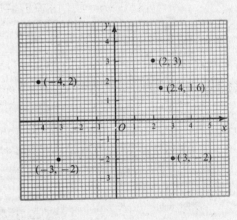

Fig. 6-4   The rectangular coordinate system.   Fig. 6-5

6.10 GRAPHING ORDERED PAIRS

An ordered pair of numbers gives the location of a point in this system.   These numbers are called the *coordinates* of the point.   The x coordinate of the point gives its distance left or right of the origin. It is also called the *abscissa* and is always written as the first of the pair of numbers.   The y coordinate (also called the *ordinate*) gives the distance of the point above or below the origin.

Example 6.36:   Plot the following pairs of numbers:

$$(2,3),\ (-4,2),\ (-3,-2),\ (3,-2),\ (2.4,1.6).$$

Solution: The locations of these points are shown in Fig. 6-5.

6.11 GRAPH OF AN EQUATION

While there are many shortcuts and special techniques for graphing equations, the simplest and most direct method is to create a table of ordered pairs and locate (plot) each pair on the coordinate axes. The line connecting these points is the graph of the equation.

Example 6.37:   Graph the equation $y = x^2 - 2x$ for values of x from -1 to 4.

Solution: We obtain a table of point pairs by substituting values of x into the equation.

x	y
-1	$1 - (-2) = 3$
0	0
1	$1 - 2 \quad = -1$
2	$4 - 4 \quad = 0$
3	$9 - 6 \quad = 3$
4	$16 - 8 \quad = 8$

The six points obtained are $(-1, 3)$, $(0, 0)$, $(1, -1)$, $(2, 0)$, $(3, 3)$, and $(4, 8)$. They are plotted in Fig. 6-6 and joined by a smooth curve. Note the use of different scales on the x and y axes to obtain a graph which is neither too cramped nor too spread out.

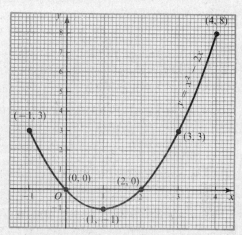

Graph of $y = x^2 - 2x$ for $-1 \le x \le 4$.

Fig. 6-6

Common Error	It is easy to make a mistake when substituting negative values of x into the equation. Be especially careful.

6.12 SOLVING EQUATIONS GRAPHICALLY

The values of x at which the graph of an equation crosses or touches the x axis are called the *x intercepts*. They are also the *roots* or *solutions* to the equation.

Example 6.38:   Find an approximate graphical solution to the equation $(x + 2)^2 = 3x + 7$.

Solution: Transpose terms to one side of the equals sign:

$$(x + 2)^2 - 3x - 7 = 0$$

Simplifying, $$x^2 + 4x + 4 - 3x - 7 = 0$$

$$x^2 + x - 3 = 0$$

The solutions to this equation are the values of x that will make both sides of the equation equal. We thus graph the function $f(x) = x^2 + x - 3$ and see which values of x make $f(x)$ equal to zero. Computing a table of point pairs,

x	-3	-2	-1	0	1	2
$f(x)$	3	-1	-3	-3	-1	3

These points are plotted in Fig. 6-7 and connected with a smooth curve. Reading the x intercepts from the graph as accurately as possible, we get for the solution

$$x \cong -2.31 \qquad \text{and} \qquad x \cong 1.31$$

The values of x that make the function $f(x)$ equal to zero are also called the *zeros* of the function.

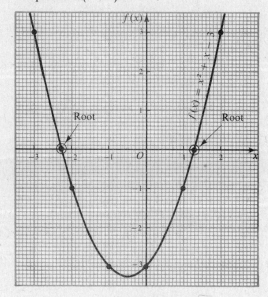

Fig. 6-7

Solved Problems

DEPENDENT AND INDEPENDENT VARIABLES

6.1   In the following equations, which are the independent variables and which are the dependent variables?

(a) $y = 3x - 5$ (b) $x = y^2 + 3y$ (c) $w^3 - 2w^2 = z$ (d) $y = 8x - 2y$

Solution

	Independent	Dependent
(a)	x	y
(b)	y	x
(c)	w	z
(d)	Neither	Neither

FUNCTIONS

6.2   Express the following statements in functional notation:

(a) The load L that a column may safely support depends upon its radius of gyration r.

(b) The compressive strength S of concrete depends upon the elapsed time T since pouring.

(c) The resistance R of a wire depends upon its length L, its cross-sectional area A, and the resistivity of the metal ρ.

(d) The moment of inertia I of a beam of rectangular cross section depends upon its width b and its depth h.

Solution

(a) $L = f(r)$ (b) $S = f(T)$ (c) $R = f(L, A, \rho)$ (d) $I = f(b, h)$

6.3   Write the area A of a square as a function of its side x.

Solution

$$A = f(x) = x^2$$

6.4   Write the area A of a circle as a function of its radius r.

Solution

$$A = f(r) = \pi r^2$$

6.5   Write the volume V of a rectangular box as a function of its length L, its width W, and its height H.

Solution

$$V = f(L, W, H) = LWH$$

IMPLICIT AND EXPLICIT FORM

6.6   Which of the following equations are explicit and which are implicit?

(a) $x^2 + y^2 = 25$ (b) $y = mx + b$ (c) $y = x^2 - xy$ (d) $2w - 3z = x$

Solution

(a) Implicit   (b) explicit   (c) implicit   (d) explicit

REARRANGING FUNCTIONS

6.7   If $y = 3x - 5$,   rewrite this equation in the form $x = f(y)$.

Solution

Solving for x, we transpose the -5,

$$3x = y + 5$$

Dividing by 3, $$x = f(y) = \frac{y + 5}{3}$$

6.8   If $x = 2y^2 z + 3$,   rewrite this equation in the form $z = f(x, y)$.

Solution

Transposing 3, $$2y^2 z = x - 3$$

Dividing by $2y^2$, $$z = f(x, y) = \frac{x - 3}{2y^2}$$

6.9   The deflection y at the end of a cantilever beam of length L, bearing a load W at one end, is

$$y = f(L, W, E, I) = \frac{WL^3}{3EI}$$

Rewrite this expression in the form $L = f(y, W, E, I)$.

Solution

Cross-multiplying, $$3EIy = WL^3$$

Dividing, $$L^3 = \frac{3EIy}{W}$$

and taking the cube root, $$L = \sqrt[3]{\frac{3EIy}{W}} = f(y, W, E, I)$$

6.10   If the extreme fiber stress for a beam is $\sigma = Mc/I$,   where M is the bending moment, c is the distance from neutral axis to extreme fiber, and I is the moment of inertia, write $M = f(c, I, \sigma)$.

Solution

Multiplying by I, $$Mc = \sigma I$$

Dividing by c, $$M = \frac{\sigma I}{c}$$

6.11   In order to make small corrections to a transit setting, the following formula is used: $D = 0.000\,004\,85SR$, where D is the distance the track is to be shifted, S is seconds (angular) of error, and R is the distance from the transit to the stake.   Rewrite this equation in the form $S = f(D, R)$.

Solution

Dividing both sides by $0.000\,004\,85R$,

$$S = \frac{D}{0.000\,004\,85R}$$

6.12   If the resistance of a piece of wire is

$$R = f(\rho, L, A) = \frac{\rho L}{A} \qquad \boxed{} \quad \mathbf{209}$$

find $\rho = f(R, L, A)$.

Solution

Cross-multiplying, $\rho L = RA$

Dividing, $\rho = \dfrac{RA}{L} = f(R, L, A)$

6.13   The elongation e of a bar of length L, cross-sectional area a, and modulus of elasticity E, when subject to a load P is

$$e = f(P, L, a, E) = \frac{PL}{aE}$$

Rewrite this equation in the form $E = f(e, L, a, P)$.

Solution

Cross-multiplying, $eaE = PL$

Dividing by ea,

$$\boxed{E = \frac{PL}{ae}} \quad \mathbf{192}$$

6.14   If the power dissipated in a resistance R when a voltage V is across its terminals is

$$\boxed{P = \frac{V^2}{R}} \quad \mathbf{204}$$

find $V = f(P, R)$.

Solution

Multiplying both sides by R, $V^2 = PR$

Taking the square root, $V = \sqrt{PR} = f(P, R)$

COMBINING EQUATIONS

6.15   If $x = w^2 + 2y$,   and $y = z - 3$,   find $x = f(w, z)$.

Solution

Substituting $z - 3$ for y in the first equation,

$$x = w^2 + 2y = w^2 + 2(z - 3)$$

So $x = f(w, z) = w^2 + 2z - 6$

6.16   If

$$x = f(y, z) = y^2 - 3z$$

and

$$y = f(w) = w + 2$$

and

$$z = f(w, y) = w^2 - 2y,$$

find $x = f(w)$.

Solution

Substituting $w + 2$ for y and $w^2 - 2y$ for z in the first equation,

$$x = (w + 2)^2 - 3(w^2 - 2y)$$

$$= w^2 + 4w + 4 - 3w^2 + 6y$$

Collecting terms and substituting $w + 2$ for y,

$$x = -2w^2 + 4w + 4 + 6(w + 2)$$

$$= -2w^2 + 4w + 4 + 6w + 12$$

So

$$x = f(w) = -2w^2 + 10w + 16$$

6.17   If stress is defined as

$$\boxed{\sigma = f(P, A) = \frac{P}{A}} \quad \mathbf{190}$$

and strain as

$$\boxed{\epsilon = f(e, L) = \frac{e}{L}} \quad \mathbf{191}$$

and the modulus of elasticity $E = f(e, L, A, P)$ as in Eq. 192, find $E = f(\sigma, \epsilon)$.

Solution

From Eq. 192, we know that $E = PL/Ae$.   Substituting σ for P/A and ϵ for e/L,

$$\boxed{E = \sigma \cdot \frac{1}{\epsilon} = \frac{\sigma}{\epsilon} = f(\sigma, \epsilon)} \quad \mathbf{193}$$

6.18   If

$$Z = f(X, R) = \sqrt{R^2 + X^2}$$

and

$$X = f(X_L, X_C) = X_L - X_C$$

and

$$X_C = f(C, \omega) = \frac{1}{\omega C}$$

and

$$X_L = f(L, \omega) = \omega L$$

find

$$Z = f(\omega, L, C, R)$$

Solution

$$Z = \sqrt{R^2 + X^2}$$

$$= \sqrt{R^2 + (X_L - X_C)^2}$$

So

$$Z = \sqrt{R^2 + \left(\omega L - \frac{1}{\omega C}\right)^2} = f(\omega, L, C, R)$$

6.19   If the power P dissipated in a resistance wire is a function of the current I through the wire and the voltage drop V across the wire, as in

$$P = f(V, I) = VI$$

and the current through the wire is a function of the resistance R of the wire and the voltage drop V across it, as in

$$I = f(V, R) = V/R$$

and the resistance of the wire is a function of its length L, cross-sectional area A, and resistivity ρ, as in Eq. 209, find $P = f(V, A, L, \rho)$.

Solution

Substituting,

$$P = VI = V\left(\frac{V}{R}\right) = \frac{V^2}{R}$$

$$= \frac{V^2}{\dfrac{\rho L}{A}} = \frac{V^2 A}{\rho L} = f(V, A, L, \rho)$$

6.20   If the moment of inertia of a circular rod is

$$I = f(d) = \frac{\pi d^4}{64}$$

and the deflection of a cantilever beam is

$$y = f(L, W, E, I) = \frac{WL^3}{3EI}$$

write

$$y = f(W, L, E, d)$$

Solution

$$y = \frac{WL^3}{3E} \cdot \frac{1}{I}$$

$$= \frac{WL^3}{3E} \cdot \frac{1}{\dfrac{\pi d^4}{64}} = \frac{WL^3}{3E}\left(\frac{64}{\pi d^4}\right)$$

$$\frac{64WL^3}{3\pi Ed^4} = f(L, W, E, d)$$

EVALUATING FUNCTIONS

6.21   If $f(x) = x - 2x^2$,   find $f(0), f(1), f(2)$, and $f(3)$.

Solution

Substituting into the given function,

$$f(0) = 0 - 0 = 0$$
$$f(1) = 1 - 2(1)^2 = -1$$
$$f(2) = 2 - 2(2)^2 = 2 - 8 = -6$$
$$f(3) = 3 - 2(3)^2 = 3 - 18 = -15$$

6.22   If $f(z) = 3z^3 + 2z - 5$,   find $f(4.83)$.

Solution

Substituting,

$$f(4.83) = 3(4.83)^3 + 2(4.83) - 5$$
$$= 338 + 9.66 - 5 = 343$$

6.23   If $g(x) = 2x + 3x^2$,   find $g(a + b)$.

Solution

Substituting $a + b$ for x in the given function,

$$g(a + b) = 2(a + b) + 3(a + b)^2$$
$$= 2a + 2b + 3(a^2 + 2ab + b^2)$$
$$= 2a + 2b + 3a^2 + 6ab + 3b^2$$

6.24   If $f(x, y) = 4x - y^2$,   find $f(3, 2)$.

Solution

Substituting 3 for x and 2 for y,

$$f(3, 2) = 4(3) - (2)^2$$
$$= 12 - 4 = 8$$

6.25   If $h(x) = x^2 + 2$,   find $h(3) - 2h(5)$.

Solution

Substituting,

$$h(3) = 3^2 + 2 = 11$$

and

$$h(5) = 5^2 + 2 = 27$$

so

$$h(3) - 2h(5) = 11 - 2(27) = -43$$

6.26   If $f_1(x) = x^2$, $f_2(x) = 3 - x$,   and $f_3(x) = x/2$, find

$$\frac{2f_3(6) - 3f_1(2)}{f_2(2)}$$

Solution

Substituting into each function,

$$f_1(2) = 2^2 = 4$$

$$f_2(2) = 3 - 2 = 1$$

$$f_3(6) = \frac{6}{2} = 3$$

Combining as indicated,

$$\frac{2(3) - 3(4)}{1} = 6 - 12 = -6$$

6.27 To find the resonant frequency F for a series resonant circuit consisting of a capacitor and inductor in series, the following formula is used:

$$F = f(L, C) = \frac{1}{2\pi\sqrt{LC}}$$

where f is in hertz

   L is in henries

   C is in farads

Find   (a) $f(6, 20)$,   (b) $f(37, 6)$,   and   (c) $f(7, 7)$.

Solution

Substituting,

(a) $f(6, 20) = \dfrac{1}{2\pi\sqrt{(6)(20)}} = 0.0145$

(b) $f(37, 6) = \dfrac{1}{2\pi\sqrt{(37)(6)}} = 0.0107$

(c) $f(7, 7) = \dfrac{1}{2\pi\sqrt{(7)(7)}} = 0.0227$

6.28 The maximum deflection of a certain cantilever beam, with a concentrated load applied r feet from the fixed end, is

$$d = f(r) = 0.000\,02 r^2 (60 - r) \qquad \text{inches}$$

Find the deflections $f(10)$ and $f(20)$.

Solution

Substituting,

$$f(10) = 0.000\,02(10)^2(60 - 10) = 0.1 \text{ in.}$$

$$f(20) = 0.000\,02(20)^2(60 - 20) = 0.32 \text{ in.}$$

6.29 When taping distances on sloping ground, a correction C must be made to the taped distance L in order to obtain the horizontal distance between the stations. This slope correction, in feet, depends upon the difference in elevation h of the two stations (feet) and the length L along the slope l (feet), as in the formula

$$C = f(h, L) = -h^2/2L$$

Find (a) $f(5, 60)$ and (b) $f(16, 200)$.

Solution

$$f(5, 60) = -\frac{5^2}{2(60)} = -\frac{25}{120} = -0.208 \text{ ft}$$

$$f(16, 200) = -\frac{16^2}{2(200)} = -\frac{256}{400} = -0.64 \text{ ft}$$

6.30 The resistance R of a conductor is a function of temperature.

$$\boxed{R = R_0(1 + \alpha t) = f(t)} \quad \textbf{208}^*$$

where R_0 is the resistance at $0°C$ and α is the temperature coefficient of resistance ($0.004\,27$ for copper). If the resistance of a copper coil is $1000\ \Omega$ at $0°C$, find $f(20)$, $f(50)$, and $f(80)$.

Solution

Substituting 1000 for R_0 and $0.004\,27$ for α,

$$R = f(t) = 1000(1 + 0.004\,27t)$$

Then

$$f(20) = 1000[1 + 0.004\,27(20)] = 1085\ \Omega$$

$$f(50) = 1000[1 + 0.004\,27(50)] = 1214\ \Omega$$

$$f(80) = 1000[1 + 0.004\,27(80)] = 1342\ \Omega$$

DOMAIN AND RANGE

6.31 What values of x are permissible in the following functions?

(a) $y = \sqrt{x - 2}$ (b) $y = \dfrac{1 - x}{3 - x}$ (c) $y = \dfrac{5}{\sqrt{x - 2}}$ (d) $y = \dfrac{2}{(x + 2)(x - 1)}$

Solution

(a) Values of x less than 2 will result in imaginary values of y, and so the domain of x is all real numbers equal to or greater than 2.
(b) All real values of x are permitted except $x = 3$, which causes division by zero.
(c) Values of x less than 2 will result in imaginary y.
Also, $x = 2$ will give division by zero. So, x must be greater than 2.
(d) All real values of x are permitted except $x = -2$ and $x = 1$, which result in division by zero.

VARIATION

6.32 For the following equations, state in words how the dependent variable varies with the independent variables.

(a) $y = 4x^2$ (b) $V = \pi \dfrac{r}{w}$ (c) $w = \dfrac{6\sqrt{y}}{x^2}$ (d) $m = \dfrac{4.93xy^4}{\sqrt[3]{z}}$

---

* This is actually a variation of Eq. 208 for the case when $t_1 = 0°$.

Solution

(a) y varies directly as the square of x.

(b) V varies directly as r and inversely as w.

(c) w varies directly as the square root of y and inversely as the square of x.

(d) m varies directly as x and the fourth power of y and inversely as the cube root *of* z.

6.33 Write the equations given in the following statements:

(a) By Hooke's law, the stress σ in a bar subjected to a tensile load is directly proportional to the strain ϵ in the bar.   Write an equation for σ.

(b) The force F needed to stretch a spring is directly proportional to the distance x.   Write the equation for F.

(c) In a vacuum diode, the plate current I_b varies as the three-halves power of the plate voltage E_b, regardless of cathode and anode geometry.   Write an equation for I_b.

Solution

(a) By Eq. 52 $\sigma = k\epsilon$ where the constant of proportionality is called the *modulus of elasticity, E.*   So

$$\boxed{\text{Hooke's Law} \quad \sigma = E\epsilon} \quad \mathbf{193}$$

(b) By Eq. 52,

$$\boxed{F = kx} \quad \mathbf{199}$$

The constant k is called the *spring constant*.

(c) From the problem statement

$$I_b = kE_b^{3/2}$$

The constant of proportionality is called the *perveance* and is usually given by the symbol G.

6.34 Fill in the missing values, if y varies directly as x.

x	2	3	
y	10		55

Solution

By Eq. 52, $y = kx$. Substituting the first pair of values,

$$k = \frac{10}{2} = 5$$

So

$$y = 5x$$

When $x = 3$,

$$y = 5(3) = 15$$

and when $y = 55$,

$$x = \frac{55}{5} = 11$$

6.35 Fill in the missing values, if y is inversely proportional to x.

x	2.74		4.73
y		9.47	15.8

Solution

By Eq. 53, $y = k/x$. Substituting the third pair of values,

$$k = xy = 4.73(15.8) = 74.7$$

So

$$y = \frac{74.7}{x}$$

When $x = 2.74$,

$$y = \frac{74.7}{2.74} = 27.3$$

and when $y = 9.47$,

$$x = \frac{74.7}{9.47} = 7.89$$

6.36 Fill in the missing values, if y varies jointly as w and x.

w	3.1		4.6	1.4
x	2.7	9.4		6.3
y		11.3	5.8	9.2

Solution

By Eq. 54,

$$y = kxw$$

Using the fourth set of values to find k,

$$9.2 = k(6.3)(1.4)$$
$$k = 1.04$$

So

$$y = 1.04xw$$

When $x = 2.7$ and $w = 3.1$,

$$y = 1.04(2.7)(3.1) = 8.7$$

When $x = 9.4$ and $y = 11.3$,

$$11.3 = 1.04(9.4)w$$
$$w = 1.2$$

and when $w = 4.6$ and $y = 5.8$,

$$5.8 = 1.04(4.6)x$$
$$x = 1.2$$

6.37 The tensile strength S of a round bar is proportional to the square of its diameter d. If a 1.25-cm-diameter bar safely supports 170 000 kg, how much will a 1.75-cm-diameter bar support?

Solution

From the problem statement,

$$S = kd^2$$

Finding k,

$$k = \frac{S}{d^2} = \frac{170\,000}{(1.25)^2} = 108\,800$$

So

$$S = 108\,800d^2$$

For the 1.75-cm-diameter bar,

$$S = 108\,800(1.75)^2 = 333\,200 \text{ kg}$$

6.38 A city of 250 000 people is supplied by a 35-in.-diameter water main. What diameter main will be needed for a population of 500 000?

Solution

The population a water main can service is proportional to its carrying capacity, which is proportional to its cross-sectional area, which is proportional to the *square* of its diameter d. So

$$P = kd^2$$

Finding k,

$$k = \frac{P}{d^2} = \frac{250\,000}{(35)^2} = 204$$

So

$$P = 204d^2$$

When $P = 500\,000$,

$$d^2 = \frac{500\,000}{204} = 2451$$

$$d = 49.5 \text{ in.}$$

6.39 By Ohm's law, the voltage drop V across a resistor is directly proportional to the current I in the resistor. If a current of 2.45 A causes a voltage drop of 55.3 V, find the drop V caused by 3.16 A.

Solution

From the problem statement,

$$V = kI$$

Finding the constant of proportionality k by substituting $I = 2.45$ and $V = 55.3$,

$$k = \frac{V}{I} = \frac{55.3}{2.45} = 22.6$$

So

$$V = 22.6I$$

When $I = 3.16$ A,

$$V = 22.6(3.16) = 71.4 \text{ V}$$

6.40 The power P dissipated in a resistor varies directly as the square of the current I. If the power is 610 W when the current is 2.5 A, find the power when the current is 3.5 A.

Solution

We are told that

$$P = kI^2$$

Finding k,

$$k = \frac{P}{I^2} = \frac{610}{(2.5)^2} = 97.6$$

and so

$$P = 97.6I^2$$

When $I = 3.5$ A,

$$P = 97.6(3.5)^2 = 1200 \text{ W}$$

6.41   If 35 transformer laminations make a stack 1.06 cm thick, how many laminations are contained in a stack 2 cm thick?

Solution

Since the thickness t of the stack is directly proportional to the number of laminations N in the stack, we write

$$N = kt$$

Finding k when $N = 35$ and $t = 1.06$,

$$k = \frac{N}{t} = \frac{35}{1.06} = 33$$

So

$$N = 33t$$

When $t = 2$ cm,

$$N = 33(2) = 66 \text{ laminations}$$

6.42   A certain screw machine can produce 98 machine screws in 7.5 min.   How many screws can it make in 7.5 h?

Solution

Since the amount produced P is directly proportional to the running time t, we write

$$P = kt$$

When $t = 7.5$ min, $P = 98$,   and so

$$k = \frac{P}{t} = \frac{98}{7.5} = 13.1$$

When $t = 7.5(60)$ min,

$$P = 13.1(7.5)(60) = 5895 \text{ screws}$$

6.43   The power P of a gasoline engine is directly proportional to the total volume swept out by the pistons, the engine *displacement d*.   If a certain 55-hp engine has a displacement of 175 in^3, what power P would you expect from a similar 215-in^3 engine?

Solution

From the problem statement

$$P = kd$$

Since $P = 55$ hp when $d = 175$ in^3,

$$k = \frac{P}{d} = \frac{55}{175} = 0.314$$

So $$P = 0.314d$$

When $d = 215 \text{ in}^3$,

$$P = 0.314(215) = 68 \text{ hp}$$

6.44 The cost for a five-man crew to do a certain 7-day job was \$2800. Estimate the cost C for a four-man crew to do a similar 12-day job.

Solution

Assuming the cost varies jointly as the number of men N and the number of days d,

$$C = kNd$$

Since $C = 2800$ when $N = 5$ and $d = 7$,

$$k = \frac{C}{Nd} = \frac{2800}{5(7)} = 80$$

and so $$C = 80Nd$$

When $N = 4$ and $d = 12$,

$$C = 80(4)(12) = \$3840$$

6.45 The average gasoline mileage that a company gets with its fleet of cars is 19 mi/gal and its annual gasoline bill is \$3592. What annual bill would you estimate if they switched to smaller cars getting an average of 32 mi/gal?

Solution

The gasoline bill G will be inversely proportional to the mileage per gallon M; so

$$G = \frac{k}{M}$$

Since $G = 3592$ when $M = 19$,

$$k = GM = 3592(19) = 68{,}248$$

and so $$G = \frac{68{,}248}{M}$$

When $M = 32$,

$$G = \frac{68{,}248}{32} = \$2132.75$$

6.46 Boyle's law states that the pressure p of a confined gas is inversely proportional to the volume V. (a) Write an equation for p. (b) Find the piston position in Fig. 6-8 that will cause the pressure in the cylinder to triple.

Solution

(a) From the problem statement,

$$p = \frac{k}{V}$$

(b) Since the volume of the gas is $\pi r^2 L$, where r is the cylinder radius and L is the distance traveled by the piston,

$$p = \frac{k}{\pi r^2 L}$$

Multiplying by L and dividing by p,

$$L = \frac{k}{\pi r^2 p}$$

Letting L_1 represent the new piston position that will cause the pressure to be $3p$, we get

$$L_1 = \frac{k}{\pi r^2 (3p)} = \frac{1}{3} \frac{k}{\pi r^2 p} = \frac{1}{3} L$$

The new length is thus 1/3 of its former value.

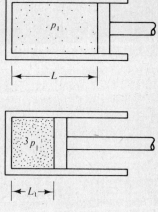

Fig. 6-8

6.47 The kinetic energy of a moving body varies directly as its weight W and as the square of its speed V. (a) Write an equation for the kinetic energy E. (b) If the weight is doubled and the speed is halved, find the ratio of the new kinetic energy E_1 to the original.

Solution

(a) From the problem statement, $E = kWV^2$.

(b) Let E_1 represent the kinetic energy when W is doubled and V is halved.

$$E_1 = k(2W)\left(\frac{V}{2}\right)^2 = \frac{1}{2} kWV^2 = \frac{1}{2} E$$

Thus, the kinetic energy is half what it was originally.

6.48 When an electric current flows through a wire, the resistance R to the flow varies directly as the length L and inversely as the cross-sectional area A of the wire. (a) Write an equation for R. (b) If the length and the diameter are both doubled, by what percentage does the resistance change?

Solution

(a) From the problem statement, $R = kL/A$. The constant in this equation is called the *resistivity* ρ.

$$\boxed{R = \frac{\rho L}{A}} \quad \textbf{209}$$

(b) Letting $A = \pi d^2/4$, where d is the wire diameter,

$$R = \frac{4\rho L}{\pi d^2}$$

Replacing L by $2L$ and d by $2d$, we get a new resistance R_1,

$$R_1 = \frac{4\rho(2L)}{\pi(2d)^2} = \frac{2\rho L}{\pi d^2} = \frac{1}{2}\left(\frac{4\rho L}{\pi d^2}\right) = \frac{1}{2} R$$

Thus, the new resistance is just half its former value, or, a 50% drop.

THE RECTANGULAR COORDINATE SYSTEM

6.49   Give the rectangular coordinates of the
six points in Fig. 6-9.

Solution

   Reading the x and then the y coordinate
of each point, we get

(a)   (2, 3)

(b) $(-1, 2)$

(c) $(-3, -1)$

(d) $(1.5, -2.3)$

(e)   (2.13, 1.58) approximately

(f) $(-1.82, 2.97)$ approximately

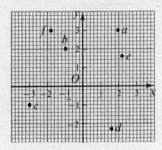

Fig. 6-9

6.50   Graph the following sets of points, connect them, and name the geometric figure formed:

(a) $(2, -0.5)$, $(3, -1.5)$, $(1.5, -3)$, and $(0.5, -2)$

(b)   (0.7, 2.1), (2.3, 2.1), (2.3, 0.5), and (0.7, 0.5)

(c) $(-1.6, 2.8)$, $(-2.2, 1)$, and $(-1, 1)$

(d) $(-2, -1)$, $(-0.8, -0.8)$, $(-1.3, -2.3)$, and $(-2.5, -2.5)$

Solution

   The points are shown plotted in Fig. 6-10.   The figures are

(a)   Rectangle   (b)   square   (c)   isosceles triangle   (d)   parallelogram

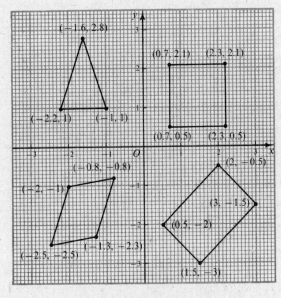

Fig. 6-10

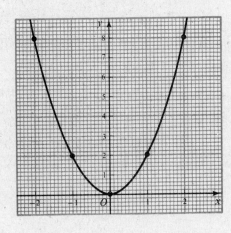

Fig. 6-11

GRAPHING EQUATIONS

6.51   Plot the equation $y = 2x^2$ from $x = -2$ to $x = 2$,   and locate any roots within that region.

Solution

Making a table of point pairs,

x	-2	-1	0	1	2
y	8	2	0	2	8

These points are plotted in Fig. 6-11, and connected with a smooth curve. There is one root, where $x = 0$. This equation, of the form $y = ax^n$,   is called a *power function*.

6.52   Graph the equation $y = 1/x - 2$ from $x = 0$ to $x = 2$,   and locate any roots within this region.

Solution

Making a table of point pairs,

x	0	1/4	1/2	3/4	1	2
y	Not defined	2	0	$-2/3$	-1	-1.5

These points are plotted in Fig. 6-12 and connected with a smooth curve, one branch of a *hyperbola* in this case. Notice from the equation that as x approaches zero in value, the curve gets closer and closer to the y axis but never touches it. Also, as x gets larger and larger, the curve gets closer to the line $y = -2$,   but never reaches it. Such lines, the y axis and $y = -2$ in this case, are called *asymptotes*. There is one root, at $x = 1/2$.

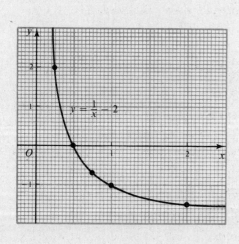

Fig. 6-12

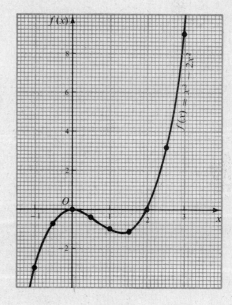

Fig. 6-13

6.53   Graph the function $f(x) = x^3 - 2x^2$ from $x = -1$ to $x = 3$,   and locate any roots within this region.

Solution

Making a table of point pairs,

x	-1	$-1/2$	0	$1/2$	1	1.5	2	2.5	3
$f(x)$	-3	-0.63	0	-0.38	-1	-1.13	0	3.12	9

Plotting the points, we get a curve which is characteristic of the graph of a *cubic* equation. We find roots at $x = 0$ and $x = 2$. See Fig. 6-13, page 153.

6.54   Plot the function $f(x) = 1.13x^2 - 1.85x - 1.44$ between $x = 0$ and $x = 3$, and locate any roots within this region.

Solution

Making a table of point pairs,

x	0	0.5	1	1.5	2	2.5	3
$f(x)$	-1.44	-2.08	-2.16	-1.67	-0.62	1.00	3.18

Plotting these points, we obtain a portion of a *parabola*, Fig. 6-14. It has a root at $x = 2.21$, approximately.

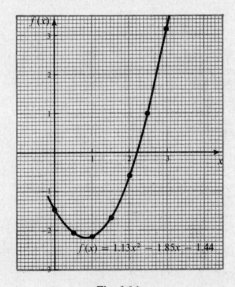

Fig. 6-14

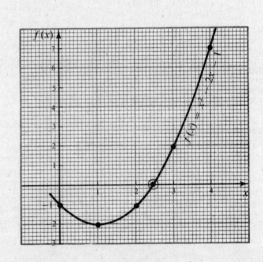

Fig. 6-15

6.55   The equation $x^2 - 1 = 2x$ has two solutions, one of which lies between $x = 0$ and $x = 3$. Graphically find an approximate value for that root.

Solution

Transposing all terms to one side of the equals sign,

$$f(x) = x^2 - 2x - 1$$

Making a table of point pairs,

x	0	1	2	3	4
$f(x)$	-1	-2	-1	2	7

Graphing this function, Fig. 6-15, we find that it crosses the x axis at approximately $x = 2.4$.

GRAPHING EXPERIMENTAL DATA

6.56   A beam, simply supported at both ends, was loaded at its midspan and the following data
was collected.   Graph this data, using the horizontal axis for strain and the vertical axis for
stress.

Strain	Stress (lb/in²)
0.005	30 000
0.010	32 500
0.015	33 000
0.025	32 000
0.040	31 000
0.100	35 000
0.150	42 000
0.200	46 300
0.250	47 100
0.300	43 000
0.350	36 000
0.400	28 000

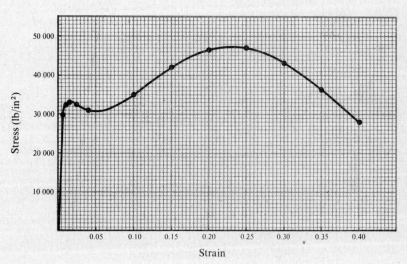

Fig. 6-16

Solution

The given point pairs are plotted in Fig. 6-16 and connected with a smooth curve.

6.57   The efficiency of a gasoline engine was measured at various power output levels and the results
are tabulated below.   Graph these values, using the horizontal axis for horsepower.

Horsepower	Efficiency, %
10	30
20	35
30	50
40	75
50	80
60	60
70	40
80	30

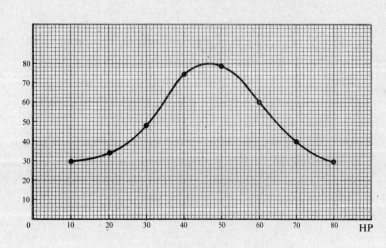

Fig. 6-17

Solution

See Fig. 6-17.

Supplementary Problems

6.58   Identify the independent variable.

   (a) $x = y^2 + 3$ (c) $f = L + w$ (e) $x = 3y$

   (b) $A = b^2 - 2$ (d) $y = 2x + 7$ (f) $a = 9c$

6.59   Represent the following equations in functional notation:

   (a) $a = 3b^2 - 2b + 2$ (b) $x = 4y^3 - 3y + 6$ (c) $c = a^5 - a^3 + 3$ (d) $N = 3z$

6.60   Express the following statements as equations:

   (a) a is directly proportional to the square of b.

   (b) x is inversely proportional to y and the cube root of z.

   (c) P is directly proportional to a and the square of c and inversely proportional to x.

6.61   Express the following statements in functional notation:

   (a) The shearing stress V in a certain beam depends upon the applied load P and the section modulus z.

   (b) The allowable stress F in a weld depends upon its throat size T and the strength F_v of the type of electrode used.

   (c) The distance D it takes to stop a car depends upon its speed S and the driver's reaction time T.

6.62   Label the following expressions as implicit or explicit:

   (a) $x = 3y + 2$ (c) $x = 6y^2 - 2x^2$ (e) $y = 6x + 4$

   (b) $2x + 7y = 5$ (d) $x + y = x - y$ (f) $z - 2 = x + bz$

6.63   (a) Rewrite the equation $y = 2x + 3$ in the form $x = f(y)$.

   (b) If $x = 4y^3z + 7$, rewrite this equation in the form $y = f(x, z)$.

   (c) The section modulus for a beam is $S = M/F_b$, where M is the bending moment and F_b is the allowable bending stress. Rewrite the equation in the form $F_b = f(S, M)$.

6.64   Combine the following equations as indicated:

   (a) If $x = y^2 + z$, and $y = a - 3$, find $x = f(a, z)$.

   (b) If $N = f(c, b) = c^2 + b^2$, and $b = f(a) = a - 3$, and $c = f(a, b) = a^2 - 3b$, find $N = f(a)$.

   (c) If the area of a circle is $A = f(r) = \pi r^2$ and the radius of a circle is $r = f(c) = c/2\pi$, write $A = f(c)$.

6.65   Evaluate the following functions:

   (a) Find $f(2)$, if $f(x) = x^2 - 3x + 1$.

   (b) Find $f(6)$, if $f(b) = b - b/2 + 2$.

   (c) Find $f(4)$, if $f(a) = a - 3a$.

   (d) Find $f(9)$, if $f(N) = N^3 - 3N^2 + 3$.

(e) If $h(x) = 2x + 7x^2$,   find $h(a - c)$.

(f) If $L(y) = 3y^2 - y + 1$,   find $L(4)$ and $L(2/3)$.

(g) If $f(x) = 5x^5 - x^4 + 2x^3 - x + 4$,   find $f(3.5), f(-2)$, and $f(1.7)$.

6.66   What values of x are permissible in the following functions?

(a) $y = \sqrt{x + 3}$ (b) $y = \dfrac{2 + x}{1 - x}$ (c) $y = \dfrac{1 - x}{\sqrt{x - 2}}$

6.67   If $y = kb^2$,   by what factor will y change if b is multiplied by the square root of 2?

6.68   If $x = k/y\sqrt[3]{z}$,   how will x change if y is doubled and z is increased by a factor of 8?

6.69   If $P = kac^2/x$,   how will P change if x is quartered?

6.70   State in words how the dependent variable varies with the independent variables.

(a) $x = 2y$ (b) $V = \dfrac{Ah}{3}$ (c) $y = \dfrac{2.7H}{3A}$

6.71   Write the equations given in the following statements:

(a) The temperature T needed to boil water is directly proportional to the altitude h above sea level.

(b) The area A of a sector of a circle is directly proportional to the square of the radius r and the internal angle θ.

6.72   Fill in the missing values, if x varies inversely as y.

x	2	3		
y	5		2	0.5

6.73   Fill in the missing values, if x varies jointly as y and z.

x	1.6		3.0	5.0	1.0
y		2.0	7.0	1.0	
z	2.3	3.0	6.0		4.4

6.74 x varies directly as the cube of y. If x is 70 when y is 2, find x when y is 3.7.

6.75   The allowable strength of a column F_a varies directly as the effective length L and inversely as the radius of gyration r. If $F_a = 21$ kg   when L is 15 m and r is 2.5 cm, find F_a when L is 12 m and r is 3.09 cm.

6.76   The intensity I of a sound varies directly as the resonant frequency H and inversely as the harmonic distortion D. If I is 50 decibels (dB) when H is 60 Hz and D is 20, find H when I is 70 dB and D is 27.

6.77 In what quadrant are the following points located? See quadrant numbering in Fig. 6-4.

  (a) $(2,3)$ (c) $(3,-2)$ (e) $(1,-2)$ (g) $(0,-1)$

  (b) $(-4,7)$ (d) $(-6,-3)$ (f) $(2,0)$ (h) $(-9,5)$

6.78 Graph the equation $y = x^2 - 2x$ from $x = -3$ to 5, and graphically find any roots within this region.

6.79 Graph the equation $y = x^3 + 3$ from $x = -2$ to 2, and graphically find any roots within this region.

6.80 Graph the equation $y = \sqrt{x^2 + 3x}$ from $x = -5$ to 4 and graphically find any roots within this region.

Answers to Supplementary Problems

6.58   (a) y (b) b (c) L, w (d) x (e) y (f) c

6.59   (a) $a = f(b)$ (b) $x = f(y)$ (c) $c = f(a)$ (d) $N = f(z)$

6.60   (a) $a = kb^2$ (b) $x = \dfrac{k}{y\sqrt[3]{z}}$ (c) $P = \dfrac{kac^2}{x}$

6.61   (a) $V = f(P, z)$ (b) $F = f(T, F_v)$ (c) $D = f(S, T)$

6.62   (a) Explicit   (b) implicit   (c) implicit   (d) implicit   (e) explicit   (f) implicit

6.63   (a) $x = f(y) = \dfrac{y-3}{2}$ (b) $y = f(x, z) = \sqrt[3]{\dfrac{x-7}{4z}}$ (c) $F_b = f(S, M) = \dfrac{M}{S}$

6.64   (a) $x = f(a, z) = a^2 - 6x + 9 + z$ (b) $N = f(a) = a^4 - 6a^3 + 28a^2 - 60a + 90$

  (c) $A = f(c) = \dfrac{c^2}{4\pi}$

6.65   (a) -1 (b) 5 (c) -8 (d) 489 (e) $h(a-c) = 7a^2 - 14ac + 7c^2 + 2a - 2c$

  (f) $L(4) = 45$, $L\left(\dfrac{2}{3}\right) = 1.667$ (g) $f(3.5) = 2562$, $f(-2) = -186$, $f(1.7) = 74.8$

6.66   (a) All real values greater than or equal to -3.   (b) All real values except 1.

  (c) All real values greater than 2.

6.67 y increases by a factor of 2.

6.68 x will be one-quarter of its original value.

6.69 P increases by a factor of 4.

6.70   (a) x varies directly as y.   (b) V varies directly as A and h.

  (c) y varies directly as H and inversely as A.

6.71   (a) $T = kh$ (b) $A = kr^2\theta$

6.72

x	2	3	5	20
y	5	3.33	2	0.5

6.73

x	1.6	0.43	3.0	5.0	1.0
y	9.7	2.0	7.0	1.0	3.2
z	2.3	3.0	6.0	71	4.4

6.74 $x = 443$

6.75 $F_a = 13.6 \text{ kg}$

6.76 $H = 113 \text{ Hz}$

6.77   (a)   1   (b)   2   (c)   4   (d)   3   (e)   4

      (f)   on x axis   (g)   on y axis   (h)   2

6.78   Root at $x \cong 0$ and $x \cong 2$.   (See Fig. 6-18.)

6.79   Root at $x \cong -1.45$.   (See Fig. 6-19.)

6.80   Roots at $x \cong -3$ and $x \cong 0$.   (See Fig. 6-20.)

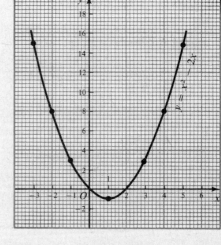

Fig. 6-18

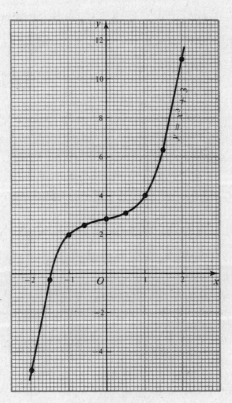

Fig. 6-19

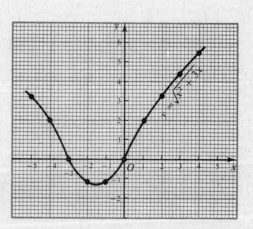

Fig. 6-20

Chapter 7

Geometry

Geometry deals with the measurement and properties of points, lines, angles, planes, and solids, and with the relationships between them. *Plane* geometry is concerned with geometric figures lying in a plane and *solid* geometry deals with figures that lie in more than one plane.

7.1 STRAIGHT LINES AND ANGLES

(a) Terminology. A *line segment* is that portion of a straight line lying between two points (the *endpoints*) on the line, Fig. 7-1*a*.

A *ray* or *half-line* is that portion of a line to one side of a point (called the endpoint) on the line, Fig. 7-1*b*.

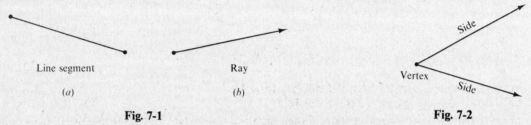

Line segment   Ray   Vertex · Side · Side

(*a*)   (*b*)

Fig. 7-1   **Fig. 7-2**

An *angle* is formed when two rays intersect at their endpoints, Fig. 7-2. The point of intersection is called the *vertex* of the angle, and the two rays are called the *sides* of the angle.

An angle can also be thought of as having been *generated* by a ray turning from some *initial position* to a *terminal position*, Fig. 7-3.

Fig. 7-3   **Fig. 7-4**

The angle shown in Fig. 7-4 can be *designated* in any of the following ways:

angle ABC   angle CBA   angle B   angle θ

The symbol $\angle$ means *angle*. So, $\angle B$ means angle *B*.

One *revolution* is the amount a ray would turn to return to its original position.

The *units of angular measure* in common use are the *degree* and the *radian*. The *measure* of an angle is the number of units of measure it contains. For brevity, we shall use statements like "add angle *A* to angle *B*" instead of the more correct "add the measure of angle *A* to the measure of angle *B*."

The *degree* is a unit of angular measure equal to 1/360 of a revolution. The symbol for degrees is °. Thus there are 360° in one complete revolution. In the SI system of measurement the unit of measure is the radian (rad). There are 2π radians in a circle.

A *right angle* is a quarter of a revolution, or 90°. Two lines at right angles to each other are said to be *perpendicular*.

An *acute angle* is smaller than 90°.

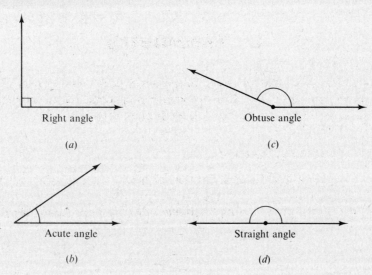

Fig. 7-5

An *obtuse angle* is greater than 90° but less than 180°.

A *straight angle* is half a revolution, or 180°.

Two angles are called *complementary* if their sum is a right angle and *supplementary* if their sum is a straight angle.

(b) Angles between Intersecting Lines. In Fig. 7-6, angles *A* and *B* are called *opposite*, or *vertical* angles, and angles *A* and *C* are *adjacent* angles.

> Opposite angles of two intersecting
> straight lines are equal.   **55**

Thus, in Fig. 7-6, angle *A* = angle *B*.

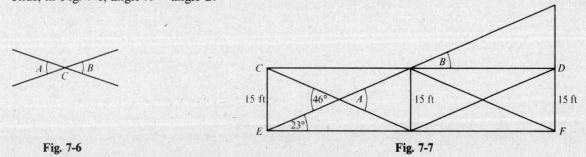

Fig. 7-6   Fig. 7-7

Example 7.1:   Find angle *A* in the truss, Fig. 7-7.

Solution: By Statement 55, angle *A* equals its opposite angle, or *A* = 46°.

In Fig. 7-8, the two parallel lines 1 and 2 are cut by line 3 (called a *transversal*). Angles *A*, *B*, *G*, and *H* are called *exterior* angles, and *C*, *D*, *E*, and *F* are called *interior* angles. Angles *A* and *E* are called *corresponding* angles. Other corresponding angles in Fig. 7-8 are *C* and *G*, *B* and *F*, and *D* and *H*.

> If two parallel straight lines are cut by a transversal, corresponding angles are equal and alternate interior angles are equal.   **56**

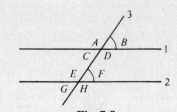

Fig. 7-8

Thus in Fig. 7-8,

$$\angle A = \angle E = \angle D = \angle H \quad \text{and} \quad \angle B = \angle F = \angle C = \angle G$$

Example 7.2:   Find angle B in the truss of Fig. 7-7.

Solution: Since girder CD is parallel to EF,

$$\text{Angle } B = 23°$$

Figure 7-9 shows two lines cut by four parallels.   The portions of the two lines lying between the same parallels are called corresponding segments (such as a and b).

> If two lines are cut by a number of parallels, the corresponding segments are proportional.   **57**

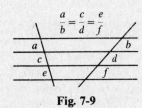

Fig. 7-9

or
$$\frac{a}{b} = \frac{c}{d} = \frac{e}{f}$$

Example 7.3:   Find the length of the roof section L, Fig. 7-10.

Solution: By Statement 57,

$$\frac{L}{12} = \frac{7}{10}$$

$$L = \frac{7}{10} \times 12$$

$$= 8.4 \text{ m}$$

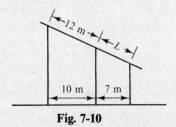

Fig. 7-10

7.2   TRIANGLES

(a)   Terminology.   A *polygon* is a plane figure formed by three or more line segments joined at their endpoints, as in Fig. 7-11a.   A *triangle* is a polygon having three *sides*.   The sides intersect at the *vertices* of the triangle.   The angles between the sides are the *interior angles* of the triangle, usually referred to as simply the angles of the triangle.

A *scalene* triangle has no equal sides, an *isosceles* triangle has two equal sides, and an *equilateral* triangle has three equal sides.

An *acute* triangle has three acute angles, an *obtuse* triangle has one obtuse angle, and a *right* triangle has one right angle.

The *altitude* of a triangle is the perpendicular distance from a vertex to the opposite side, called the *base*, or an extension of that side.

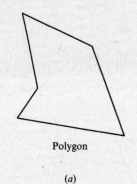

Polygon

(a)

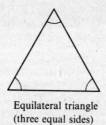

Equilateral triangle
(three equal sides)

(d)

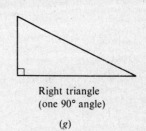

Right triangle
(one 90° angle)

(g)

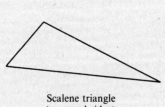

Scalene triangle
(no equal sides)

(b)

Acute triangle
(three acute angles)

(e)

Altitude

Base

(h)

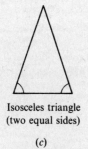

Isosceles triangle
(two equal sides)

(c)

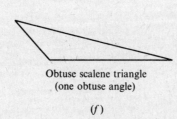

Obtuse scalene triangle
(one obtuse angle)

(f)

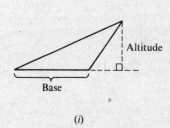

Altitude

Base

(i)

Fig. 7-11

(b) Any Triangle

Area equals one half the product of the base and the altitude
to that base.

$$A = \frac{bh}{2}$$

58

Example 7.4:   Find the area of the triangular window in Fig. 7-12.

Solution: By Eq. 58,

$$\text{Area} = \frac{5(6)}{2}$$

$$= 15 \text{ ft}^2$$

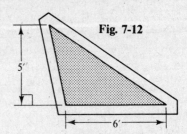

Fig. 7-12

Another formula for the area of a triangle having sides of lengths a, b, and c is

$$\boxed{\text{Area} = \sqrt{s(s-a)(s-b)(s-c)} \quad \text{where } s = \frac{a+b+c}{2}} \quad \mathbf{59}$$

Example 7.5: Find the area of a triangle having sides of lengths 85 mm, 47 mm, and 62 mm.

Solution: From Eq. 59,

$$s = \frac{85 + 47 + 62}{2} = 97$$

So,

$$\text{Area} = \sqrt{97(97-85)(97-47)(97-62)}$$

$$= 1427 \text{ mm}^2$$

$$\boxed{\begin{array}{l}\text{The sum of the three interior angles of any triangle is } 180°. \\ A + B + C = 180°\end{array}} \quad \mathbf{60}$$

Example 7.6 Find angle A in Fig. 7-13.

Solution: By Eq. 60,

$$A + 43 + 105 = 180$$

$$A = 180 - 43 - 105 = 32°$$

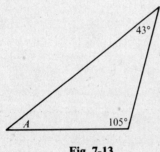

Fig. 7-13

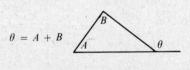

Fig. 7-14

In Fig. 7-14, angle θ is called an *exterior* angle.

$$\boxed{\begin{array}{l}\text{An exterior angle equals the sum of the} \\ \text{two opposite interior angles.} \\ \theta = A + B\end{array}} \quad \mathbf{63}$$

Thus in Fig. 7-14, $\theta = A + B$.

Example 7.7: Find angle A in Fig. 7-15.

Solution: By Eq. 63,

$$A = 101 + 29 = 130°$$

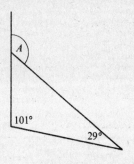

101°

29°

Fig. 7-15

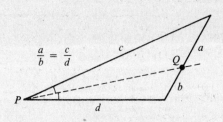

Fig. 7-16

An *angle bisector* is a line that divides an angle into two equal angles, such as line *PQ* in Fig. 7-16.

> An angle bisector of a triangle divides the opposite side in proportion to the two other sides.

67

Example 7.8:   Find the distance between holes *A* and *B* in Fig. 7-17.

Solution: From Statement 67,

$$\frac{AB}{1.895} = \frac{2.475}{4.338}$$

$$A = \frac{1.895(2.475)}{4.338}$$

$$= 1.081 \text{ cm}$$

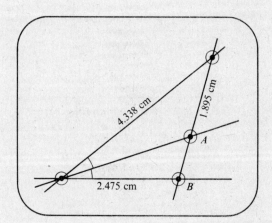

Fig. 7-17

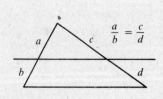

Fig. 7-18

> A line parallel to one side of a triangle divides the other two sides proportionately.

68

Thus in Fig. 7-18,

$$\frac{a}{b} = \frac{c}{d}$$

Example 7.9:   In the truss of Fig. 7-19, beam AB is parallel to beam CD.   Find distance BD.

Solution: By Statement 68,

$$\frac{BD}{14.8} = \frac{15.5}{12.3}$$

$$BD = \frac{14.8(15.5)}{12.3}$$

$$= 18.7 \text{ ft}$$

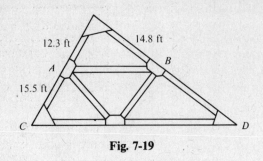

Fig. 7-19

> A line which joins the midpoints of two sides of a triangle is parallel to the third side and is equal to one-half the third side.
>
> **69**

This is illustrated in Fig. 7-20.

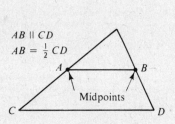

$AB \parallel CD$
$AB = \frac{1}{2} CD$

Midpoints

Fig. 7-20

Fig. 7-21

6.24 m

2.45 m | 2.45 m

Example 7.10:   Find the length of vertical brace AB in Fig. 7-21.

Solution: We assume the wall of the house to be vertical, so AB is parallel to DE.   Since the brace AB bisects side CD and is parallel to DE, it must be half of DE, or 3.12 m.

(c) Similar Triangles.   Two triangles are *similar* if the angles of one triangle equal the angles of the other triangle, as in Fig. 7-22.   Sides a and d are called *corresponding sides*, because they lie opposite corresponding equal angles in each triangle.   Sides b and e and sides c and f are also corresponding sides.

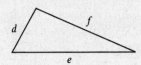

$$\frac{a}{d} = \frac{b}{e} = \frac{c}{f}$$

Fig. 7-22

| If two angles of a triangle equal two angles of another triangle, their third angles must be equal, and the triangles are similar. | **70** |

and

| Corresponding sides of similar triangles are in proportion. | **71** |

Example 7.11:   Two vertical posts are connected by cables as in Fig. 7-23.   Find the distance *BC*.

Solution: By Statement 55, opposite angles *ACB* and *DCE* are equal.   By Statement 56, since the posts are parallel, angle *BAC* equals angle *CED*.   Triangle *ABC* is thus similar to triangle *CED*, and so, by Statement 71,

$$\frac{BC}{4} = \frac{8}{12}$$

Therefore,

$$BC = \frac{8}{3} = 2\frac{2}{3} \text{ ft}$$

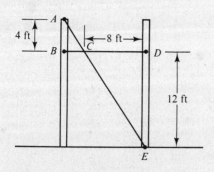

Fig. 7-23

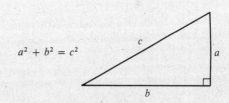

Fig. 7-24

(d) Right Triangles.   In a right triangle, the side opposite the right angle is called the *hypotenuse* and the other two sides are called *legs*.

| Pythagorean Theorem | The square of the hypotenuse is equal to the sum of the squares of the other two sides (legs). $$a^2 + b^2 = c^2$$ | **72** |

See Fig. 7-24.

Example 7.12:   A right triangle has legs of length 3 units and 4 units.   Find the length of the hypotenuse.

Solution: By Eq. 72, letting *C* = the length of the hypotenuse,

$$C^2 = 3^2 + 4^2 = 9 + 16 = 25$$

$$C = 5$$

(The 3-4-5 right triangle is a useful one to remember.)

In a right triangle, the altitude to the
hypotenuse forms two right triangles
which are similar to each other and to the
original triangle.

84

See Fig. 7.25.

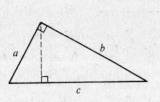

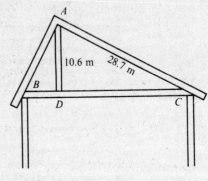

Fig. 7-25   **Fig. 7-26**

Example 7.13:   In the roof of Fig. 7-26, angles *BAC* and *ADB* are right angles.   Find the lengths of *AB*, *BD*, and *CD*.

Solution: From the Pythagorean Theorem,

$$(CD)^2 = AC^2 - AD^2$$

$$(CD)^2 = (28.7)^2 - (10.6)^2 = 711$$

$$CD = 26.7 \text{ m}$$

By Statement 84, triangle *ABD* is similar to triangle *ACD*; so, by Statement 71,

$$\frac{BD}{10.6} = \frac{10.6}{26.7}$$

$$BD = \frac{(10.6)^2}{26.7} = 4.21 \text{ m}$$

By the Pythagorean Theorem, in triangle *ABD*,

$$(AB)^2 = (10.6)^2 + (4.21)^2 = 130$$

$$AB = 11.4 \text{ m}$$

(e) Congruent Triangles. If all three angles and three sides of one triangle equal the sides and angles of another triangle, the two triangles are equal, or *congruent*, as in Fig. 7-27.

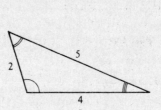

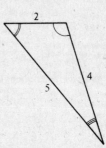

Fig. 7-27

Two Triangles Are Congruent if	Two angles and a side of one are equal to two angles and the corresponding side of the other (ASA), (AAS).	**86**
	Two sides and the included angle of one are equal, respectively, to two sides and the included angle of the other (SAS).	**87**
	Three sides of one are equal to the three sides of the other (SSS).	**88**

Example 7.14:　　Two parallel beams are connected by cables as in Fig. 7-28.　Point C bisects cable BD.　Find distance AB.

Solution:　　Since, by Statement 55, $\angle ACB$ equals $\angle DCE$, and, by Statement 56, $\angle BAC$ equals $\angle CED$, and BC equals CD, triangle ABC is congruent to triangle CDE (AAS).　Therefore,

$$AB = DE = 15 \text{ m}$$

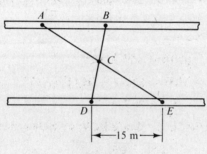

Fig. 7-28

(f)　Some Special Triangles.　In a 30-60-90° right triangle, Fig. 7-29, the side opposite the 30° angle is half the length of the hypotenuse.

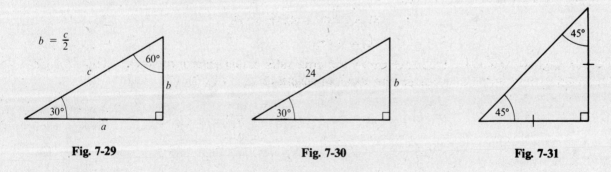

Fig. 7-29　　　　　　　　　**Fig. 7-30**　　　　　　　　　**Fig. 7-31**

Example 7.15:　　The length of side b in Fig. 7-30 is 12 units.

A *45° right triangle*, Fig. 7-31, is also isosceles and the hypotenuse is $\sqrt{2}$ times the length of either side.

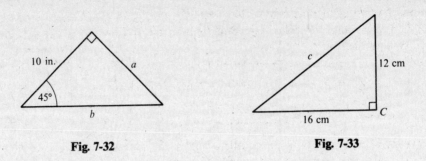

Fig. 7-32   Fig. 7-33

Example 7.16:   In the triangle of Fig. 7-32, side a is 10 in. and the hypotenuse is $10\sqrt{2}$ or 14.1 in.

A *3-4-5 triangle* is a right triangle in which the sides are in the ratio of 3 to 4 to 5.

Example 7.17:   In Fig. 7-33, side c is 20 cm.

7.3 QUADRILATERALS

A four-sided polygon is called a *quadrilateral*, Fig. 7-34a.

(a) Square.   The *square*, Fig. 7-34b, has four equal sides and four right angles.

$$\boxed{\begin{array}{c} \text{Area of square} = (\text{length of side})^2 \\ \text{Area} = a^2 \end{array}}\quad \mathbf{89}$$

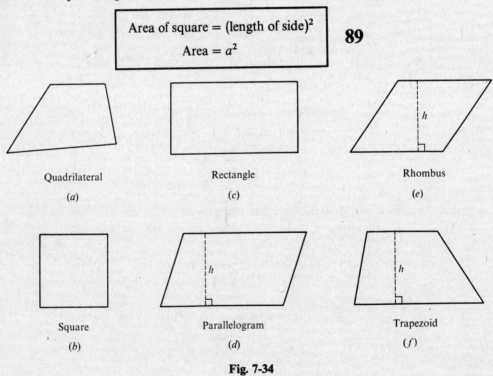

Fig. 7-34

(b) Rectangle.   A *rectangle*, Fig. 7-34c, has four right angles and its opposite sides are parallel and equal.

$$\boxed{\begin{array}{c} \text{Area of rectangle} = \text{length} \times \text{width} \\ \text{Area} = ab \end{array}}\quad \mathbf{90}$$

(c) Parallelogram. A *parallelogram*, Fig. 7-34d, has opposite sides which are parallel and equal. Opposite angles are equal and the diagonals bisect each other. The *altitude* is the perpendicular distance between two parallel sides, the bases.

$$\boxed{\begin{array}{c} \text{Area of parallelogram} = \text{base} \times \text{altitude} \\ \text{Area} = bh \end{array}} \quad \mathbf{91}$$

(d) Rhombus. A *rhombus*, Fig. 7-34e, is a parallelogram with four equal sides. Its diagonals bisect each other at right angles and bisect the angles of the rhombus.

$$\boxed{\begin{array}{c} \text{Area of rhombus} = \text{base} \times \text{altitude} \\ \text{Area} = bh \end{array}} \quad \mathbf{92}$$

(e) Trapezoid. A *trapezoid*, Fig. 7-34f, has only two parallel sides, called the *bases*. The *altitude* is the distance between the bases.

$$\boxed{\begin{array}{c} \text{Area of trapezoid} = (\text{altitude})(\text{average of the bases}) \\ \text{Area} = \dfrac{(a+b)h}{2} \end{array}} \quad \mathbf{93}$$

7.4 THE CIRCLE

(a) Circumference and Area. For a circle of radius r and diameter d,

$$\boxed{\text{Circumference} = 2\pi r = \pi d} \quad \mathbf{94}$$

$$\boxed{\text{Area} = \pi r^2 = \dfrac{\pi d^2}{4}} \quad \mathbf{95}$$

where $\pi \approx 3.1416$.

Example 7.18: A flat, circular mirror, 75.0 cm in diameter, is to be fabricated. The cost of grinding the edge is $2.25/cm and the cost of polishing the face is $0.55/cm². Find the cost of the two operations.

Solution: The circumference is, by Eq. 94,

$$C = 75\pi$$
$$= 236 \text{ cm}$$

So the cost of edge grinding is

$$236(2.25) = \$531$$

The area is, by Eq. 95,

$$A = \pi(37.5)^2$$
$$= 4418 \text{ cm}^2$$

So the polishing cost is

$$4418(0.55) = \$2430$$

(b) Angles and Arcs. A *central angle* is one whose vertex is at the center of the circle. An *arc* is a portion of the circle. An *inscribed angle* is one whose vertex is on the circle, Fig. 7-35.

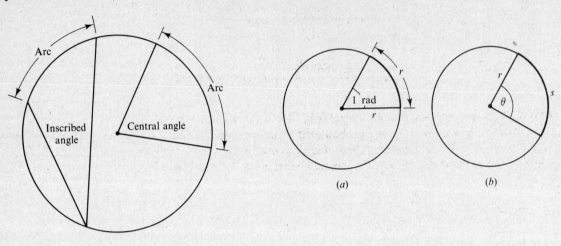

Fig. 7-35   **Fig. 7-36**

(c) Radian Measure. If an arc is laid off along a circle Fig. 7-36*a*, with a length equal to the radius of the circle, the central angle subtended by this arc is defined as 1 *rad*. There are 2π rad in one complete revolution.

$$\boxed{1 \text{ rev} = 2\pi \text{ rad} = 360°} \quad \mathbf{98}$$

Example 7.19:   Convert 2.00 rad to degrees.

Solution

$$2.00 \text{ rad}\left(\frac{360°}{2\pi \text{ rad}}\right) = \frac{360°}{\pi} = 115°$$

Example 7.20:   Convert 74.8° to radians.

Solution

$$74.8°\left(\frac{2\pi \text{ rad}}{360°}\right) = 1.31 \text{ rad}$$

If an arc of length s is intercepted by a central angle θ (in radians), Fig. 7-36*b*, then,

$$\boxed{\theta = \frac{s}{r}} \quad \mathbf{96}$$

Example 7.21:   An arc of length 3 in. on a circle having a 4-in. radius will intercept a central angle of

$$\theta = \frac{3}{4} \text{ rad}$$

(d) Sector. A *sector* of a circle is the plane figure bounded by an arc and two radii, as in Fig. 7-36.

$$\boxed{\text{Area of sector} = \frac{rs}{2} = \frac{r^2\theta}{2}} \quad \mathbf{97}$$

where r is the radius, s is the arc length, and θ is the central angle in radians.

Example 7.22: Find the area of a sector having a radius of 3.0 and a central angle of 28°.

Solution: Converting the central angle to radians, by Eq. 98,

$$\theta = 28° \left(\frac{2\pi \text{ rad}}{360°} \right) = 0.489 \text{ rad}$$

Then by Eq. 97,

$$\text{Area} = \frac{(3.0)^2(0.489)}{2} = 2.2 \text{ square units}$$

(e) Chord, Segment, and Sagitta. A *chord* (Fig. 7-37) is a line segment connecting two points on a circle. A *segment of a circle* is the region bounded by an arc and the chord connecting the ends of the arc, as in Fig. 7-38. (Do not confuse these with line segments.) The *sagitta* is a line connecting the midpoints of the arc and the chord of a segment, such as line *AB* in Fig. 7-38.

Fig. 7-37

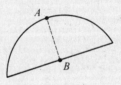

Fig. 7-38

> If two chords in a circle intersect, the product of the parts of one chord is equal to the product of the parts of the other chord.

102

Thus, in Fig. 7-39, $ab = cd$.

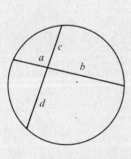

Fig. 7-39

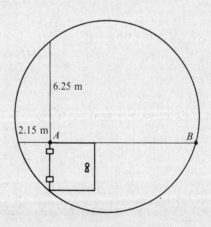

Fig. 7-40

Example 7.23: A 2.94-m square door in the end of a circular oven is shown in Fig. 7-40. Find dimension *AB*.

Solution: By Statement 102,

$$2.15(AB) = 2.94(6.25)$$

$$AB = \frac{2.94(6.25)}{2.15} = 8.55 \text{ m}$$

(f) Semicircle. A *semicircle* is a half circle. An *inscribed angle* is one whose vertex is on the circle.

| Any angle inscribed in a semicircle is a right angle. | **99** |

Thus, in Fig. 7-41, since AC is a diameter, angle ABC is a right angle.

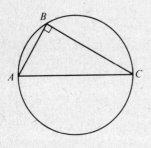

Fig. 7-41

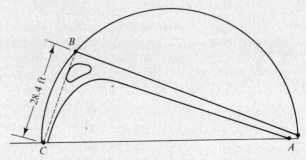

Fig. 7-42

Example 7.24:   A hemispherical roof 80 ft in diameter is supported by concrete members, one of which is shown in Fig. 7-42. Find the dimension AB.

Solution: By Statement 99, we know that angle ABC is a right angle. Then by the Pythagorean Theorem,

$$(AB)^2 = (80)^2 - (28.4)^2 = 5593$$
$$AB = 74.8 \text{ ft}$$

(g) Tangents and Secants. A *tangent* is a straight line, such as AB in Fig. 7-43, that touches a circle in just one point. A *secant*, line CD, touches in two points.

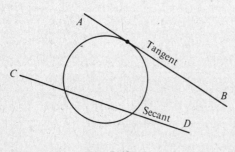

Fig. 7-43

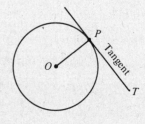

Fig. 7-44

| A tangent is perpendicular to the radius drawn through the point of contact. | **100** |

In Fig. 7-44, the tangent T touches the circle at P and makes a right angle with the radius OP.

Example 7.25:   Sheet steel in a rolling mill passes over a 22.6-cm-diameter roller, then passes between two smaller rollers, as in Fig. 7-45. Find dimension BC.

Solution: From Statement 100, we know that angle *ABC* is a right angle.   Then by the Pythagorean Theorem,

$$(BC)^2 = (39.5)^2 - (11.3)^2 = 1433$$
$$BC = 37.8 \text{ cm}$$

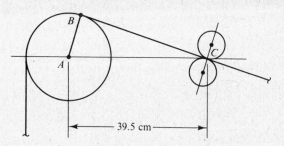

Fig. 7-45

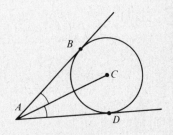

Fig. 7-46

> Two tangents drawn to a circle from a
> point outside the circle are equal, and they
> make equal angles with a line from the
> point to the center of the circle.

101

Thus in Fig. 7-46, $AB = AD$,   and $\angle BAC = \angle CAD$.

Example 7.26:   A 1.000-in.-diameter rod is placed in a dovetail, as in Fig. 7-47.   Find the distance *AC*.
Solution: Finding *AB*,

$$AB = 1.565 - 0.500 = 1.065 \text{ in.}$$

Since, by Statement 100, triangle *ACB* is a right triangle,

$$(AC)^2 = (1.065)^2 + (0.500)^2 = 1.384$$
$$AC = 1.177 \text{ in.}$$

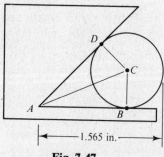

Fig. 7-47

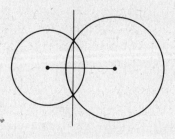

Fig. 7-48

(h)   Intersecting Circles

> If two circles intersect, their line
> of centers is perpendicular to their
> common chord at its midpoint.

103

See Fig. 7-48.

| If two circles are tangent to each other, the line of centers passes through the point of contact. | **104** |

See Fig. 7-49.

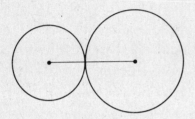

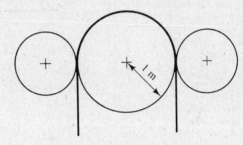

Fig. 7-49   Fig. 7-50

Example 7.27:   Three rollers have their centers on a common line, as in Fig. 7-50.   Find the length of sheet material between the two contact points.

Solution: By Statement 104, we know that the contact points lie on the line of centers.   The distance between them is simply half the circumference of the middle roller.

$$D = \frac{\pi d}{2} = \pi r = 3.14 \text{ m}$$

7.5 POLYHEDRONS

A *polyhedron* is a solid bounded entirely by planes.   The bounding planes are called *faces*.   The faces intersect at the *edges* of the polyhedron and the edges meet at the *vertices*.

A *prism* is a polyhedron two faces of which are equal (congruent) polygons lying in parallel planes, called *bases*, and whose other faces, called *lateral* faces, are parallelograms, as in Fig. 7-51. A *lateral* edge is the line of intersection of two lateral faces.   The *altitude* of a prism is the perpendicular distance between the bases.   A *right prism* is one whose bases are perpendicular to its lateral faces, which are now rectangles.   For a right prism,

| Volume = (area of base)(altitude) | **109** |

| Lateral area (not including the bases) = (perimeter of base)(altitude) | **110** |

A *rectangular parallelepiped* is a prism with six faces, all of which are rectangles, with adjoining faces meeting at right angles, as in Fig. 7-52.

For a rectangular parallelepiped of length l, width w, and height h,

| Volume = lwh | **107** |
| Surface area = $2(lw + hw + lh)$ | **108** |

If the six faces are squares, the rectangular parallelepiped becomes a *cube*.
For a cube of side *a*,

$$\text{Volume} = a^3 \qquad \textbf{105}$$

$$\text{Surface area} = 6a^2 \qquad \textbf{106}$$

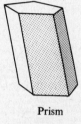

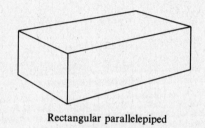

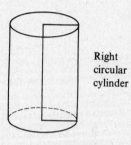

Right circular cylinder

Prism

Fig. 7-51

Rectangular parallelepiped

Fig. 7-52

Fig. 7-53

7.6 RIGHT CIRCULAR CYLINDER

A right circular cylinder is generated by revolving a rectangle about one of its sides, Fig. 7-53.   The cross section of a right circular cylinder is a circle.

$$\text{Volume} = (\text{area of base})(\text{altitude}) \qquad \textbf{109}$$

The *lateral area* is the area of the curved surface of the cylinder and does not include the area of the bases.

$$\text{Lateral area} = (\text{perimeter of base})(\text{altitude}) \qquad \textbf{110}$$

Formulas 109 and 110 apply also to cylinders that do not have a circular cross section and to prisms as well.

Example 7.28:   A cylindrical kiln of circular cross section is to have a diameter of 15.0 ft and a volume of 3500 ft³.
(*a*) Find its length.   (*b*) How many square feet of steel are needed to construct the kiln?

Solution: From Eq. 95,

$$\text{Area of base} = \pi(7.5)^2 = 177 \text{ ft}^2$$

From Eq. 109,

$$\text{Volume} = 177L = 3500$$

$$L = 19.8 \text{ ft}$$

From Eq. 110,

$$\text{Lateral area} = 2\pi(7.5)(19.8) = 933 \text{ ft}^2$$

to which we add the area of the ends.

$$\text{Total area} = 933 + 2(177) = 1287 \text{ ft}^2$$

7.7 CONES

A right circular cone, Fig. 7-54, is generated by rotating a right triangle about one of its legs. Point A is called the *vertex*, line AC is the *axis* of the cone, distance AC is the altitude, distance AB is the *slant height*, and the circle generated by line CB is the *base*.

The volume equals the product of one-third the area of the base A and the altitude h.

$$V = \frac{Ah}{3}$$

113

The lateral area equals the product of one-half the circumference (perimeter) of the base P and the slant height s.

$$\text{Lateral area} = \frac{Ps}{2}$$

114

Example 7.29:   A right circular cone has a base radius of 3 and an altitude of 4. Find its volume and lateral area.

Solution: The area of the base is $\pi r^2 = 9\pi$, so by Eq. 113,

$$\text{Volume} = \frac{9\pi(4)}{3} = 12\pi$$

The perimeter of the base is $2\pi r = 6\pi$. Since the altitude, a base radius, and the cone edge form a 3-4-5 right triangle, the slant height is 5. So, by Eq. 114,

$$\text{Lateral area} = \frac{6\pi(5)}{2} = 15\pi$$

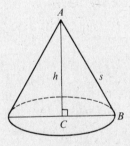

Right circular cone

Fig. 7-54

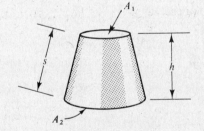

Frustum of a cone

Fig. 7-55

The *frustum* of a cone is that portion intercepted between the base and a plane parallel to the base, as in Fig. 7-55. For a frustum of altitude h, slant height s, and whose base areas are A_1 and A_2,

$$\text{Volume} = \frac{h}{3}\left(A_1 + A_2 + \sqrt{A_1 A_2}\right)$$

115

and

$$\boxed{\begin{aligned} \text{Lateral area} &= \frac{s}{2} \text{ (sum of base perimeters)} \\ &= \frac{s}{2} (P_1 + P_2) \end{aligned}} \quad \mathbf{116}$$

7.8 SPHERE

For a sphere of radius r,

$$\boxed{\begin{array}{l} \text{Volume} = \dfrac{4}{3} \pi r^3 \quad \mathbf{111} \\ \hline \text{Surface area} = 4\pi r^2 \quad \mathbf{112} \end{array}}$$

Example 7.30:   The hemispherical dome on a certain statehouse has an outside radius of 61.5 ft. Find the cost of gilding the dome, if the rate is \$1.55 ft^2.

Solution: By Eq. 112,

$$\text{Surface area} = \frac{4\pi r^2}{2} = 2\pi(61.5)^2 = 23\,765 \text{ ft}^2$$

So the cost is

$$23\,765(\$1.55) = \$36,836$$

7.9 SIMILAR FIGURES

Similar geometric figures, either plane or solid, are those whose angles are respectively equal. Corresponding sides of similar figures are those which are included between equal angles in the respective figures. Thus, in Fig. 7-56, the two figures shown are similar because the angles of the smaller figure equal the angles of the larger, and sides AB and EF, for example, are corresponding sides.

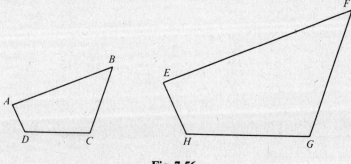

Fig. 7-56

(a)   Dimensions

$$\boxed{\text{Corresponding dimensions of plane or solid similar figures are in proportion.}} \quad \mathbf{117}$$

Example 7.31:   Two similar solids are shown in Fig. 7-57.   Find dimension AB.

Solution: From Statement 117,

$$\frac{AB}{4.5} = \frac{3.7}{1.8}$$

$$AB = \frac{4.5(3.7)}{1.8} = 9.3 \text{ cm}$$

(b)   Areas

> Areas of similar plane or solid figures are proportional to the *squares* of any two corresponding dimensions.

118

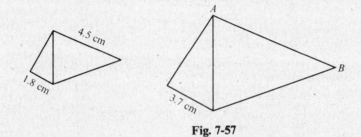

Fig. 7-57

Example 7.32:   If the surface area of the smaller solid in Fig. 7-57 is 29 cm², find the surface area s of the larger solid.

Solution: By Statement 118,

$$\frac{s}{29} = \left(\frac{3.7}{1.8}\right)^2 = 4.23$$

$$s = 29(4.23) = 123 \text{ cm}^2$$

(c)   Volumes

> Volumes of similar solid figures are proportional to the *cubes* of any two corresponding dimensions.

119

Example 7.33:   If the volume of the larger solid in Fig. 7-57 is 65 cm³, find the volume V of the smaller solid.

Solution: By Statement 119,

$$\frac{V}{65} = \left(\frac{1.8}{3.7}\right)^3 = 0.115$$

$$V = 65(0.115) = 7.5 \text{ cm}^3$$

Common Error	Students often forget to *square* corresponding dimensions when finding areas and to *cube* corresponding dimensions when finding volumes.

(d)   Scale Drawings.   Engineering and architectural drawings, surveys, and maps are examples of scale drawings.   The *scale* relates distances on the drawing to distances on the actual object and is often given as a ratio.   Realize that a scale drawing, and the object it represents, are similar figures.

Example 7.34:   A geological survey map has a scale of $1:62\,500$.   How many miles are represented by a map distance of 2.4 in?

Solution: The distance on the land is $2.4 \times 62\,500$ in.   Converting to miles,

$$2.4(62\,500) \text{ in.} \left(\frac{1 \text{ ft}}{12 \text{ in.}} \right) \left(\frac{1 \text{ mi}}{5280 \text{ ft}} \right) = 2.4 \text{ mi}$$

Solved Problems

TRIANGLES

7.1   Two angles of a triangle are 57° and 84°.   Find the third angle.

Solution

Let $x =$ the third angle.   By Eq. 60,

$$x + 57° + 84° = 180°$$

Solving for x,

$$x = 180° - 57° - 84° = 39°$$

7.2   The three sides of a triangle have lengths 3.46, 2.13, and 4.17.   Find the area of the triangle.

Solution

By Eq. 59,

$$s = \frac{3.46 + 2.13 + 4.17}{2} = 4.88$$

So,
$$\text{Area} = \sqrt{4.88(4.88 - 3.46)(4.88 - 2.13)(4.88 - 4.17)}$$
$$= \sqrt{13.5} = 3.67$$

7.3   One side of a triangle is 15.8 cm long and the altitude drawn to that side is 9.54 cm.   Find the area of the triangle.

Solution

By Eq. 58,

$$\text{Area} = \frac{15.8(9.54)}{2} = 75.4 \text{ cm}^2$$

7.4   Find x in Fig. 7-58.   Line AD bisects angle A.

Solution

Since line AD is an angle bisector, by Statement 67,

$$\frac{x}{4.86} = \frac{4.24}{2.98}$$

Solving for x,

$$x = \frac{4.86(4.24)}{2.98} = 6.91$$

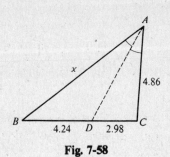

Fig. 7-58

7.5   In Fig. 7-59, line DE is parallel to AC.   Find x.

Solution

By Statement 68,

$$\frac{x}{182} = \frac{852}{212}$$

Solving for x,

$$x = \frac{182(852)}{212} = 731 \text{ mm}$$

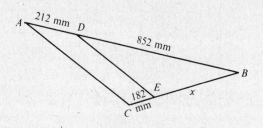

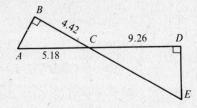

Fig. 7-59   **Fig. 7-60**

7.6   Find sides AB, CE, and DE in Fig. 7-60.

Solution

Applying the Pythagorean Theorem to right triangle ABC,

$$(AB)^2 + (4.42)^2 = (5.18)^2$$
$$(AB)^2 = (5.18)^2 - (4.42)^2 = 7.30$$
$$AB = 2.70$$

Triangle ABC is similar to triangle CDE since each triangle contains a right angle and the vertical angles at C are equal.   By Statement 71,

$$\frac{CE}{5.18} = \frac{9.26}{4.42}$$

Solving for CE,

$$CE = \frac{5.18(9.26)}{4.42} = 10.9$$

Similarly,

$$\frac{DE}{2.70} = \frac{9.26}{4.42}$$
$$DE = \frac{2.70(9.26)}{4.42} = 5.66$$

7.7   Find the distance AB between the centers of the two rollers in Fig. 7-61.

Solution

The vertical distance AC between the centers is equal to the difference in the radii of the rollers. Since the diameters are 48.4 cm and 27.6 cm, the corresponding radii are 24.2 cm and 13.8 cm.   So,

$$24.2 - 13.8 = 10.4 \text{ cm}$$

By the Pythagorean Theorem in right triangle ABC,

$$(AB)^2 = (10.4)^2 + (64.3)^2 = 4243$$
$$AB = 65.1 \text{ cm}$$

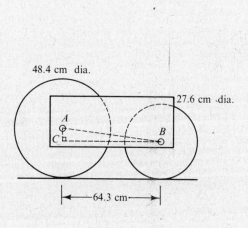

Fig. 7-61

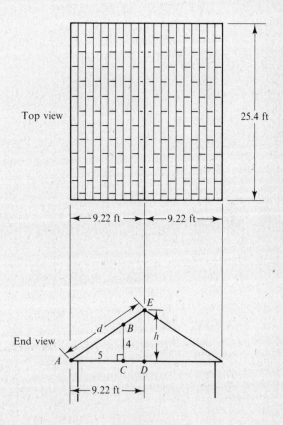

Fig. 7-62

7.8 The top and end views of a roof to be shingled are shown in Fig. 7-62. The slope or pitch of the roof is 4/5, which means that it rises 4 units every 5 horizontal units. Find the total shingle area.

Solution

Since triangle ABC is similar to triangle ADE, we may write the proportion

$$\frac{h}{9.22} = \frac{4}{5}$$

and so

$$h = \frac{4(9.22)}{5} = 7.38 \text{ ft}$$

Using the Pythagorean Theorem in triangle ADE,

$$d^2 = (7.38)^2 + (9.22)^2 = 139.5$$
$$d = 11.8 \text{ ft}$$

Multiplying this value by the length of the roof will give half the shingled area. So the total shingled area is

$$A = 2(11.8)(25.4) = 599 \text{ ft}^2$$

7.9   An antenna, 75 m high, is supported by two cables of length L which reach the ground at a distance D from the base of the antenna, Fig. 7-63.   The distance between the cable anchor points is one and one-half times the length of each cable.   Find L and D.

Solution

By the Pythagorean Theorem,

$$L^2 = D^2 + (75)^2 = D^2 + 5625$$

From the problem statement,

$$2D = 1.5L$$

or

$$D = 0.75L$$

Substituting,

$$L^2 = (0.75L)^2 + 5625 = 0.5625L^2 + 5625$$
$$0.4375L^2 = 5625$$
$$L^2 = 12857$$
$$L = 113 \text{ m}$$

and

$$D = 0.75L = 85 \text{ m}$$

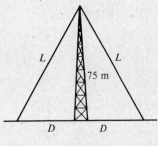

Fig. 7-63

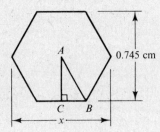

Fig. 7-64

7.10   A hex head bolt, Fig. 7-64, measures 0.745 cm across the flats.   Find the distance x across the corners.

Solution

The six flats each subtend an angle of

$$\frac{360}{6} = 60°$$

at the center.   So, angle BAC is 30°, making triangle ABC a 30-60-90 right triangle.   The side AC is equal to half the distance across the flats.

$$AC = \frac{1}{2}(0.745) = 0.3725$$

The side AB is equal to half the distance across the corners.

$$AB = \frac{x}{2}$$

In a 30-60-90 triangle, the side opposite the 30° angle equals half the hypotenuse.   So,

$$BC = \frac{1}{2}AB = \frac{x}{4}$$

By the Pythagorean Theorem,

$$(AB)^2 = (BC)^2 + (AC)^2$$

So,

$$\frac{x^2}{4} = \frac{x^2}{16} + (0.3725)^2$$

Solving for x,

$$\frac{3x^2}{16} = 0.1388$$

$$x^2 = \frac{(0.1388)(16)}{3} = 0.740$$

$$x = 0.860 \text{ cm}$$

7.11   A 24.0-in.-long piece of steel hexagonal stock measuring 1.125 in. across the flats has a 0.5000-in.-diameter hole running lengthwise from end to end.   Find the volume and weight of the bar. Assume the density of steel to be 450 lb/ft³.

Solution

As in Problem 7.10, triangle ABC, Fig. 7-65, is a 30-60-90 right triangle; so $AB = 2(BC)$. AC equals half the distance across the flats or 0.5625.   By the Pythagorean Theorem,

$$(2d)^2 = d^2 + (0.5625)^2$$
$$3d^2 = 0.316$$
$$d = 0.325$$

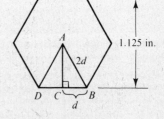

The area of triangle ABD is, by Eq. 58,

$$\text{Area} = 0.325(0.5625) = 0.183 \text{ in}^2$$

Since there are six such areas in the hexagon, for the hexagon,

$$\text{Area} = 6(0.183) = 1.10 \text{ in}^2$$

Fig. 7-65

The area of the hole is, by Eq. 95,

$$\frac{\pi(0.5000)^2}{4} = 0.196 \text{ in}^2$$

Subtracting, $1.10 - 0.196 = 0.904 \text{ in}^2 = \text{area of end}$

By Eq. 109,

$$\text{Volume} = 0.904(24.0) = 21.7 \text{ in}^3$$

By Eq. 184,

$$\text{Weight} = \text{volume} \times \text{density}$$

Note that $1 \text{ ft}^3 = 1728 \text{ in}^3$.   So,

$$\text{Weight} = 21.7 \text{ in}^3 \left(\frac{\text{ft}^3}{1728 \text{ in}^3} \right) \left(\frac{450 \text{ lb}}{\text{ft}^3} \right) = 5.65 \text{ lb}$$

7.12   A 60° screw thread is measured by placing three wires on the thread and measuring the distance T as in Fig. 7-66.   Find the distance D if the wire diameters are 0.125 cm and the distance T is 1.250 cm, assuming the root of the thread is a sharp V shape.   See Fig. 7-67 for an enlarged version of the root.

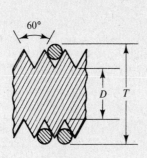

Fig. 7-66

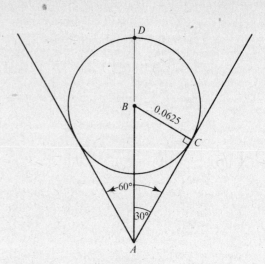

Fig. 7-67

Solution

Line AB bisects the 60° thread angle; so $\angle BAC = 30°$. By Statement 100, AC is perpendicular to BC, so $\angle ACB = 90°$. By Eq. 60,

$$\text{Angle } ABC = 180 - 90 - 30 = 60°$$

and so triangle ABC is a 30-60-90 triangle. Its hypotenuse AB is twice the shortest side, BC.

$$AB = 2(0.0625) = 0.125 \text{ cm}$$

So, $$D = T - 2(AB + BD)$$
$$= 1.250 - 2(0.125 + 0.0625) = 1.250 - 0.375 = 0.875 \text{ cm}$$

AVERAGE ORDINATE

7.13   Find the average ordinate for the triangular waveform shown in Fig. 7-68.

Solution

The *average ordinate* (or mean value) is the average height of the waveform. If the original waveform is replaced by a rectangle having the same width and same area, the average ordinate, y_{avg}, is the height of that rectangle. So, since the area of a rectangle equals width × height,

$$y_{avg} = \frac{\text{area under the waveform}}{\text{width}}$$

By Eq. 58, the area is

$$A = \frac{4(10)}{2} = 20$$

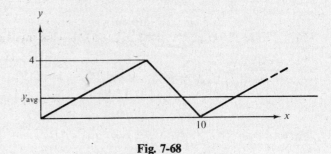

So,

$$y_{avg} = \frac{20}{10} = 2 \text{ units}$$

Fig. 7-68

7.14   Find the average voltage for the triangular wave in Fig. 7-69.

Solution

By Eq. 58,

$$\text{Area} = \frac{28(2)}{2} = 28$$

So,

$$y_{\text{avg}} = \frac{28}{2} = 14 \text{ V}$$

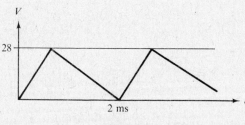

Fig. 7-69

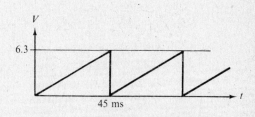

Fig. 7-70

7.15   Find the average ordinate for the sawtooth wave in Fig. 7-70.

Solution

By Eq. 58,

$$\text{Area} = \frac{6.3(45)}{2} = 142$$

So,

$$y_{\text{avg}} = \frac{142}{45} = 3.2 \text{ V}$$

THE CIRCLE

7.16   Find the circumference and area of a circle of radius 8.74 cm.

Solution

By Eq. 94,

$$\text{Circumference} = 2\pi(8.74) = 54.9 \text{ cm}$$

By Eq. 95,

$$\text{Area} = \pi(8.74)^2 = 240 \text{ cm}^2$$

7.17   Find the circumference of a circle having an area of 846 in^2.

Solution

By Eq. 95,

$$\text{Area} = \pi r^2 = 846$$

$$r^2 = \frac{846}{\pi} = 269$$

$$r = 16.4 \text{ in.}$$

By Eq. 94,

$$\text{Circumference} = 2\pi(16.4) = 103 \text{ in.}$$

7.18   Find the central angle in degrees subtended by a 16.3-m-long arc in a circle having a 10.3-m radius.

Solution

By Eq. 96,

$$\theta = \frac{16.3}{10.3} = 1.58 \text{ rad}$$

Using Eq. 98 to convert to degrees,

$$\theta = 1.58 \text{ rad} \left(\frac{180°}{\pi \text{ rad}} \right) = 90.5°$$

7.19   Find the area of the sector of the previous problem.

Solution

By Eq. 97,

$$\text{Area} = \frac{10.3(16.3)}{2} = 83.9 \text{ m}^2$$

7.20   Find x in Fig. 7-71.

Solution

By Eq. 102,

$$8.42x = 4.21(9.78)$$
$$x = \frac{4.21(9.78)}{8.42}$$
$$= 4.89$$

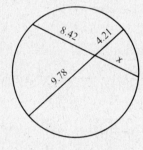

Fig. 7-71

7.21   A circular tile walk is 50.0 in. wide and surrounds a circular swimming pool 32.0 ft in diameter. Find the area of the walk.

Solution

The area of the pool is

$$A_1 = \frac{\pi(32.0)^2}{4} = 804 \text{ ft}^2$$

The outside diameter of the circular walk is found by adding the diameter of the pool to twice the width of the walk.   So,

$$d = 32.0 + \frac{2(50.0)}{12} = 40.3 \text{ ft}$$

The area of the walk and pool combined is

$$A_2 = \frac{\pi(40.3)^2}{4} = 1276 \text{ ft}^2$$

Subtracting the pool area,

$$\text{Walk area} = 1276 - 804 = 472 \text{ ft}^2$$

7.22 What must be the diameter d of a cylindrical piston so that a pressure of 1250 lb/in² on its circular end will result in a total force of 12 500 lb? *Hint:* Use Eq. 187 from the Summary of Formulas at the beginning of the book.

Solution

By Eq. 187, the piston area is

$$A = \frac{\text{force}}{\text{pressure}} = \frac{12\,500 \text{ lb}}{1250 \text{ lb/in}^2} = 10.0 \text{ in}^2$$

By Eq. 95,

$$\text{Area} = \frac{\pi d^2}{4} = 10.0$$

$$d^2 = \frac{10.0(4)}{\pi} = 12.7$$

$$d = 3.57 \text{ in.}$$

7.23 Find the length of belt needed to connect the three 1.25-cm-diameter tuning pulleys in Fig. 7-72.

Solution

By the Pythagorean Theorem,

$$(AB)^2 + (1.44)^2 = (2.88)^2$$
$$(AB)^2 = 8.29 - 2.07 = 6.22$$
$$AB = 2.49 \text{ cm}$$

The straight sections of belt equal

$$2.49 + 1.44 + 2.88 = 6.81 \text{ cm}$$

The curved portions of the belt equal the circumference of one 1.25-cm-diameter pulley, or, by Eq. 94,

$$1.25\pi = 3.93 \text{ cm}$$

So the total length is

$$L = 6.81 + 3.93 = 10.74 \text{ cm}$$

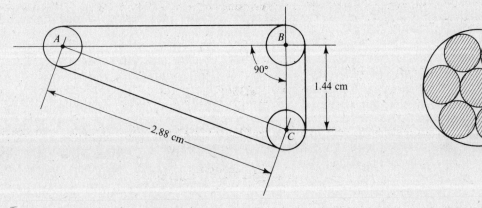

Fig. 7-72   Fig. 7-73

7.24 Seven cables are contained within a circular conduit, as in Fig. 7-73, and a coolant is to be circulated in the spaces between the cables. If the inside diameter of the conduit is 3.840 cm, find the cross-section area *not* occupied by the cables.

Solution

Let D = the diameter of the conduit.   The diameter of each cable is then $D/3$.   By Eq. 95,

$$\text{Area of conduit} = \frac{\pi D^2}{4}$$

and $$\text{Area of each cable} = \frac{\pi\left(\dfrac{D}{3}\right)^2}{4}$$

The space between the conduit and the cables is then

$$A = \frac{\pi D^2}{4} - 7\left[\frac{\pi\left(\dfrac{D}{3}\right)^2}{4}\right] = \frac{\pi(3.840)^2}{4} - \frac{7\pi}{4}\left(\frac{3.840}{3}\right)^2 = 11.58 - 9.008 = 2.57\ \text{cm}^2$$

7.25   In laying out the circular highway curve of Fig. 7-74, show that the deflection angle D is half the central angle Δ.

Solution

Since all radii of a circle must be equal, $OQ = OR$ and so triangle OQR is isosceles.   Therefore the base angles are equal.   Angle A = angle ORQ.

By Eq. 60, the sum of the angles of a triangle equals $180°$.   So,

$$2A + \Delta = 180°$$

$$A = \frac{180 - \Delta}{2} = 90 - \frac{\Delta}{2}$$

$$D = 90 - A = 90 - \left(90 - \frac{\Delta}{2}\right)$$

So, $$D = \frac{\Delta}{2}$$

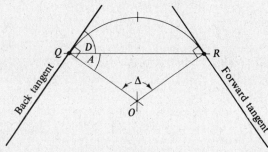

Fig. 7-74

SIGHT DISTANCE

7.26   A circular highway curve of radius R has some visual obstacle beside the roadway, at a distance m from the white line, Fig. 7-75.   Find the sight distance C.

Solution

Draw line PO, which bisects C.   Then, by Eq. 102, noting that one of the intersecting chords is the diameter, or $2R$,

$$\left(\frac{C}{2}\right)\left(\frac{C}{2}\right) = m(2R - m)$$

$$\frac{C^2}{4} = 2Rm - m^2$$

$$C^2 = 8Rm - 4m^2$$

$$C = \sqrt{8Rm - 4m^2}$$

As m is usually quite small compared with R, the quantity $4m^2$ will be much smaller than $8Rm$ and is often omitted, leaving

$$C = \sqrt{8Rm}$$

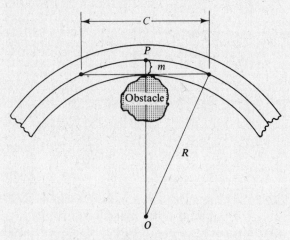

Fig. 7-75

SOLIDS

7.27　The base of a certain right prism is a right triangle having sides of 6.00 cm, 8.00 cm, and 10.0 cm, and the length of the prism is 21.6 cm.　Find the volume of the prism and the total area, including the ends.

Solution

By Eq. 58, the area of the base is

$$\text{Base area} = \frac{6.00(8.00)}{2} = 24.0 \text{ cm}^2$$

By Eq. 109,

$$\text{Volume} = 24.0(21.6) = 518 \text{ cm}^3$$

The perimeter of the base is

$$6.00 + 8.00 + 10.0 = 24.0 \text{ cm}$$

By Eq. 110,

$$\text{Lateral area} = 24.0(21.6) = 518 \text{ cm}^2$$

Adding to this the areas of the ends,

$$\text{Total area} = 518 + 24.0 + 24.0 = 566 \text{ cm}^2$$

7.28　Find the volume and lateral area of a right circular cylinder 12.8 ft in diameter and 21.4 ft long.

Solution

By Eq. 95, the base area is

$$\text{Base area} = \pi(6.40)^2 = 129 \text{ ft}^2$$

By Eq. 109,

$$\text{Volume} = 129(21.4) = 2760 \text{ ft}^3$$

By Eq. 94, the base circumference is

$$\text{Circumference} = 12.8\pi = 40.2 \text{ ft}$$

By Eq. 110,

$$\text{Lateral area} = 40.2(21.4) = 86\overline{0} \text{ ft}^2$$

7.29　Find the volume and lateral area of a right circular cone having a base radius of 11.4 and a height of 18.6.

Solution

By Eq. 95, the area of the base is

$$\text{Base area} = \pi(11.4)^2 = 408 \text{ square units}$$

By Eq. 113,

$$\text{Volume} = \frac{408(18.6)}{3} = 2530 \text{ cubic units}$$

By Eq. 72, the slant height is

$$\text{Slant height} = \sqrt{(11.4)^2 + (18.6)^2} = 21.8$$

By Eq. 94, the circumference of the base is

$$\text{Circumference} = 2\pi(11.4) = 71.6$$

By Eq. 114,

$$\text{Lateral area} = \frac{71.6(21.8)}{2} = 78\bar{0} \text{ square units}$$

7.30 Find the volume and surface area of a sphere having a radius of 53.8 cm.

Solution

By Eq. 111,

$$\text{Volume} = \frac{4\pi(53.8)^3}{3} = 652\,000 \text{ cm}^3$$

By Eq. 112,

$$\text{Surface area} = 4\pi(53.8)^2 = 36\,400 \text{ cm}^2$$

7.31 A truckload of gravel is spread on a 120-ft section of roadbed and found to have an average depth of 1.0 in. (see Fig. 7-76). The truck carries 216 ft^3.

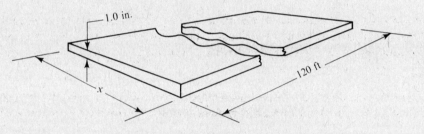

Fig. 7-76

(*a*) How wide is the roadbed?

(*b*) How many loads of gravel will be needed to cover 2.0 mi of roadbed of the same width to a depth of 0.25 ft?

Solution

(*a*) Let x = width of roadbed, ft. By Eq. 107, the volume of gravel in the roadbed is

$$\text{Volume} = lwh = (120 \text{ ft})(1.0 \text{ in.})\left(\frac{\text{ft}}{12 \text{ in.}}\right)(x \text{ ft}) = 10x \text{ ft}$$

Since this volume is equal to the truck's capacity (216 ft^3),

$$10x = 216$$

$$x = 21.6 \text{ ft}$$

(*b*) Let V = the required volume of gravel, ft^3. By Eq. 107,

$$V = lwh = (2.0 \text{ mi})\left(\frac{5280 \text{ ft}}{\text{mi}}\right)(21.6 \text{ ft})(0.25 \text{ ft}) = 57\,000 \text{ ft}^3$$

Dividing by the capacity of one truck,

$$V = \frac{57\,000}{216} = 264 \text{ loads}$$

7.32 A rectangular settling tank is being filled with liquid, with each 50.0 gal of liquid increasing the depth by 0.250 in.   The length of the tank is 30.0 ft.   (*a*) What is the width of the tank? (*b*) How many gallons will be required to fill the tank to a depth of 10.0 ft?

Solution

(*a*) Let W = width of tank, ft.   Converting gallons to cubic feet, with   1 ft^3 = 7.48 gal,

$$(50.0 \text{ gal})\left(\frac{\text{ft}^3}{7.48 \text{ gal}}\right) = 6.68 \text{ ft}^3$$

By Eq. 107,

$$\text{Volume} = lwh = (0.250 \text{ in.})\left(\frac{\text{ft}}{12 \text{ in.}}\right)(30.0 \text{ ft})(W) = 6.68 \text{ ft}^3$$

Solving for W,

$$W = \frac{12(6.68)}{0.25(30.0)} = 10.7 \text{ ft}$$

(*b*) Let V = gallons required to fill tank to 10.0-ft depth.   By Eq. 107,

$$V = 10.0(10.7)(30.0) \text{ ft}^3 = 3210 \text{ ft}^3$$

Converting to gallons,

$$V = (3210 \text{ ft}^3)\left(\frac{7.48 \text{ gal}}{\text{ft}^3}\right) = 24\,000 \text{ gal}$$

7.33 A 2-in. cube of steel is placed in a surface grinding machine, and the vertical feed set so that 0.0005 in. of metal is removed at each cut.   How many cuts will be needed to reduce the weight of the cube by 10 g?   Assume the density of steel to be 450 lb/ft^3.   There are 454 g/lb.

Solution

Let x = number of cuts.   Each cut removes a slab of steel measuring $2 \times 2 \times 0.0005$, with a volume of 0.002 in^3.   By Eq. 184, the weight of material removed with each cut = volume × density, or

$$(0.002 \text{ in}^3)\left(\frac{450 \text{ lb}}{\text{ft}^3}\right)\left(\frac{1 \text{ ft}^3}{1728 \text{ in}^3}\right)\left(\frac{454 \text{ g}}{\text{lb}}\right) = 0.236 \text{ g/cut}$$

So, $0.236x = 10$

$$x = 42.4 \text{ cuts}$$

So 42 cuts will remove slightly less than the required 10 g, and 43 cuts will remove slightly more.

7.34 A spherical antenna dish, Fig. 7-77, has a diameter of 4.16 m and a depth (sagitta) of 26.8 cm. Find the radius of curvature r.

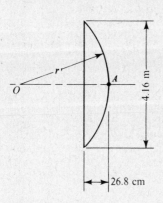

Fig. 7-77

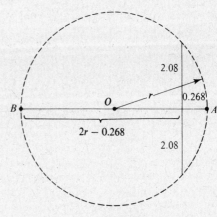

Fig. 7-78

Solution

We complete the circle, Fig. 7-78, and draw line AOB from the center A of the dish through the center of curvature O. By Eq. 102,

$$(2r - 0.268)(0.268) = (2.08)(2.08)$$

$$0.536r - 0.0718 = 4.33$$

$$0.536r = 4.40$$

$$r = 8.21 \text{ m}$$

7.35   A spherical radome encloses a volume of $40\,800 \text{ ft}^3$.

(a)   Find the radome radius r.

(b)   If constructed of a material weighing 20.5 lb per square yard (lb/yd^2), find the weight of the radome.

Solution

(a)   By Eq. 111,

$$\text{Volume} = \frac{4}{3}\pi r^3 = 40\,800$$

$$r^3 = \frac{3(40\,800)}{4\pi} = 9740$$

$$r = 21.4 \text{ ft}$$

(b)   By Eq. 112,

$$\text{Surface area} = 4\pi(21.4)^2 = 5750 \text{ ft}^2 = 639 \text{ yd}^2$$

So, $$\text{Weight} = (639 \text{ yd}^2)\left(\frac{20.5 \text{ lb}}{\text{yd}^2}\right) = 13\,100 \text{ lb} = 6.55 \text{ tons}$$

7.36   Find the weight in grams of 1000 steel balls, 4.22 mm in diameter.   Assume the density of steel to be 7.85 g/cm^3.

Solution

Their volume, by Eq. 111, is

$$V = \frac{4\pi(2.11)^3}{3} \times 1000 = 3.93 \times 10^4 \text{ mm}^3$$

Converting to cubic centimeters,

$$1 \text{ cm} = 10 \text{ mm}$$

Cubing both sides,

$$1 \text{ cm}^3 = 10^3 \text{ mm}^3$$

and so the volume is

$$V = (3.93 \times 10^4 \text{ mm}^3)\left(\frac{\text{cm}^3}{10^3 \text{ mm}^3}\right) = 39.3 \text{ cm}^3$$

By Eq. 184,

$$\text{Weight} = \text{volume} \times \text{density}$$

$$= (39.3 \text{ cm}^3)\left(\frac{7.85 \text{ g}}{\text{cm}^3}\right) = 309 \text{ g}$$

7.37   A steel gear (density = 450 lb/ft³) is to be lightened by drilling holes through the gear.   The gear is 0.25 in. thick.   Find the diameter d of the holes if each is to remove 1.0 oz.

Solution

The volume of steel to be removed is, by Eq. 184,

$$V = \frac{(1.0 \text{ oz})(\text{ft}^3)}{450 \text{ lb}}\left(\frac{1 \text{ lb}}{16 \text{ oz}}\right)\left(\frac{1728 \text{ in}^3}{\text{ft}^3}\right) = 0.24 \text{ in}^3$$

Since the steel removed is a cylinder of diameter d, 0.25 in. in length, we get, by Eq. 109,

$$\frac{\pi d^2}{4}(0.25) = 0.24$$

$$d^2 = \frac{0.24(4)}{0.25\pi} = 1.2$$

$$d = 1.1 \text{ in.}$$

7.38   In order to balance a flywheel, it is determined that 0.0276 lb of steel (density = 450 lb/ft³) must be removed.   This is done by drilling a blind hole 0.750 in. in diameter with a twist drill having a conical tip, as in Fig. 7-79.   Find the required hole depth d.

Solution

The volume of steel to be removed is, by Eq. 184,

$$V = \frac{(0.0276 \text{ lb})(\text{ft}^3)}{450 \text{ lb}}\left(\frac{1728 \text{ in}^3}{\text{ft}^3}\right) = 0.106 \text{ in}^3$$

The cross-sectional area of the hole is, by Eq. 95,

$$A = \frac{\pi(0.750)^2}{4} = 0.442 \text{ in}^2$$

The volume of the conical tip is, by Eq. 113,

$$V_1 = \frac{0.442(0.225)}{3} = 0.0331 \text{ in}^3$$

So the volume of the cylindrical portion of the hole is

$$V_2 = 0.106 - 0.0331 = 0.073 \text{ in}^3$$

By Eq. 109,

$$0.442d = 0.073$$

$$d = 0.17 \text{ in.}$$

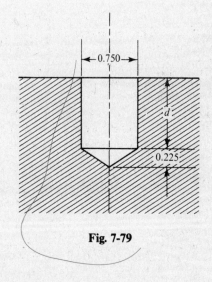

Fig. 7-79

7.39   One element in a core memory unit is in the shape of a hollow cylinder 4.24 mm in diameter and 2.54 mm long, with an axial hole 2.17 mm in diameter.   If the density of the material from which they are made is 6.86 g/cm³, find the weight of half a million such elements.

Solution

The volume of the device is, by Eq. 109,

$$V = \text{length} \times \text{base area} = 2.54\left[\frac{\pi(4.24)^2}{4} - \frac{\pi(2.17)^2}{4}\right]$$

$$= \frac{2.54\pi}{4}[(4.24)^2 - (2.17)^2] = 26.5 \text{ mm}^3$$

and for 500 000 units

$$\text{Volume} = 26.5(5 \times 10^5) = 1.32 \times 10^7 \text{ mm}^3$$

$$= (1.32 \times 10^7 \text{ mm}^3)\left(\frac{\text{cm}^3}{10^3 \text{ mm}^3}\right) = 13\,200 \text{ cm}^3$$

By Eq. 184,

$$\text{Weight} = \text{volume} \times \text{density}$$

$$= 13\,200 \text{ cm}^3\left(\frac{6.86 \text{ g}}{\text{cm}^3}\right) = 90\,600 \text{ g} = 90.6 \text{ kg}$$

7.40 A 10.0-ft-long piece of iron pipe (density = 450 lb/ft^3) has an outside diameter of 15.3 in. and weighs 1255 lb.   Find the wall thickness.

Solution

The outside radius is

$$\frac{15.3}{12(2)} = 0.638 \text{ ft}$$

Letting the inside radius be r, the volume of the pipe is

$$V = \pi[(0.638)^2 - r^2](10.0) \text{ ft}^3$$

but, by Eq. 184,

$$\text{Volume (density)} = \text{weight} = 1255 \text{ lb}$$

So

$$\pi(0.407 - r^2)(10.0)(450) = 1255$$

Solving for r,

$$0.407 - r^2 = \frac{1255}{4500\pi} = 0.0888$$

$$r^2 = 0.318$$

$$r = 0.564 \text{ ft}$$

The wall thickness is then

$$0.638 - 0.564 = 0.074 \text{ ft} = 0.888 \text{ in.}$$

7.41 The slump test for concrete requires that the concrete mixture be packed into a conical form, which is later removed.   This *slump cone* is in the shape of a frustum of a cone, with open ends 4.00 and 8.00 in. in diameter and height 12.0 in.   See Fig. 7-80.

(*a*) Find the volume of concrete that will fit in the slump cone.

(*b*) Find the surface area of the slump cone.

Solution

(*a*) The area of the top is, by Eq. 95,

$$A_1 = \pi(2.00)^2 = 4.00\pi$$

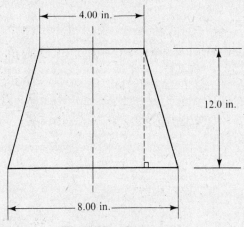

Fig. 7-80   Slump cone.

and of the bottom,

$$A_2 = \pi(4.00)^2 = 16.0\pi$$

By Eq. 115,

$$\text{Volume} = \frac{1}{3}[4.00\pi + 16.0\pi + \sqrt{4.00(16.0)\pi^2}](12.0) = 352 \text{ in}^3$$

(b) The slant height is, by the Pythagorean Theorem,

$$s^2 = (2.00)^2 + (12.0)^2 = 148$$

$$s = 12.2 \text{ in.}$$

The perimeters of the top and bottom are, by Eq. 94,

$$P_1 = \pi 4.00 \qquad P_2 = \pi 8.00$$

By Eq. 116,

$$\text{Area} = \frac{12.2}{2}(\pi 4.00 + \pi 8.00) = 230 \text{ in}^2$$

7.42 Find the surface area and weight of a tapered steel roller (density = 450 lb/ft^3) 4 ft 8 in. long and having end diameters of 5.25 and 2.47 in.

Solution

By Eq. 95, the end areas are

$$\frac{\pi(5.25)^2}{4} = 21.6 \text{ in}^2$$

and

$$\frac{\pi(2.47)^2}{4} = 4.79 \text{ in}^2$$

Converting the length to inches, 4 ft 8 in. equals 56 in., and by Eq. 115,

$$\text{Volume} = \frac{56}{3}[21.6 + 4.79 + \sqrt{(21.6)(4.79)}] = 682 \text{ in}^3 = 0.395 \text{ ft}^3$$

By Eq. 184,

$$\text{Weight} = 0.395(450) = 178 \text{ lb}$$

The slant height S is, by Eq. 72,

$$S^2 = (56)^2 + \left(\frac{5.25}{2} - \frac{2.47}{2}\right)^2$$

that is, S^2 equals the square of the length plus the square of the difference between the two end radii. So,

$$S^2 = (56)^2 + (2.625 - 1.235)^2 = 3138$$

$$S = 56.0 \text{ in.}$$

The base perimeters, by Eq. 94, are

$$5.25\pi = 16.5 \text{ in.}$$

and

$$2.47\pi = 7.76 \text{ in.}$$

By Eq. 116,

$$\text{Lateral area} = \frac{56}{2}(16.5 + 7.76) = 679 \text{ in}^2$$

7.43   A sector-shaped piece of sheet metal, Fig. 7-81, is to be rolled into a cone 15.5 cm high and having a volume of 1780 cm³, Fig. 7-82.   Find the central angle A and the radius r of the required sector.

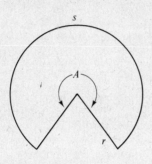

Fig. 7-81

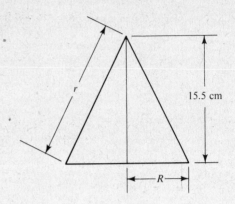

Fig. 7-82

Solution

By Eq. 113,

$$\text{Volume} = 1780 = \frac{Bh}{3} = \frac{15.5B}{3}$$

where B is the area of the base.   Solving for B,

$$B = \frac{3(1780)}{15.5} = 345 \text{ cm}^3$$

The base radius R is found from Eq. 95.

$$\pi R^2 = 345$$

$$R = \sqrt{\frac{345}{\pi}} = 10.5 \text{ cm}$$

The slant height r is, by Eq. 72,

$$r^2 = R^2 + (15.5)^2 = (10.5)^2 + (15.5)^2 = 350$$

$$r = 18.7 \text{ cm}$$

The circumference s of the base is, by Eq. 94,

$$s = 2\pi(10.5) = 66.0 \text{ cm}$$

By Eq. 96,

$$A = \frac{s}{r} = \frac{66.0}{18.7} = 3.53 \text{ rad} = 202°$$

SIMILAR FIGURES

7.44   The shortest side of one triangle is 13.8 cm, its perimeter is 59.4 cm, and its area is 455 cm². Find the perimeter P and area A of a second triangle, similar to the first, whose shortest side is 37.4 cm long.

Solution

The perimeters of the two triangles will be proportional to the ratio of corresponding sides; so

$$\frac{P}{59.4} = \frac{37.4}{13.8}$$

Solving for P,

$$P = \frac{59.4(37.4)}{13.8} = 161 \text{ cm}$$

By Statement 118, the areas will be proportional to the square of corresponding dimensions; so

$$\frac{A}{455} = \left(\frac{37.4}{13.8}\right)^2 = 7.34$$

Thus, $A = 455(7.34) = 3340 \text{ cm}^2$

7.45 A certain pyramid has a height of 11.3 in. and a volume of 257 in³.   Another pyramid, of similar shape, has a height of 21.8 in.   Find the volume of the second pyramid.

Solution

By Statement 119,

$$\frac{V}{257} = \left(\frac{21.8}{11.3}\right)^3 = 7.18$$

where V is the volume of the larger pyramid.   Solving for V,

$$V = 257(7.18) = 1850 \text{ in}^3$$

7.46 The floor plan of a certain factory has a scale of   1/4 in. = 1 ft,   and it shows a room having an area of 15.2 in².   What is the actual room area A in square feet?

Solution

By Statement 118,

$$\frac{A}{15.2} = \left(\frac{12}{0.25}\right)^2 = 2304$$

$$A = 15.2(2304) = 35\,000 \text{ in}^2 = 243 \text{ ft}^2$$

7.47 The surface area of a one-quarter scale model of a radar antenna measures 5.24 square meters (m²).   Find the surface area A of the full-sized antenna.

Solution

By Statement 118,

$$\frac{A}{5.24} = \left(\frac{4}{1}\right)^2$$

$$A = 5.24(4)^2 = 83.8 \text{ m}^2$$

7.48 The area of one electrode in a certain cathode ray tube (CRT) measures 1.45 cm² on a drawing having a scale of 1 : 4.   Find the actual electrode area A in square centimeters.

Solution

By Statement 118,

$$\frac{A}{1.45} = \left(\frac{4}{1}\right)^2 = 16$$

$$A = 16(1.45) = 23.2 \text{ cm}^2$$

7.49   Each side of a square semiconductor chip is increased by 1.0 mm, and the area is seen to increase by 8.6 mm^2. What were the dimensions of the original chip?

Solution

Let x = length of side of original chip, mm. From the problem statement,

$$\text{New area} = \text{old area} + 8.6$$

$$(x + 1.0)^2 = x^2 + 8.6$$

Solving for x,

$$x^2 + 2x + 1.0 = x^2 + 8.6$$

$$2x = 7.6$$

$$x = 3.8 \text{ mm}$$

7.50   The outline of a field was traced onto sheet metal, cut out, and weighed at 46.8 g. A 10.0-in^2 piece of the same metal weighed 24.3 g. If the scale of the map from which the outline was obtained is   1 in. = 60 ft,   find the acreage of the field.

Solution

Since the weight of a sheet metal shape is proportional to its area A,

$$\frac{A}{46.8} = \frac{10.0}{24.3}$$

$$A = \frac{46.8(10.0)}{24.3} = 19.3 \text{ in}^2$$

The map scale is   1 in. = 60 ft.   Squaring both sides,   1 in^2 = 3600 ft^2.

So, $\text{Area of the field} = (19.3 \text{ in}^2)\left(\frac{3600 \text{ ft}^2}{\text{in}^2}\right) = 69\,500 \text{ ft}^2$

Since   1 acre = 43 560 ft^2,

$$\text{Area of the field} = \frac{69\,500}{43\,560} = 1.60 \text{ acres}$$

7.51   A certain water tank is 28 ft high and holds 150 000 gal. How much would a 35-ft-high tank of similar shape hold?

Solution

By Statement 119,

$$\frac{V}{150\,000} = \left(\frac{35}{28}\right)^3$$

$$V = \frac{150\,000(35)^3}{(28)^3} = 290\,000 \text{ gal}$$

7.52   A semiconductor chip is viewed under a magnification of 40×, and the apparent volume is 554 cm^3. Find the actual volume V of the chip.

Solution

$$\frac{V}{554} = \left(\frac{1}{40}\right)^3$$

$$V = \frac{554}{(40)^3} = 8.66 \times 10^{-3} \text{ cm}^3 = 8.66 \text{ mm}^3$$

7.53   A certain lathe weighs 1380 lb.   If all dimensions were scaled up by a factor of 1.50, what would the larger lathe be expected to weigh?

Solution

Since the weight is proportional to the volume, we get, by Statement 119,

$$\frac{W}{1380} = (1.50)^3$$

$$W = 1380(1.50)^3 = 4660 \text{ lb}$$

Supplementary Problems

7.54   Find the area of the following:

(a)   A 3 cm by 7 cm rectangle

(b)   A 2 ft by 2 ft square

(c)   The triangle in Fig. 7-83

(d)   The triangle in Fig. 7-84

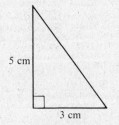

Fig. 7-83

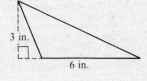

Fig. 7-84

7.55   Find θ in the three triangles of Fig. 7-85.

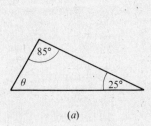

(a)

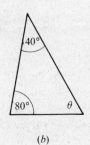

(b)

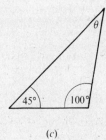

(c)

Fig. 7-85

7.56   Find the missing side in Fig. 7-86.

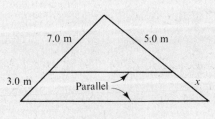

Fig. 7-86

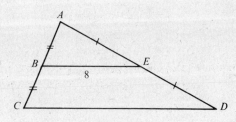

Fig. 7-87

7.57   Find length CD in Fig. 7-87.

7.58   Find the missing side x in each of the triangles in Fig. 7-88.

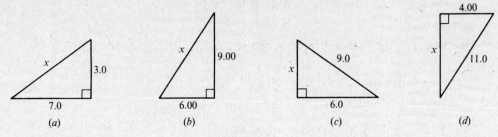

(a)   (b)   (c)   (d)

Fig. 7-88

7.59   State which pairs of triangles are congruent in Fig. 7-89.

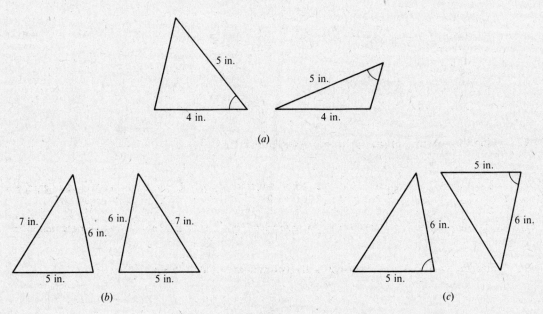

(a)

(b)   (c)

Fig. 7-89

7.60   Find the area of the trapezoid in Fig. 7-90.

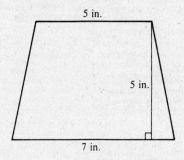

Fig. 7-90

7.61   Find the area and circumference of the following circles:

(a) $r = 6.00$ (b) $d = 3.0$ (c) $d = 5.00$ (d) $r = 7.00$

7.62   How large a central angle will an arc length of 2 in. intercept in a circle with a 4-in. radius?

7.63   Convert   (a) 6.00 rad to degrees   and   (b) 75° to radians.

7.64   Solve for the unknown line segments in Fig. 7-91.

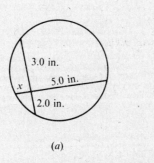
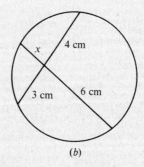

3.0 in.

5.0 in.

x

2.0 in.

x 4 cm

3 cm   6 cm

(a)   (b)

Fig. 7-91

7.65   Find the volume and surface area of a 4.00-m-radius sphere.

7.66   Find the volume and lateral area of a right circular cylinder of base radius 4.00 in. and height 6.00 in.

7.67   Find the volume and lateral area of a cone with a 6.00-cm altitude and a 9.00-cm base perimeter.

7.68   Find the volume V of the smaller of the two similar cylinders in Fig. 7-92.

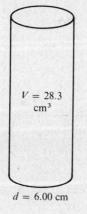
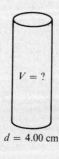

$V = 28.3$ cm^3

$V = ?$

$d = 4.00$ cm

$d = 6.00$ cm

Fig. 7-92

7.69   On a map of scale 1 : 550, a distance of how many feet is represented by a map distance of 3.00 in.?

Answers to Supplementary Problems

7.54   (a) 21 cm^2 (b) 4 ft^2 (c) $7\frac{1}{2}$ cm^2 (d) 9 in^2

7.55   (a) 70°   (b) 60°   (c) 35°

7.56 $x = 2.1$ m

7.57   16

7.58   (a) $x = 7.6$ (b) $x = 10.8$ (c) $x = 6.7$ (d) $x = 10.2$

7.59   (b), (c)

7.60   30 in^2

7.61   (a) $A = 113$ (b) $A = 7.1$ (c) $A = 19.6$ (d) $A = 154$
             $C = 37.7$ $C = 9.4$ $C = 15.7$ $C = 44.0$

7.62 $\theta = \frac{1}{2}$ rad

7.63   (a) 344°   (b) 1.3 rad

7.64   (a) $x = 1.2$ in.   (b) $x = 2$ cm

7.65 $V = 268$ m^3, surface area $= 201$ m^2

7.66 $V = 302$ in^3, lateral area $= 151$ in^2

7.67 $V = 12.8$ cm^3, lateral area $= 28$ cm^2

7.68 $V = 8.39$ cm^3

7.69   138 ft

Chapter 8

Introduction to Trigonometry

Trigonometry is the branch of mathematics that deals with, but is not limited to, the solution of triangles.

8.1 ANGLES AND THEIR MEASURES

(a) Definitions

1. In Chapter 7 we defined an *angle*, Fig. 8-1, as being formed when a ray turns from some *initial position OA* to a *terminal position OB*. When the turning is counterclockwise (CCW) as in Fig. 8-1, the angle is called *positive*. Clockwise (CW) turning of the ray, as in Fig. 8-2, generates a *negative* angle.

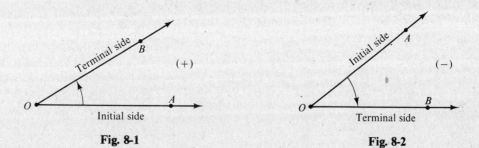

Fig. 8-1   **Fig. 8-2**

2. A *degree* (°) has already been defined as 1/360 of a revolution.

Example 8.1

$$\text{Half a revolution (rev)} = \frac{360°}{2} = 180°$$

$$\frac{2}{3} \text{ rev} = \frac{2}{3}(360°) = 240°$$

$$\frac{1}{48} \text{ rev} = \frac{360°}{48} = 7\frac{1}{2}° = 7.5°$$

Note that the fractional part of a degree can be expressed as a common fraction or as a decimal; angles written in decimal form are sometimes called *decimal degrees*.

Example 8.2: Express 1/96 rev in decimal degrees.

Solution

$$\frac{360°}{96} = 3.75°$$

205

Another way of expressing the fractional part of a degree is with *minutes* and *seconds*.

3. A *minute* (') is defined as 1/60 of a degree, or

$$60' = 1°$$

Example 8.3

$$0.8° \ (60'/\text{degree}) = 48'$$

4. A *second* (") is 1/60 of a minute, or

$$60'' = 1'$$

and

$$3600'' = 1°$$

Example 8.4

$$0.01° \ (3600''/\text{degree}) = 36''$$

5. A *radian* is the central angle subtended by an arc equal to the radius of the circle. Since an arc equal to the full circumference subtends an angle of 360°, and since the circumference equals 2π times the radius r,

$$\boxed{2\pi \text{ radians} = 360° = 1 \text{ revolution}} \quad \mathbf{98}$$

Example 8.5:   Express 2.41 rad in decimal degrees.

Solution

$$2.41 \text{ rad} \left(\frac{180°}{\pi \text{ rad}} \right) = 138°$$

Example 8.6:   Express 81.8° in radians.

Solution

$$81.8° \left(\frac{\pi \text{ rad}}{180°} \right) = 1.43 \text{ rad}$$

8.2 ANGLE CONVERSIONS

Angular units are converted by the same method (see Chapter 1) used for converting other units. Write the conversion factor as a fraction with the units which are to cancel located in the denominator. Multiply the given angle by this fraction and round the result to the same number of significant figures as in the given angle.

Example 8.7:   Convert 0.7350 rev to degrees.

Solution: Our conversion factor is

$$1 \text{ rev} = 360°$$

and so,

$$0.7350 \text{ rev} \left(\frac{360°}{\text{rev}} \right) = 264.6°$$

Example 8.8:   Convert 0.7350 rev to radians.

Solution: Our conversion factor is

$$1 \text{ rev} = 2\pi \text{ rad}$$

Therefore,

$$0.7350 \text{ rev} \left(\frac{2\pi \text{ rad}}{\text{rev}} \right) = 4.618 \text{ rad}$$

Example 8.9:   Convert 81.583° to (*a*) revolutions and (*b*) radians.

Solution

(*a*)
$$81.583° \left(\frac{1 \text{ rev}}{360°} \right) = 0.226\,62 \text{ rev}$$

(*b*)
$$81.583° \left(\frac{\pi \text{ rad}}{180°} \right) = 1.4239 \text{ rad}$$

Example 8.10: Convert 1.4822 rad to (*a*) degrees and (*b*) revolutions.

Solution

(*a*)
$$1.4822 \text{ rad} \left(\frac{180°}{\pi \text{ rad}} \right) = 84.924°$$

(*b*)
$$1.4822 \text{ rad} \left(\frac{1 \text{ rev}}{2\pi \text{ rad}} \right) = 0.235\,90 \text{ rev}$$

Example 8.11:   Convert 146.3° to revolutions.

Solution: Our conversion factor is

$$1 \text{ rev} = 360°$$

so
$$146.3° \left(\frac{1 \text{ rev}}{360°} \right) = 0.4064 \text{ rev}$$

8.3 OPERATIONS WITH ANGLES EXPRESSED IN DEGREES, MINUTES, AND SECONDS (DMS)

(a) Angle Conversions. To convert an angle in DMS to any other units, first convert to decimal degrees. The angle in decimal degrees can then be converted to other units.

Example 8.12:   Convert 38°17′26″ to decimal degrees.

Solution: We first convert the *seconds* to degrees.

$$26″ \left(\frac{1°}{3600″} \right) = 0.0072°$$

Since there are two significant digits in "26," we retain just two significant digits after converting to degrees.

We now convert the *minutes*, treating 17′ as an *exact number*. We round to four decimal places after converting, however, as there is no point in carrying more places than contained in "0.0072."

$$17′ \left(\frac{1°}{60′} \right) = 0.2833°$$

Adding, we get

$$38.0000 + 0.0072 + 0.2833 = 38.2905°$$

Example 8.13: Convert 3°45′18.6″ to radians.

Solution: We first convert to decimal degrees.

$$18.6″ \left(\frac{1°}{3600″} \right) = 0.005\,17°$$

$$45′ \left(\frac{1°}{60′} \right) = 0.750\,00°$$

$$\begin{array}{r} 3.000\,00° \\ \hline \text{Adding} \quad 3.755\,17° \end{array}$$

Converting the decimal degrees to radians,

$$3.755\,17°\left(\frac{\pi\ \text{rad}}{180°}\right) = 0.065\,540\,1\ \text{rad}$$

Convert decimal degrees to DMS as shown in the following example.

Example 8.14:   Convert 47.3941° to DMS.

Solution: Converting the decimal portion to minutes,

$$0.3941°\left(\frac{60'}{\text{degree}}\right) = 23.65'$$

Converting the decimal portion of 23.65 to seconds,

$$0.65'\left(\frac{60''}{\text{minute}}\right) = 39''$$

so, $47.3941° = 47°23'39''$

(b) Addition and Subtraction.   Add or subtract degrees, minutes, and seconds separately. Remove multiples of 60 from the *minutes* and *seconds* sums and "carry."

Example 8.15:   Add 25°32'18'' to 47°21'59''.

Solution

	°	′	″
	25	32	18
	47	21	59
Adding	72°	53′	77″
Removing multiples of 60	72°	54′	17″

Example 8.16:   Subtract 15°32'41'' from 26°12'27''.

Solution: "Borrowing" from minutes and degrees,

	°	′	″
	25	71	87
	−15	32	41
Subtracting	10°	39′	46″

(c) Multiplication and Division.   Convert the given angle to decimal degrees, perform the multiplication or division as required, and convert back to DMS.

Example 8.17:   Multiply 12°27'36'' by 2.6435.

Solution: Converting to decimal degrees,

$$27'\left(\frac{1°}{60'}\right) = 0.450°$$

$$36''\left(\frac{1°}{3600''}\right) = 0.010°$$

$$\underline{\hphantom{0000000}12.000°}$$

$$\text{Adding} \quad 12.460°$$

Multiplying,

$$12.460(2.6435) = 32.938°$$

Converting back to DMS,

$$0.938° \left(\frac{60'}{\text{degree}}\right) = 56.28'$$

$$0.28' \left(\frac{60''}{\text{minute}}\right) = 17''$$

so,

$$(12°27'36'')(2.6435) = 32°56'17''$$

Example 8.18:   Divide 68°30′ by 2.625.

Solution: Converting to decimal degrees,

$$68°30' = 68.50°$$

Dividing,

$$\frac{68.50}{2.625} = 26.10°$$

Converting back to DMS,

$$26.10° = 26°6'$$

8.4 THE TRIGONOMETRIC RATIOS

(a) Angles in Standard Position. If an angle is placed on coordinate axes with the vertex of the angle at the origin and its initial side along the positive *x* axis, the angle is said to be in *standard position*, Fig. 8-3.

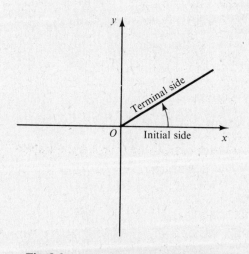

Fig. 8-3 An angle in standard position.

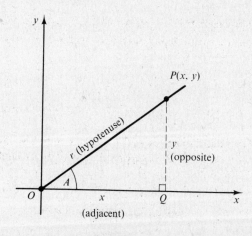

Fig. 8-4

(b) Trigonometric Ratios. Let *A* be some angle in standard position, Fig. 8-4. From any point *P* on the terminal side of the angle we draw a perpendicular to the *x* axis, forming a right triangle *OPQ*.

Note that side *OP* is the *hypotenuse* of the triangle and has a length *r*; *PQ* is the side *opposite* angle *A* and has a length *y*; and *OQ* is *adjacent* to angle *A* and has a length *x*.

The six *trigonometric ratios* are now defined, both in terms of the coordinates of point *P*, as well as in terms of the sides of right triangle *OPQ*.

$\text{sine } A = \sin A = \dfrac{y}{r} = \dfrac{\text{opposite side}}{\text{hypotenuse}}$	**73**
$\text{cosine } A = \cos A = \dfrac{x}{r} = \dfrac{\text{adjacent side}}{\text{hypotenuse}}$	**74**
$\text{tangent } A = \tan A = \dfrac{y}{x} = \dfrac{\text{opposite side}}{\text{adjacent side}}$	**75**
$\text{cotangent } A = \cot A = \text{ctn } A = \dfrac{x}{y} = \dfrac{\text{adjacent side}}{\text{opposite side}}$	**76**
$\text{secant } A = \sec A = \dfrac{r}{x} = \dfrac{\text{hypotenuse}}{\text{adjacent side}}$	**77**
$\text{cosecant } A = \csc A = \dfrac{r}{y} = \dfrac{\text{hypotenuse}}{\text{opposite side}}$	**78**

Example 8.19:   A point on the terminal side of angle θ in standard position has the coordinates $(7.1, 4.8)$, Fig. 8-5. Write the six trigonometric functions of θ.

Solution: Computing the distance OP by the Pythagorean Theorem,

$$OP = \sqrt{(7.1)^2 + (4.8)^2} = 8.6$$

Then by Eqs. 73 to 78,

$$\sin \theta = \frac{4.8}{8.6} = 0.56 \qquad \cot \theta = \frac{7.1}{4.8} = 1.5$$

$$\cos \theta = \frac{7.1}{8.6} = 0.83 \qquad \sec \theta = \frac{8.6}{7.1} = 1.2$$

$$\tan \theta = \frac{4.8}{7.1} = 0.68 \qquad \csc \theta = \frac{8.6}{4.8} = 1.8$$

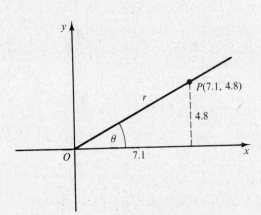

Fig. 8-5

Example 8.20:   A point Q on the terminal side of angle A has the coordinates $(-5.72, -3.15)$, as in Fig. 8-6. Write the six trigonometric functions of A.

Solution: The distance OQ is

$$OQ = \sqrt{(-5.72)^2 + (-3.15)^2} = 6.53$$

Then by Eqs. 73 to 78,

$$\sin A = \frac{-3.15}{6.53} = -0.482 \qquad \cot A = \frac{-5.72}{-3.15} = 1.82$$

$$\cos A = \frac{-5.72}{6.53} = -0.876 \qquad \sec A = \frac{6.53}{-5.72} = -1.14$$

$$\tan A = \frac{-3.15}{-5.72} = 0.551 \qquad \csc A = \frac{6.53}{-3.15} = -2.07$$

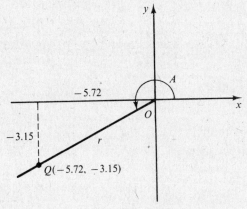

Fig. 8-6

8.5 ALGEBRAIC SIGNS OF THE TRIGONOMETRIC FUNCTIONS

As shown by the two previous examples, the signs of the trigonometric functions are positive for first quadrant angles and depend upon the signs of x and y (r is always positive) for angles in the other quadrants.   To determine the sign of a particular trigonometric function, sketch the angle in the proper quadrant, and draw x and y.   Note the signs of x and y, remembering that x is negative to the left of the origin and y is negative below the origin.

Example 8.21:   What is the algebraic sign of sec 110°?

Solution: We sketch the angle in standard position, Fig. 8-7, in the second quadrant.   Draw a perpendicular to the x axis from any point P on the terminal side of the angle.   Note that x is negative and y is positive.   Then,

$$\sec \theta = \frac{(+)}{(-)} = \text{negative}$$

Trigonometric tables rarely give the algebraic signs of the functions.   Determine them by the above method, or refer to Fig. 8-8.

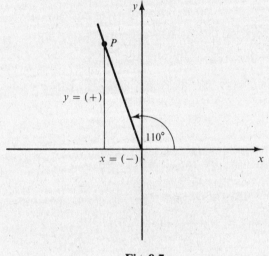

Fig. 8-7

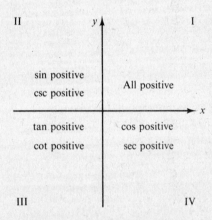

Algebraic signs of the trigonometric functions in all quadrants.

Fig. 8-8

8.6 RECIPROCAL RELATIONSHIPS

When we defined the six trigonometric ratios, we saw that

$$\sin A = \frac{\text{opposite side}}{\text{hypotenuse}} \quad \text{and} \quad \csc A = \frac{\text{hypotenuse}}{\text{opposite side}}$$

We see that the cosecant is the reciprocal of the sine; so,

Similarly,

$$\sin A = \frac{1}{\csc A}$$

$$\cos A = \frac{1}{\sec A} \qquad \textbf{79}$$

$$\tan A = \frac{1}{\cot A}$$

These are the three *reciprocal relationships*.

Example 8.22:   If the sine of a certain angle θ is 0.486, its cosecant is, from Eq. 79,

$$\csc \theta = \frac{1}{\sin \theta} = \frac{1}{0.486} = 2.06$$

8.7   REFERENCE ANGLE

When an angle is drawn in standard position, the *acute* angle that the terminal side makes with the x axis is called the *reference angle*, or working angle.

Example 8.23:   Find the reference angle m for an angle of 105°.

Solution: The reference angle is shown in Fig. 8-9 and is

$$m = 180° - 105° = 75°$$

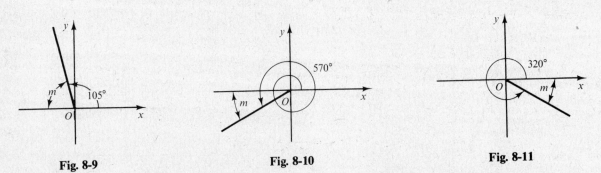

Fig. 8-9   Fig. 8-10   Fig. 8-11

Example 8.24:   Find the reference angle for an angle of 570°.

Solution: In Fig. 8-10, we see that 570° will have the same terminal side as $570° - 360° = 210°$,   a third quadrant angle.   The reference angle is then

$$m = 210° - 180° = 30°$$

Example 8.25:   Find the reference angle m for an angle of 320°.

Solution: From Fig. 8-11,

$$m = 360° - 320° = 40°$$

Common Error	The reference angle is always the acute angle between the terminal side and the x axis, never the y axis.

Reference angles will be needed in Section 8.9, when looking up the trigonometric functions of angles greater than 90° in the tables.

8.8   TRIGONOMETRIC FUNCTIONS ON THE CALCULATOR

(a)   Sine, Cosine, and Tangent.   Simply enter the angle and depress the *sin*, *cos*, or *tan* key.

Example 8.26:   Find sin 36.8° to four decimal places.

Solution: If your calculator has a *Degree/Radian* switch, be sure it is in the *Degree* position.

Enter 36.8

Depress the *sin* key

Read 0.5990

So, $\sin 36.8° = 0.5990$

Example 8.27:   Find the tangent of 1.45 rad to four significant figures.

Solution: Put the *Degree/Radian* switch into the *Radian* position.

>   Enter 1.45
>   Depress the *tan* key
>   Read 8.238

So, $\tan 1.45 \text{ rad} = 8.238$

(b) Cotangent, Secant, and Cosecant. Since calculators do not usually have keys for these three functions, we make use of the reciprocal relationships, Eq. 79.   We obtain reciprocals on the calculator with the *1/x* key.

Example 8.28:   Find sec 18.4° to four significant figures.

Solution: Put the calculator in the *Degree* mode.

>   Enter 18.4
>   Depress the *cos* key
>   Depress the *1/x* key
>   Read 1.054

So,

$$\sec 18.4° = \frac{1}{\cos 18.4°} = 1.054$$

8.9   TABLES OF TRIGONOMETRIC FUNCTIONS

A simple table of trigonometric ratios is located in Appendix C.   The sine, cosine, and tangent are given, to four decimal places, for angles from 0 to 90°.

(a) Angles Less than 90°.   The sine, cosine, and tangent are found in the table.   Use the reciprocal relationships to compute the cotangent, secant, and cosecant.

Example 8.29:   Find the six trigonometric functions of 26°.

Solution: From the table, we read

$$\sin 26° = 0.4384 \qquad \cos 26° = 0.8988 \qquad \tan 26° = 0.4877$$

By the reciprocal relationships, Eq. 79,

$$\cot 26° = \frac{1}{0.4877} = 2.050$$

$$\sec 26° = \frac{1}{0.8988} = 1.113$$

$$\csc 26° = \frac{1}{0.4384} = 2.281$$

(b) Angles Greater than 90°
   1. Compute the reference angle.
   2. Find the required trigonometric function of the reference angle.
   3. Determine the algebraic sign of the function.

Example 8.30:　　Find cos 152°.

Solution:　　The reference angle, Fig. 8-12, is

$$m = 180° - 152° = 28°$$

From the table of trigonometric functions,

$$\cos 28° = 0.8829$$

Since the cosine is negative in the second quadrant,

$$\cos 152° = -0.8829$$

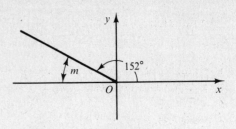

Fig. 8-12

Example 8.31:　　Find tan 304°.

Solution:　　The reference angle is

$$m = 360° - 304° = 56°$$

From the trigonometric table,

$$\tan 56° = 1.483$$

Since the tangent is negative in the third quadrant,

$$\tan 304° = -1.483$$

Example 8.32:　　Find sin 420°.

Solution:　　The reference angle is

$$m = 420° - 360° = 60°$$

So,　　　　　　　　$$\sin 420° = \sin 60° = 0.866$$

8.10　FINDING THE ANGLE WHEN THE FUNCTION IS GIVEN

(a)　By Calculator. Locate the *arc* or *inv* key on your calculator. This is to be depressed before pressing the *sin, cos,* or *tan* key.

Example 8.33:　If　sin $A = 0.455$,　find A, in degrees, to three significant figures.

Solution:　　Put the calculator into the *Degree* mode.

　　　　Enter .455
　　　　Depress the *arc* or *inv* key
　　　　Depress the *sin* key
　　　　Read 27.1°

However, 27.1° is not the only angle that has a sine of 0.455. A second quadrant angle having a reference angle of 27.1° will have the same sine. This angle is

$$180° - 27.1° = 152.9°$$

Angles greater than 360° having the same terminal side as 27.1° or 152.9° will also have a sine of 0.455. It is customary, however, to give only the two angles less than 360°.

The notation $\theta = \arcsin 0.455$

or $\theta = \sin^{-1} 0.455$

is read, "θ is the angle whose sine is 0.455." These are referred to as the *arc functions*, or the *inverse trigonometric functions*.

Example 8.34:   Find the two angles less than 360° that have a cotangent of 1.21. Find the angles in decimal degrees to three significant figures.

Solution: With the calculator in the *Degree* mode,

      Enter 1.21
      Depress $1/x$
      Depress *arc* (or *inv*)
      Depress *tan*
      Read 39.6

The third quadrant angle having a reference angle of 39.6° will also have a cotangent of 1.21. This angle is

$$180° + 39.6° = 219.6°$$

(b) From the Tables. Use the tables in reverse, first finding the function in the proper column; then reading the angle at the left edge of the table.

Example 8.35:   Find the two angles less than 360° whose sine is 0.5000.

Solution: Finding 0.5000 in the *sin* column, we read across to

$$\theta = 30°$$

However, 30° is not the only angle having a sine of 0.5000. A second quadrant angle having a reference angle of 30° will also have the same sine. This angle is

$$\theta = 180° - 30° = 150°$$

Example 8.36:   Find the two angles less than 360° that have a cotangent of 0.4452.

Solution: Let A = the unknown angle(s); so,

$$\cot A = 0.4452$$

Since our trigonometric table does not list cotangents, we convert to the tangent. By Eq. 79,

$$\tan A = \frac{1}{\cot A} = \frac{1}{0.4452} = 2.246$$

So instead of finding the angle that has a cotangent of 0.4452, we find the angle that has a tangent of 2.246. From the table,

$$A = 66°$$

Since the cotangent is positive in the first and third quadrants, we get another angle,

$$A = 180° + 66° = 246°$$

which also has a cotangent of 0.4452. So,

$$A = \text{arccot} \, (0.4452) = 66° \quad \text{and} \quad 246°$$

Example 8.37:   Find the two angles less than 360° that have a cosine of −0.3420.

Solution: Since the cosine is negative, our angles are in the second and third quadrants. We first find the reference angle from the trigonometric table,

$$\cos m = 0.3420$$

$$m = 70°$$

The second quadrant angle is then

$$\theta = 180° - 70° = 110°$$

and the third quadrant angle is

$$\theta = 180° + 70° = 250°$$

Thus, $\arccos(-0.3420) = 110°$ and $250°$

8.11   INTERPOLATING IN THE TRIGONOMETRIC TABLES

1. If the angle for which we need a trigonometric function is not listed in the table, but is located somewhere *between* two table values, we find the intermediate value by a process called *interpolation*, as in the following examples.

Example 8.38:   Find sin 21.4°.

Solution: From the trigonometric table, we find the sines of 21.0° and 22.0°.

$$\sin 21.0 = 0.3584$$

and $$\sin 22.0 = 0.3746$$

We now assume that *differences* in the values of the sines are directly proportional to differences in the measures of the angles themselves. We compute the differences between the angles and the differences between the sines of the angles,

	Angle	sin
	21°	0.3584
1.0 0.4	21.4°	Unknown x 0.0162
	22°	0.3746

letting x represent the difference between the sine of 21.4°, which is the value we seek, and the sine of 21.0° (0.3584). Writing a proportion,

$$\frac{0.4}{1} = \frac{x}{0.0162}$$

Solving for x,

$$x = 0.0065$$

This is the difference between sin 21° and sin 21.4°. We note that the sine is increasing with the angle and *add* x to 0.3584,

$$\sin 21.4° = 0.3584 + 0.0065 = 0.3649$$

By writing a proportion as we did in the previous example, we were making the assumption that the graph of the *sine* is a straight line between the two table values to either side of our given angle: This method is therefore called *linear* interpolation.

Example 8.39:   Find cos 56.7°.

Solution: The tabular difference between cos 56° and cos 57° is

$$0.5592 - 0.5446 = 0.0146$$

so,

$$\frac{0.7}{1} = \frac{x}{0.0146}$$

$$x = 0.7(0.0146) = 0.0102$$

Since the cosine is decreasing in this region, we *subtract* x from 0.5592,

$$\cos 57.6° = 0.5592 - 0.0102 = 0.5490$$

Common Error	Do not forget to *subtract* the amount x when the table values are *decreasing*, as in the previous example.

2. To find the angle when the trigonometric function is given, we simply reverse the procedure.

Example 8.40:   Find arctan 0.6355.

Solution: We see in the table that the angle lies between 32° and 33°.

$$
\begin{array}{llll}
 & \tan 32° & 0.6249 & \\
1\;\lbrack\; x\;\lbrack & \tan \theta & 0.6355 \;\rbrack 0.0106 & \rbrack 0.0245 \\
 & \tan 33° & 0.6494 &
\end{array}
$$

Forming a proportion,

$$\frac{x}{1} = \frac{0.0106}{0.0245}$$

$$x = 0.43$$

So,

$$\text{arctan } 0.6355 = 32.43°$$

which rounds to 32.4°.   Since the tangent is positive in the third quadrant also,

$$\text{arctan } 0.6355 = 180° + 32.4° = 212.4°$$

Example 8.41:   Find arcsec (-1.3294).

Solution: Let θ = the angle we seek.   Then,

$$\sec \theta = -1.3294$$

$$\cos \theta = -\frac{1}{1.3294} = -0.7522$$

The minus sign tells us that θ is in the second and third quadrants.   Find the reference angle m,

$$\cos m = 0.7522$$

From the trigonometric table, m is seen to lie between 41° and 42°.   Taking tabular differences and writing the proportion,

$$\frac{x}{1} = \frac{0.7547 - 0.7522}{0.7547 - 0.7431} = 0.22$$

$$m = \text{arccos } 0.7522 = 41.2°$$

$$\theta = \text{arccos } (-0.7522) = \text{arcsec } (-1.3294)$$

$$= 180° + 41.2° = 221.2° \quad \text{and} \quad 180° - 41.2° = 138.8°$$

8.12 SPECIAL ANGLES

Some angles appear so often in technical problems that it is convenient to be able to write their trigonometric functions without consulting the tables or using a calculator.   Try to do so for the angles given in the following table.

Angle	sin	cos	tan
0°	0	1	0
30°	0.5	0.866	0.577
45°	0.707	0.707	1
60°	0.866	0.5	1.732
90°	1	0	Undefined
180°	0	−1	0
270°	−1	0	Undefined
360°	Same as for 0°		

The angles 0°, 90°, 180°, and 360° are called *quadrantal* angles because the terminal side of each of them lies along one of the coordinate axes.

8.13 RADIAN MEASURE AND ARC LENGTH

In a circle of radius r, let s be the length of the arc intercepted by a central angle θ, Fig. 8-13.   If θ is expressed in radians, then s, r, and θ are related by the expression

$$\boxed{\theta = \frac{s}{r}} \qquad \textbf{96}$$

(θ must be in radians)

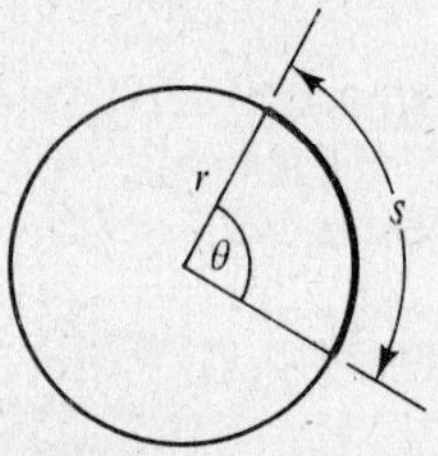

Relationship between arc length, radius, and central angle.
Fig. 8-13

This is because an angle θ is to 2π, the total number of radians in a circle, as s is to the circumference $2\pi r$.   Making a proportion,

$$\frac{\theta}{2\pi} = \frac{s}{2\pi r}$$

multiplying both sides by 2π, $\theta = s/r$.

Example 8.42:   Find the angle that would intercept an arc of 8.6 in. in a circle of radius 14 in.

Solution: From Eq. 96,

$$\theta = \frac{s}{r} = \frac{8.6}{14} = 0.61 \text{ rad}$$

Example 8.43:   Find the arc length intercepted by a central angle of 48.3° in a 25.0-cm-radius circle.

Solution: Converting the angle to radians,

$$48.3° \left(\frac{\pi \text{ rad}}{180°} \right) = 0.843 \text{ rad}$$

By Eq. 96,

$$s = r\theta = 25.0(0.843) = 21.1 \text{ cm}$$

Example 8.44:   Find the radius of a circle in which an angle of 1.25 rad intercepts a 12.5-in. arc.

Solution: By Eq. 96,

$$r = \frac{s}{\theta} = \frac{12.5}{1.25} = 10.0 \text{ in.}$$

Common Error	Remember, when using Eq. 96, that the angle must be in *radians*.

8.14 ROTATION

The following definitions apply to the object in Fig. 8-14, which has rotated about point O from an initial position A to position B.

(a) Angular Displacement. The *angular displacement* is the angle θ through which the object turns.

(b) Angular Velocity. The *angular velocity* ω is a measure of how *fast* the object turns. If the object turns with a constant speed through an angle θ in time t,

Angular Velocity	$\omega = \dfrac{\theta}{t}$	**183**

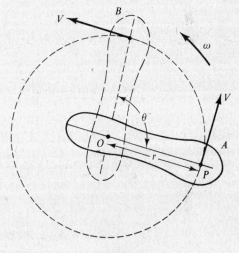

Fig. 8-14

Example 8.45:   Find the angular velocity in rev/min of a wheel that rotates 40 revolutions in one second.

Solution: By Eq. 183,

$$\omega = \frac{40 \text{ rev}}{1 \text{ s}} = 40 \text{ rev/s} = 2400 \text{ rev/min}$$

Example 8.46:   Find the angular displacement in degrees of a wheel rotating at 2.48 rad/s for 36.2 milliseconds (ms).

Solution: By Eq. 183,

$$\theta = \omega t = \left(\frac{2.48 \text{ rad}}{\text{s}}\right)(0.0362 \text{ s}) = 0.0898 \text{ rad} = (0.0898 \text{ rad})\left(\frac{180°}{\pi \text{ rad}}\right) = 5.14°.$$

Example 8.47:   How long will it take a drum rotating with an angular velocity of 1480°/s to make 10 complete revolutions?

Solution: By Eq. 183,

$$t = \frac{\theta}{\omega} = \left(\frac{10 \text{ rev}}{1480°/\text{s}}\right)\left(\frac{360°}{\text{rev}}\right) = 2.43 \text{ s}$$

(c) Linear Speed. The *linear speed* V of any point P on the rotating object, Fig. 8-14, moving in a circular path about O, is

Linear Speed	$V = \omega r$	**182**

where r is the distance from the center O to the point P. In order for the units to cancel properly, the angular velocity ω must be expressed in radians per unit time.

Example 8.48:   A 2.00-m-diameter wheel rotates at $20\overline{0}0$ rev/min. Find the linear speed of a point on the rim.

Solution: Converting the angular velocity to rad/min,

$$\omega = \left(\frac{20\overline{0}0 \text{ rev}}{\text{min}}\right)\left(\frac{2.00\pi \text{ rad}}{\text{rev}}\right) = 40\overline{0}0\pi \text{ rad/min}$$

From Eq. 182,

$$V = \omega r = 40\overline{0}0\pi(1.00) = 12\,600 \text{ m/min}$$

Example 8.49:   Find the angular velocity in rev/min of a 30.0-in.-diameter tire on a car traveling 50.0 mi/h.

Solution: Converting 50.0 mi/h to in./min,

$$\left(\frac{50.0 \text{ mi}}{\text{h}}\right)\left(\frac{5280 \text{ ft}}{\text{mi}}\right)\left(\frac{12 \text{ in.}}{\text{ft}}\right)\left(\frac{\text{h}}{60 \text{ min}}\right) = 52\,800 \text{ in./min}$$

From Eq. 182,

$$\omega = \frac{V}{r} = \frac{52\,800 \text{ in./min}}{15 \text{ in.}} = 3520 \text{ rad/min}$$

To convert to revolutions, we divide by 2π, the number of radians in one revolution. Thus,

$$\omega = 56\overline{0} \text{ rev/min}$$

Common Error	Remember, when using Eq. 182, that the angular velocity must be expressed in *radians* per unit time.

Solved Problems

ANGLE CONVERSIONS

8.1   Express the following angles in revolutions, degrees (decimal), and radians:

(*a*) 0.4553 rev   (*b*) 29.24°   (*c*) 1.577 rad

Solution

(*a*) $(0.4553 \text{ rev})\left(\dfrac{360°}{\text{rev}}\right) = 163.9°$ (*c*) $(1.577 \text{ rad})\left(\dfrac{1 \text{ rev}}{2\pi \text{ rad}}\right) = 0.2510 \text{ rev}$

  $(0.4553 \text{ rev})\left(\dfrac{2\pi \text{ rad}}{\text{rev}}\right) = 2.861 \text{ rad}$ $(1.577 \text{ rad})\left(\dfrac{180°}{\pi \text{ rad}}\right) = 90.36°$

(*b*) $29.24°\left(\dfrac{1 \text{ rev}}{360°}\right) = 0.081\,22 \text{ rev}$

  $29.24°\left(\dfrac{\pi \text{ rad}}{180°}\right) = 0.5103 \text{ rad}$

OPERATIONS WITH ANGLES IN DEGREES, MINUTES, AND SECONDS

8.2   Convert the following angles to decimal degrees.

(a) $15°36'43''$ (b) $68°55'$ (c) $83°54'28.7''$

Solution

(a) $43''\left(\dfrac{1°}{3600''}\right) = 0.012°$ (b) $55'\left(\dfrac{1°}{60'}\right) = 0.92°$ (c) $28.7''\left(\dfrac{1°}{3600''}\right) = 0.007\,97°$

$36'\left(\dfrac{1°}{60'}\right) = 0.600°$ Adding $\dfrac{68.00}{68.92°}$ $54'\left(\dfrac{1°}{60'}\right) = 0.900\,00$

Adding $\dfrac{15.000}{15.612°}$ Adding $\dfrac{83.000\,00}{83.907\,97°}$

8.3   Convert the following angles to degrees, minutes, and seconds.

(a) $51.5836°$ (b) $31.473°$ (c) $79.26°$

Solution

(a) $0.5836°\left(\dfrac{60'}{\text{degree}}\right) = 35.02'$ (c) $0.26°\left(\dfrac{60'}{\text{degree}}\right) = 16'$

$0.02'\left(\dfrac{60''}{\text{minute}}\right) = 1''$ (rounded)   So, $79.26° = 79°16'$

So, $51.5836° = 51°35'01''$

(b) $0.473°\left(\dfrac{60'}{\text{degree}}\right) = 28.38'$

$0.38'\left(\dfrac{60''}{\text{minute}}\right) = 23''$

So, $31.473° = 31°28'23''$

8.4   A circular highway curve is to be laid out with a central angle of 0.7500 rad. Convert this angle to degrees, minutes, and seconds.

Solution

Converting,

$$(0.7500 \text{ rad})\left(\frac{180°}{\pi \text{ rad}}\right) = 42.97°$$

$$0.97°\left(\frac{60'}{\text{degree}}\right) = 58'$$

So, $0.7500 \text{ rad} = 42°58'$

Note that the given angle, 0.7500 rad, does not contain enough significant figures to justify the inclusion of *seconds* in our answer.

8.5   The following deflection angles are to be used to lay out a curved roadway. (a) $2.6250°$, (b) $5.2500°$, (c) $7.875°$, and (d) $10.5000°$. Convert these angles so that they may be laid out using a theodolite (a surveying instrument) having scales graduated in degrees, minutes, and seconds.

Solution

(a)

$$0.6250° \left(\frac{60'}{\text{degree}} \right) = 37.50'$$

$$0.50' \left(\frac{60''}{\text{minute}} \right) = 30''$$

Thus the required angle is 2°37′30″.

(b)

$$0.2500° \left(\frac{60'}{\text{degree}} \right) = 15.00$$

Thus the required angle is 5°15′00″.

(c)

$$0.8750° \left(\frac{60'}{\text{degree}} \right) = 52.50'$$

$$0.50' \left(\frac{60''}{\text{minute}} \right) = 30''$$

Thus the required angle is 7°52′30″.

(d)

$$0.5000° \left(\frac{60'}{\text{degree}} \right) = 30.00'$$

Thus the required angle is 10°30′00″.

8.6   Add 46°38′41″ to 12°51′44″.

Solution

°	′	″
46	38	41
12	51	44
58	89	85

Removing multiples of 60,

$$59° \quad 30' \quad 25''$$

8.7   Subtract 37°44′58″ from 59°14′22″.

Solution

Borrowing from degrees and minutes,

°	′	″
58	73	82
37	44	58
21°	29′	24″

8.8   Multiply 27°35′52″ by 1.5724.

Solution

Converting to decimal degrees,

$$\frac{35'}{60} = 0.583\,33°$$

$$\frac{52''}{3600} = 0.014\,44$$

$$\underline{27.000\,00}$$
$$27.597\,77°$$

or 27.5978°, rounded.

Multiplying, $27.5978 \times 1.5724 = 43.395°$

Converting to DMS,

$$0.395° \times 60 = 23.70'$$
$$0.70' \times 60 = 42''$$

So, $27°35′52″ \times 1.5724 = 43°23′42″$

8.9   Divide 94°37′53″ by 2.75.

Solution

Converting to decimal degrees,

$$\frac{37'}{60} = 0.616\,67°$$

$$\frac{53''}{3600} = 0.014\,72$$

$$\underline{94.000\,00}$$
$$94.631\,39°$$

or 94.6314°, rounded.   Dividing,

$$\frac{94.6314}{2.75} = 34.4°$$ when rounded to three significant figures

TRIGONOMETRIC RATIOS

8.10   A point on the terminal side of an angle A in standard position has the coordinates $(5.2, 3.9)$. Write the six trigonometric functions of A.

Solution

Letting $x = 5.2$, $y = 3.9$,   and, from Eq. 72,

$$r^2 = (5.2)^2 + (3.9)^2 = 42.3$$
$$r = 6.5$$

Then, from Eqs. 73 to 78,

$$\sin A = \frac{3.9}{6.5} = 0.60$$

$$\cos A = \frac{5.2}{6.5} = 0.80$$

$$\tan A = \frac{3.9}{5.2} = 0.75$$

$$\cot A = \frac{5.2}{3.9} = 1.3$$

$$\sec A = \frac{6.5}{5.2} = 1.2$$

$$\csc A = \frac{6.5}{3.9} = 1.7$$

8.11   Find the algebraic signs of (a) sin 155°, (b) sec 200°, and (c) cot 300°.

Solution

Sketch each angle in the appropriate quadrant, as in Fig. 8-15. Drop a perpendicular to the x axis from any point P on each terminal side. Indicate the algebraic signs of the x and y coordinates of each of these points. The distance r from the origin to P is always positive.

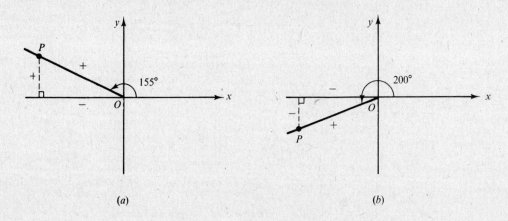

(a)   (b)

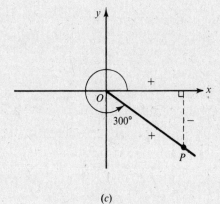

(c)

Fig. 8-15

(a) By Eq. 73, $\sin 155° = \dfrac{y}{r} = \dfrac{(+)}{(+)} = (+)$

(b) By Eq. 77, $\sec 200° = \dfrac{r}{x} = \dfrac{(+)}{(-)} = (-)$

(c) By Eq. 76, $\cot 300° = \dfrac{x}{y} = \dfrac{(+)}{(-)} = (-)$

8.12   The tangent of a certain angle is 1.554.   Find its cotangent.

Solution

By Eq. 79,

$$\cot \theta = \frac{1}{\tan \theta} = \frac{1}{1.554} = 0.6435$$

8.13   Find the sine of an angle having a cosecant of 2.47.

Solution

By Eq. 79,

$$\sin \theta = \frac{1}{\csc \theta} = \frac{1}{2.47} = 0.405$$

8.14   Find the reference angles for (a) 125°, (b) 250°, (c) 310°, and (d) 620°.

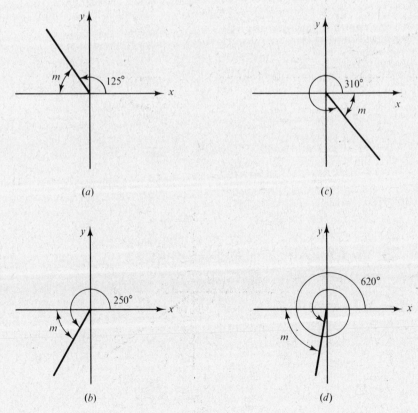

(a)

(c)

(b)

(d)

Fig. 8-16

Solution

(a) In Fig. 8-16a, we see that 125° will have a reference angle of

$$m = 180° - 125° = 55°$$

Similarly

(b) 250° has a reference angle, Fig. 8-16b, of

$$m = 250° - 180° = 70°$$

(c) and 310° has a reference angle, Fig. 8-16c, of

$$m = 360° - 310° = 50°$$

(d) In Fig. 8-16d, we see that 620° will have the same terminal side as $620° - 360° = 260°$, a third quadrant angle. The reference angle is then

$$m = 260° - 180° = 80°$$

TRIGONOMETRIC FUNCTIONS

8.15 Find the six trigonometric functions, to four significant figures, of (a) 32°, (b) 57°, (c) 162°, and (d) 310°. Use a calculator or the table.

Solution

(a) The signs of all functions are positive in the first quadrant. From the calculator or table we read

$$\sin 32° = 0.5299$$

$$\cos 32° = 0.8480$$

$$\tan 32° = 0.6249$$

By the reciprocal relationships, Eq. 79,

$$\cot 32° = \frac{1}{0.6249} = 1.600$$

$$\sec 32° = \frac{1}{0.8480} = 1.179$$

$$\csc 32° = \frac{1}{0.5299} = 1.887$$

(b) The signs of all functions are positive in the first quadrant. From the calculator or table we read

$$\sin 57° = 0.8387$$
$$\cos 57° = 0.5446$$
$$\tan 57° = 1.540$$

By the reciprocal relationships, Eq. 79,

$$\cot 57° = \frac{1}{1.540} = 0.6494$$

$$\sec 57° = \frac{1}{0.5446} = 1.836$$

$$\csc 57° = \frac{1}{0.8387} = 1.192$$

(c) The reference angle for 162° is

$$m = 180° - 162° = 18°$$

From the trigonometric tables or calculator,

$$\sin 18° = 0.3090$$
$$\cos 18° = 0.9511$$
$$\tan 18° = 0.3249$$

As 162° is in the second quadrant, the cosine and tangent are negative, so,

$$\sin 162° = 0.3090$$
$$\cos 162° = -0.9511$$
$$\tan 162° = -0.3249$$

By the reciprocal relationships, Eq. 79,

$$\cot 162° = -3.078$$
$$\sec 162° = -1.051$$
$$\csc 162° = 3.236$$

(d) The reference angle for 310° is

$$m = 360° - 310° = 50°$$

From the trigonometric tables or calculator,

$$\sin 50° = 0.7660$$
$$\cos 50° = 0.6428$$
$$\tan 50° = 1.192$$

For a fourth-quadrant angle, the sine and tangent are negative; so,

$$\sin 310° = -0.7660$$
$$\cos 310° = 0.6428$$
$$\tan 310° = -1.192$$

By the reciprocal relationships, Eq. 79,

$$\cot 310° = -0.8391$$
$$\sec 310° = 1.556$$
$$\csc 310° = -1.305$$

8.16 Find to the nearest degree the two angles less than 360° whose trigonometric functions are:

(a) $\tan \theta = 0.3640$ (c) $\cos \theta = 0.3746$ (e) $\csc \theta = 4.444$

(b) $\sin \theta = 0.8192$ (d) $\tan \theta = -2.145$ (f) $\cot \theta = 0.4878$

Use a calculator or the table.

Solution

By calculator, simply depress the *arc* or *inv* key prior to the appropriate trigonometric function key. Check your answers against the table solutions that follow.

(a) Find 0.3640 in the *tan* column, then read across to

$$\theta = 20°$$

As the tangent is positive in the third quadrant also, another angle is

$$\theta = 180° + 20° = 200°$$

(b) Find 0.8192 in the *sin* column, then read across to

$$\theta = 55°$$

The sine is positive in the second quadrant also, so another angle is

$$\theta = 180° - 55° = 125°$$

(c) Find 0.3746 in the *cos* column, then read across to

$$\theta = 68°$$

The cosine is positive in the fourth quadrant also; the angle is

$$\theta = 360° - 68° = 292°$$

(d) Find 2.145 in the *tan* column, then read across to

$$\theta = 65°$$

The tangent is negative in the second and fourth quadrants. We take 65° as our reference angle and compute the second-quadrant angle.

$$\theta = 180° - 65° = 115°$$

and the fourth-quadrant angle

$$\theta = 360° - 65° = 295°$$

(e) Since the table does not include the cosecant, we convert to the sine.

$$\sin \theta = \frac{1}{\csc \theta} = \frac{1}{4.444} = 0.2250$$

From the table,

$$\theta = 13°$$

The sine is positive in the second quadrant also, so,

$$\theta = 180° - 13° = 167°$$

(f) Since the table does not include the cotangent, we convert to the tangent.

$$\tan \theta = \frac{1}{\cot \theta} = \frac{1}{0.4878} = 2.050$$

From the table,

$$\theta = 64°$$

The tangent is positive in the third quadrant also; so,

$$\theta = 180° + 64° = 244°$$

INTERPOLATING IN THE TRIGONOMETRIC TABLES

8.17   Find the following functions by interpolating:

(a) sin 31.6°   (c)   sec 16°30′   (e)   sin 110.9°
(b) tan 67.8°   (d)   csc 59.3°   (f)   cos 210.4°

Solution

(a) The tabular difference between sin 31° and sin 32° is

$$0.5299 - 0.5150 = 0.0149$$

So,
$$\frac{0.6}{1} = \frac{x}{0.0149}$$

$$x = 0.0089$$

Since the sine is increasing, we add x to 0.5150,

$$\sin 31.6° = 0.5239$$

(b) The tabular difference between tan 67° and tan 68° is

$$2.475 - 2.356 = 0.119$$

So,
$$\frac{0.8}{1} = \frac{x}{0.119}$$

$$x = 0.0952$$

Since the tangent is increasing, we add x to 2.356,

$$\tan 67.8° = 2.451$$

(c) The tabular difference between cos 16° and cos 17° is

$$0.9613 - 0.9563 = 0.0050$$

Then,
$$\frac{30}{60} = \frac{x}{0.0050}$$

$$x = 0.0025$$

Since the cosine is decreasing, we subtract x from 0.9613,

$$\cos 16°30' = 0.9588$$

And by Eq. 79,

$$\sec 16°30' = \frac{1}{0.9588} = 1.043$$

(d) The tabular difference between sin 59° and sin 60° is

$$0.8660 - 0.8572 = 0.0088$$

Then
$$\frac{0.3}{1} = \frac{x}{0.0088}$$

$$x = 0.0026$$

Since the sine is increasing, we add x to 0.8572,

$$\sin 59.3° = 0.8598$$

Then by Eq. 79,

$$\csc 59.3° = \frac{1}{0.8598} = 1.163$$

(e) The reference angle is

$$m = 180° - 110.9° = 69.1°$$

The tabular difference between sin 69° and sin 70° is

$$0.9397 - 0.9336 = 0.0061$$

So,

$$\frac{0.1}{1} = \frac{x}{0.0061}$$

$$x = 0.0006$$

Since the sine is increasing, x is added to 0.9336.

$$\sin 69.1° = 0.9342$$

As the sine is positive in the second quadrant,

$$\sin 110.9° = +0.9342$$

(f)   The reference angle is

$$m = 210.4° - 180° = 30.4°$$

The tabular difference between cos 30° and cos 31° is

$$0.8660 - 0.8572 = 0.0088$$

So,

$$\frac{0.4}{1} = \frac{x}{0.0088}$$

$$x = 0.0035$$

Since the cosine is decreasing, x is subtracted from 0.8660.

$$\cos 30.4° = 0.8625$$

As the cosine is negative in the third quadrant,

$$\cos 210.4° = -0.8625$$

8.18   Find two angles less than 360°, having the following trigonometric functions:

(a) $\tan \theta = 0.5821$ (b) $\sin \theta = 0.9969$ (c) $\cot \theta = -1.460$

Use a calculator or interpolate in the trigonometric table.

Solution

(a)   From the table, θ is seen to lie between 30° and 31°.   The tabular differences are

$$0.6009 - 0.5774 = 0.0235$$

and

$$0.5821 - 0.5774 = 0.0047$$

Writing a proportion,

$$\frac{x}{1} = \frac{0.0047}{0.0235}$$

$$x = 0.2$$

So,

$$\arctan 0.5821 = 30.2°$$

The tangent is also positive in the third quadrant; so,

$$\arctan 0.5821 = 180° + 30.2° = 210.2°$$

(b)   From the table, arcsin 0.9969 is between 85° and 86°.   The tabular differences are

$$0.9976 - 0.9962 = 0.0014$$

$$0.9969 - 0.9962 = 0.0007$$

Writing a proportion,

$$\frac{x}{1} = \frac{0.0007}{0.0014}$$

$$x = 0.5$$

Therefore,   arcsin $0.9969 = 85.5°$

Also, the sine is positive in the second quadrant; so,

$$\text{arcsin } 0.9969 = 180° - 85.5° = 94.5°$$

(c) By the reciprocal relationships,

$$\tan \theta = \frac{1}{-1.460} = -0.6849$$

The minus sign tells us that θ is in the second and fourth quadrants.
    Finding the reference angle m,

$$\tan m = 0.6849$$

From the table, m is seen to lie between 34° and 35°. Taking tabular differences and writing the proportion,

$$\frac{x}{1} = \frac{0.6849 - 0.6745}{0.7002 - 0.6745}$$

$$x = 0.40$$

So, $m = 34.0° + 0.4° = 34.4°$

Our second-quadrant angle is then

$$\theta = 180° - 34.4° = 145.6°$$

And in the fourth quadrant,

$$\theta = 360° - 34.4° = 325.6°$$

RADIAN MEASURE AND ARC LENGTH

8.19   What angle would intercept a 15-cm arc in a circle of radius 20 cm?

Solution

By Eq. 96,

$$\theta = \frac{15 \text{ cm}}{20 \text{ cm}} = \frac{3}{4} \text{ rad}$$

8.20   Find the angle that would intercept an arc of 2.5 ft in a 50-in.-radius circle.

Solution

As in Eq. 96, s and r must have the same units. Converting,

$$2.5 \text{ ft} = 30 \text{ in.}$$

By Eq. 96,

$$\theta = \frac{30}{50} = 0.60 \text{ rad}$$

8.21 Find the arc length intercepted in a 48.3-in.-radius circle by a central angle of 1.20 rad.

Solution

By Eq. 96,

$$s = 48.3(1.20) = 58.0 \text{ in.}$$

8.22 What arc length would be intercepted by a central angle of 113° in a 355-cm-radius circle?

Solution

Converting degrees to radians,

$$113° \left(\frac{2\pi \text{ rad}}{360°} \right) = 1.97 \text{ rad}$$

By Eq. 96,

$$s = 355(1.97) = 699 \text{ cm}$$

8.23 Find the radius of the circle in which an angle of 0.644 rad intercepts an arc of 34.5 in.

Solution

By Eq. 96,

$$r = \frac{34.5}{0.644} = 53.6 \text{ in.}$$

8.24 What is the radius of a circle in which an arc of 21.4 m is intercepted by an angle of 32.6°?

Solution

Converting the angle to radians,

$$32.6° \left(\frac{\pi \text{ rad}}{180°} \right) = 0.569 \text{ rad}$$

By Eq. 96,

$$r = \frac{21.4}{0.569} = 37.6 \text{ m}$$

8.25 A *slump cone* for testing concrete, Fig. 8-17, has the shape of a frustum of a cone, with open ends 4.00 and 8.00 in. in diameter, and with a height of 12.0 in. It is made from sheet metal cut according to the pattern of Fig. 8-18, rolled and joined. Find the pattern dimensions r, R, and θ.

Solution

Since in Fig. 8-17, triangle ABC is similar to triangle ADE, we can say, by Statement 117,

$$\frac{AD}{AB} = \frac{DE}{BC}$$

So,

$$\frac{AD}{12 + AD} = \frac{2}{4} = \frac{1}{2}$$

Cross-multiplying,

$$2AD = 12 + AD$$

So,

$$AD = 12.0 \text{ in.}$$

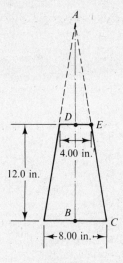

Fig. 8-17

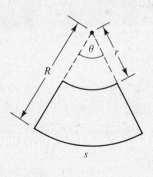

Fig. 8-18

By the Pythagorean Theorem,

$$(AC)^2 = R^2 = (AB)^2 + (BC)^2$$

So,

$$R^2 = (24.0)^2 + (4.00)^2 = 592$$

$$R = 24.3 \text{ in.}$$

Since the diameter is 8.00 in., the base of the slump cone has a circumference of

$$8.00\pi = 25.1 \text{ in.}$$

which is equal to the arc length s. We now find θ by Eq. 96,

$$\theta = \frac{s}{R} = \frac{25.1}{24.3} = 1.03 \text{ rad} = 59.2°$$

Also since triangle ADE is similar to triangle ABC, AE must be half AC, or

$$r = \frac{R}{2} = 12.2 \text{ in.}$$

8.26　Find the length of a circular highway curve having a radius of 260.098 ft and a central angle of 12°35′46″.

Solution

Converting the central angle to decimal degrees,

$$
\begin{array}{r}
12.0000° \\
35' = 0.5833° \\
46'' = 0.0128° \\
\hline
12.5961°
\end{array}
$$

Converting to radians,

$$12.5961° \left(\frac{\pi \text{ rad}}{180°} \right) = 0.219\,843 \text{ rad}$$

So, by Eq. 96,

$$s = r\theta = 260.098(0.219\,843) = 57.181 \text{ ft}$$

8.27 A circular railroad curve has an arc length of 310.00 ft and a radius of 618.56 ft. Find the central angle in degrees, minutes, and seconds.

Solution

By Eq. 96,

$$\theta = \frac{s}{r} = \frac{310.00}{618.56} = 0.501\ 16 \text{ rad}$$

Converting to decimal degrees,

$$(0.501\ 16 \text{ rad})\left(\frac{180°}{\pi \text{ rad}}\right) = 28.714°$$

Now converting 0.714° to minutes and seconds,

$$0.714°\left(\frac{60'}{\text{degree}}\right) = 42.8'$$

$$0.8'\left(\frac{60''}{\text{minute}}\right) = 48''$$

So,

$$\theta = 28°42'48''$$

8.28 Find the radius of a circular railroad track which will cause a train to change direction by 22° in a distance of 250 m.

Solution

The central angle θ is

$$\theta = 22° = 0.384 \text{ rad}$$

By Eq. 96,

$$r = \frac{s}{\theta} = \frac{250 \text{ m}}{0.384} = 651 \text{ m}$$

8.29 Find the length of contact between the cable and the 25.0-cm-diameter pulley of Fig. 8-19.

Solution

Since the interior angles of the quadrilateral $ABCD$ must add up to 360° and the two right angles total 180°,

$$A = 360° - 180° - 52.4° = 127.6° = 2.23 \text{ rad}$$

Then by Eq. 96, the length of contact s is

$$s = r\theta = (12.5 \text{ cm})(2.23) = 27.9 \text{ cm}$$

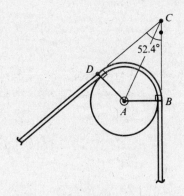

Fig. 8-19

8.30 What distance on the surface of the earth corresponds to 10° of latitude? Assume the earth is a sphere, 7920 mi in diameter.

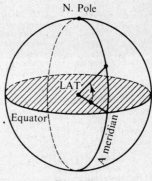

Fig. 8-20

Solution

Latitude is the angular distance north or south of the equator, measured along a *meridian* (a circle on the earth's surface passing through the poles), as in Fig. 8-20.

If the latitude is 10° (=0.175 rad), we get by Eq. 96,

$$s = r\theta$$
$$= (3960 \text{ mi})(0.175)$$
$$= 693 \text{ mi}$$

8.31 Two towns are 755 miles apart and lie on the same meridian. Find their difference in latitude.

Solution

The latitude difference is, by Eq. 96,

$$\theta = \frac{s}{r} = \frac{755 \text{ mi}}{3960 \text{ mi}} = 0.191 \text{ rad} = 10.9°$$

8.32 Certain topographic maps span 15' of latitude. How many kilometers on the earth's surface does this correspond to?

Solution

$$\theta = 15' = 0.25° = 0.004\,36 \text{ rad}$$

So, by Eq. 96,

$$s = r\theta = (3960 \text{ mi})(0.004\,36) = 17.3 \text{ mi} = 27.8 \text{ km}$$

8.33 A voltmeter pointer moves along a circular scale 7.00 in. long. Find the length of the pointer if 40.0° of rotation produces a full-scale deflection.

Solution

The angular displacement is

$$\theta = 40.0° = 0.698 \text{ rad}$$

By Eq. 96,

$$r = \frac{s}{\theta} = \frac{7.00 \text{ in.}}{0.698} = 10.0 \text{ in.}$$

8.34 A servomotor drives a rack by means of a pinion that has a pitch diameter of 0.545 cm. How far will the rack move for a 150° rotation of the motor shaft?

Solution

$$\theta = 150° = 2.62 \text{ rad}$$

By Eq. 96,

$$s = r\theta = \frac{0.545 \text{ cm}}{2}(2.62) = 0.714 \text{ cm}$$

8.35 The pointer on a "slide rule" type dial for an FM tuner is driven by a belt and pulley arrangement as in Fig. 8-21. Find the pulley radius if 1/8 revolution of the control knob is to move the pointer 12.0 mm.

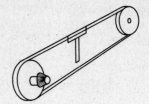

Fig. 8-21

Solution

$$\theta = \left(\frac{1}{8}\,\text{rev}\right)\left(\frac{2\pi\,\text{rad}}{\text{rev}}\right) = \frac{\pi}{4}\,\text{rad}$$

By Eq. 96,

$$r = \frac{s}{\theta} = \frac{12.0\,\text{mm}}{\dfrac{\pi}{4}} = \frac{48.0}{\pi} = 15.3\,\text{mm}$$

8.36 A 3.754-in.-radius sector gear, Fig. 8-22, is to have an arc length of 2.250 in. along the pitch circle. Find the central angle θ.

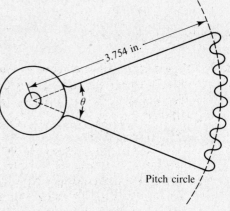

Solution

By Eq. 96,

$$\theta = \frac{s}{r} = \frac{2.250}{3.754} = 0.5994\,\text{rad} = 34.34°$$

Fig. 8-22

8.37 The beam AB in the *beam-and-crank* mechanism of Fig. 8-23 rotates 47.2° for each rotation of the crank CD. Find the length of the arc swept out by point A.

Solution

The angular displacement is

$$\theta = 47.2°\left(\frac{2\pi\,\text{rad}}{360°}\right) = 0.824\,\text{rad}$$

By Eq. 96,

$$s = r\theta = (8.75\,\text{m})(0.824) = 7.21\,\text{m}$$

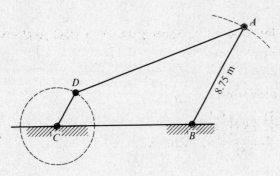

Fig. 8-23

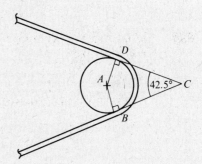

Fig. 8-24

8.38 A 1.75-in.-wide brake band is wrapped around a 12.5-in.-diameter drum, as in Fig. 8-24. Find the area of contact between the band and the drum.

Solution

Since the interior angles of quadrilateral $ABCD$ must add up to $360°$,

$$A = 360° - 180° - 42.5° = 137.5° = 2.40 \text{ rad}$$

By Eq. 96, the length of contact s is

$$s = r\theta = (6.25 \text{ in.})(2.40) = 15.0 \text{ in.}$$

So the contact area is

$$\text{Area} = 15.0(1.75) = 26.2 \text{ in}^2$$

BENDING ALLOWANCE

8.39   In Fig. 8-25 a sheet of metal of thickness t is bent through an angle A. The length of material s that must be allowed for the bend is called the *bending allowance*.

Write an equation for the bending allowance s, in terms of the metal thickness t, the angle of bend A, and the bend radius r. Assume that the neutral axis (the line at which there is no stretching or compression of the metal) is at a distance from the inside of the bend equal to four-tenths of the metal thickness t.

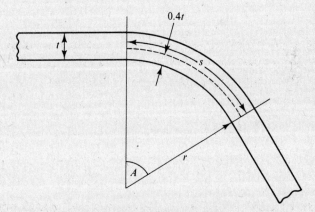

Fig. 8-25   Bending allowance.

Solution

By Eq. 96,

$$s = AR = A(r + 0.4t)$$

where A is in radians. For A in degrees,

$$s = \frac{A(r + 0.4t)\pi}{180}$$

8.40   What bending allowance is required for 1/8-in.-thick sheet steel, for a right angle bend of 1/2-in. radius?

Solution

From the equation derived in the previous problem,

$$s = \frac{90\left[\frac{1}{2} + 0.4\left(\frac{1}{8}\right)\right]\pi}{180} = 0.864 \text{ in.}$$

ROTATION

8.41   A wheel is rotating with an angular velocity of 655 rev/min. Convert this to (*a*) rad/s and (*b*) degrees/s.

Solution

(a) $\left(\dfrac{655 \text{ rev}}{\text{min}}\right)\left(\dfrac{2\pi \text{ rad}}{\text{rev}}\right)\left(\dfrac{\text{min}}{60 \text{ s}}\right) = 68.6 \text{ rad/s}$ (b) $\left(\dfrac{655 \text{ rev}}{\text{min}}\right)\left(\dfrac{360°}{\text{rev}}\right)\left(\dfrac{\text{min}}{60 \text{ s}}\right) = 3.93 \times 10^3 \text{ degrees/s}$

8.42 A 1.55-ft-diameter flywheel rotates with an angular velocity of 1500 rev/min. Find the linear speed, in inches per second, of a point on the circumference.

Solution

Converting to rad/s,

$$\omega = \left(\frac{1500 \text{ rev}}{\text{min}}\right)\left(\frac{2\pi \text{ rad}}{\text{rev}}\right)\left(\frac{\text{min}}{60 \text{ s}}\right) = 157 \text{ rad/s}$$

By Eq. 182,

$$V = 157\left(\frac{1.55}{2}\right) = 122 \text{ ft/s} = 1.46 \times 10^3 \text{ in./s}$$

8.43 Find the angular velocity, in rev/min, of a 25.0-cm-radius tire on a car traveling 50.0 km/h.

Solution

From Eq. 182, with conversion of units,

$$\omega = \left(\frac{50.0 \text{ km}}{\text{h}}\right)\left(\frac{1}{25.0 \text{ cm}}\right)\left(\frac{10^5 \text{ cm}}{\text{km}}\right)\left(\frac{\text{h}}{60 \text{ min}}\right) = 3.33 \times 10^3 \text{ rad/min} = 530 \text{ rev/min}$$

8.44 A drum, rotating at 20.0 rev/min, has a steel cable wound around it which is used to lift a construction elevator at a rate of 3.00 m/s. Find the radius of the drum.

Solution

The angular velocity is

$$\omega = \left(\frac{20.0 \text{ rev}}{\text{min}}\right)\left(\frac{2\pi \text{ rad}}{\text{rev}}\right) = 126 \text{ rad/min}$$

By Eq. 182,

$$r = \frac{V}{\omega} = \left(\frac{3.00 \text{ m}}{\text{s}}\right)\left(\frac{\text{min}}{126 \text{ rad}}\right)\left(\frac{60 \text{ s}}{\text{min}}\right) = 1.43 \text{ m}$$

8.45 Find the linear speed in miles per hour of a point on the equator due to the rotation of the earth about its axis. Assume the earth to be a sphere, 7920 mi in diameter.

Solution

Since the earth rotates once every 24 h,

$$\omega = \left(\frac{1 \text{ rev}}{24 \text{ h}}\right)\left(\frac{2\pi \text{ rad}}{\text{rev}}\right) = 0.262 \text{ rad/h}$$

Then by Eq. 182,

$$V = \omega r = \left(\frac{0.262 \text{ rad}}{\text{h}}\right)(3960 \text{ mi}) = 1040 \text{ mi/h}$$

8.46 Find the maximum diameter for the commutator of a generator armature rotating at 2250 rev/min if the surface speed of the brushes is not to exceed 6000 ft/min.

Solution

The angular velocity is

$$\omega = \left(\frac{2250 \text{ rev}}{\text{min}}\right)\left(\frac{2\pi \text{ rad}}{\text{rev}}\right) = 14\,100 \text{ rad/min}$$

By Eq. 182,

$$r = \frac{V}{\omega} = \frac{6000 \text{ ft/min}}{14\,100 \text{ rad/min}} = 0.426 \text{ ft}$$

The diameter is

$$d = 2(0.426) = 0.852 \text{ ft} = 10.2 \text{ in.}$$

8.47 A certain capstan is to rotate at 655 rev/min and drive a magnetic tape at a speed of 30.4 in./s. Find the capstan diameter in millimeters.

Solution

The angular velocity is

$$\omega = \left(\frac{655 \text{ rev}}{\text{min}}\right)\left(\frac{2\pi \text{ rad}}{\text{rev}}\right) = 4115 \text{ rad/min}$$

By Eq. 182,

$$r = \frac{V}{\omega} = \left(\frac{30.4 \text{ in.}}{\text{s}}\right)\left(\frac{\text{min}}{4115 \text{ rad}}\right)\left(\frac{60 \text{ s}}{\text{min}}\right) = 0.443 \text{ in.}$$

The diameter is

$$d = 0.886 \text{ in.}$$

To convert to millimeters, we multiply by 25.4; so

$$d = 0.886 \times 25.4 = 22.5 \text{ mm}$$

8.48 At how many inches from the center of a phonograph record turning at $33\frac{1}{3}$ rev/min will the surface speed be 1.50 ft/s?

Solution

The angular velocity is

$$\omega = \left(\frac{33.3 \text{ rev}}{\text{min}}\right)\left(\frac{2\pi \text{ rad}}{\text{rev}}\right) = 209 \text{ rad/min}$$

By Eq. 182,

$$r = \frac{V}{\omega} = \left(\frac{1.50 \text{ ft}}{\text{s}}\right)\left(\frac{\text{min}}{209 \text{ rad}}\right)\left(\frac{60 \text{ s}}{\text{min}}\right) = 0.431 \text{ ft} = 5.17 \text{ in.}$$

8.49 A conveyor belt is driven at a speed of 65.0 ft/min by a 28.0-in.-diameter wheel. How fast is the wheel rotating?

Solution

By Eq. 182,

$$\omega = \frac{V}{r} = \left(\frac{65.0 \text{ ft/min}}{14.0 \text{ in.}}\right)\left(\frac{12 \text{ in.}}{\text{ft}}\right) = \frac{55.7 \text{ rad}}{\text{min}}$$

Dividing by 2π,

$$\omega = 8.86 \text{ rev/min}$$

8.50 A 4.00-in.-diameter bar is to be turned in a lathe. Find the maximum spindle speed so that a cutting speed of 80 ft/min at the surface of the bar is not exceeded.

Solution

By Eq. 182,

$$\omega = \frac{V}{r} = \left(\frac{80 \text{ ft/min}}{2.00 \text{ in.}}\right)\left(\frac{12 \text{ in.}}{\text{ft}}\right) = \frac{480 \text{ rad}}{\text{min}} = 76.4 \text{ rev/min}$$

8.51 Find the surface speed of a 10.0-cm-diameter grinding wheel rotating at $18\overline{0}0$ rev/min.

Solution

The angular velocity is

$$\omega = \left(\frac{1800 \text{ rev}}{\text{min}}\right)\left(\frac{2\pi \text{ rad}}{\text{rev}}\right) = 11\,300 \text{ rad/min}$$

By Eq. 182,

$$V = \omega r = \left(\frac{11\,300 \text{ rad}}{\text{min}}\right)(5 \text{ cm}) = 56\,500 \text{ cm/min} = 942 \text{ cm/s}$$

Supplementary Problems

8.52 Convert the following angles to revolutions, degrees, and radians:

(a) 0.7219 rad   (b) 155.7°   (c) 0.1010 rev

8.53 Convert the following angles to decimal degrees: (a) 27°29′13″   (b) 64°13′27″

8.54 Convert the following angles to degrees, minutes, and seconds: (a) 79.376 21°   (b) 22.732 18°

8.55 Perform the indicated operations.

(a) 36°21′12″ + 12°13′14″   (c) 14°39′57″ × 1.739

(b) 76°12′26″ − 39°27′37″   (d) 87°53′37″ ÷ 2.937

8.56 Find the angle in radians that would intercept an arc of 3.9 ft in a 27-in.-radius circle.

8.57 What arc length would be intercepted by a central angle of 27.9° in a 222-cm-radius circle.

8.58 Find the radius of a circle in which an angle of 56° intercepts an arc of 29.7 in.

8.59 A cam is rotating with an angular velocity of 237 rev/min. Convert this to

(a) rad/s   (b) degrees/s

8.60 Find the angular velocity in rad/s of a 26-in.-radius wheel traveling 60 mi/h.

8.61 The two legs of a right triangle are 31.6 and 26.0. Write the six trigonometric functions of the smallest angle, x, of the triangle.

8.62 Find the algebraic signs of the following: (a) sin 135°   (b) csc 210°

8.63 Find the reference angles for (a) 112°   (b) 237°   (c) 729°

8.64   Find the six trigonometric functions of

(a) 13°   (b) 93°   (c) 145°   (d) 327°

to four significant figures.

8.65   Find two angles less than 360° whose trigonometric functions are

(a) tan θ = 2.3559   (c) cos θ = 0.017 45   (e) cot θ = −57.2899

(b) sin θ = 0.5150   (d) csc θ = 1.0576   (f) sec θ = 1.0125

8.66   Interpolate to find the functions of the following angles:

(a) sin 25.7°   (b) csc 61.2°   (c) tan 90.9°   (d) cos 137.1°

8.67   Find two angles less than 360°, having the following trigonometric functions:

(a) tan θ = 1.5282   (b) csc θ = 1.3311

Answers to Supplementary Problems

8.52   (a) 0.1149 rev, 41.36°   (b) 0.4325 rev, 2.717 rad   (c) 36.36°, 0.6346 rad

8.53   (a) 27.4869°   (b) 64.2242°

8.54   (a) 79°22′34″   (b) 22°43′56″

8.55   (a) 48°34′26″   (b) 36°44′49″   (c) 25°30′14″   (d) 29°55′35″

8.56   1.7 rad

8.57   108 cm

8.58   30.4 in.

8.59   (a) 24.82 rad/s   (b) 1422 degrees/s

8.60   40.6 rad/s

8.61   sin x = 0.636, cos x = 0.773, tan x = 0.823, cot x = 1.22, sec x = 1.29, csc x = 1.57

8.62   (a) (+)   (b) (−)

8.63   (a) 68°   (b) 57°   (c) 9°

8.64

	sin	cos	tan	cot	sec	csc
(a)	0.2250	0.9744	0.2309	4.331	1.026	4.445
(b)	0.9986	−0.0523	−19.08	−0.0524	−19.11	1.001
(c)	0.5736	−0.8192	−0.7002	−1.428	−1.221	1.743
(d)	−0.5446	0.8317	−0.6494	−1.540	1.192	−1.836

8.65   (a) 67°, 247°   (b) 31°, 149°   (c) 89°, 271°   (d) 71°, 109°   (e) 1°, 181°   (f) 9°, 351°

8.66   (a) 0.4337   (b) 1.141   (c) −63.657   (d) −0.7325

8.67   (a) 56.8°, 236.8°   (b) 48.7°, 131.3°

Chapter 9

Solution of Right Triangles

9.1 RIGHT TRIANGLES

A *right triangle* is one containing a right angle. The other two angles are always acute.

The side of the triangle opposite the right angle is called the *hypotenuse*. It is always the longest side.

In trigonometry, right triangles are commonly labeled as in Fig. 9-1. Capital letters A, B, and C indicate the angles, with C usually representing the right angle. The lowercase letters a, b, and c represent the sides, with side a opposite angle A, side b opposite angle B, and side c (the hypotenuse) opposite the right angle C.

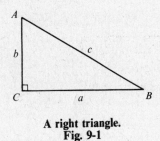

A right triangle.
Fig. 9-1

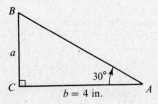

Drawing a right triangle.
Fig. 9-2

Example 9.1: In right triangle ABC, $A = 30°$ and $b = 4$ in. Draw the triangle.

Solution: In Fig. 9-2, we first draw a right angle C. Then measure 4 in. along one of the legs, and label this side b. At the other end of this same leg, draw an angle of 30°, and label it A. Extend the hypotenuse and leg a to where they meet at B.

9.2 SUM OF THE ANGLES

The sum of the three angles of any triangle must be 180°.

$$\boxed{A + B + C = 180°} \quad \textbf{60}$$

Therefore *the sum of the acute angles of a right triangle is 90°*, or

$$A + B = 90°$$

Two angles whose sum is 90° are called *complementary* angles.

Example 9.2 In right triangle ABC, $B = 42.5°$. Find A.

Solution: By Eq. 60,

$$A = 90° - 42.5° = 47.5°$$

9.3 THE PYTHAGOREAN THEOREM

In a right triangle, the square of the hypotenuse equals the sum of the squares of the other two sides.

Pythagorean Theorem	$c^2 = a^2 + b^2$	**72**

Example 9.3:   Two legs of a right triangle are 3.45 in. and 4.68 in. long.   Find the length of the hypotenuse.

Solution: By Eq. 72,

$$c^2 = (3.45)^2 + (4.68)^2 = 33.8$$

So, $c = 5.81$ in.

As a rough check, see that the length of the hypotenuse is greater than either of the legs but is less than their sum.

Example 9.4:   The hypotenuse of a right triangle is 25.8 cm and leg a is 15.3 cm.   Find the length of leg b.

Solution: By Eq. 72.

$$b^2 = (25.8)^2 - (15.3)^2 = 431.6$$

$$b = 20.8 \text{ cm}$$

Common Error	Do not use the Pythagorean Theorem for triangles that are not *right* triangles.

9.4 SOLUTION OF RIGHT TRIANGLES

To *solve* a triangle means to find all the unknown sides and angles.

(a) One Side and One Angle Known

1. Sketch the triangle.
2. Find the missing angle by subtracting the known angle from 90°.
3. Find the missing sides by using the trigonometric functions of the given angle.
4. Check your work with the Pythagorean Theorem, Eq. 72.

Example 9.5:   In right triangle ABC, $A = 25°$ and $c = 10.0$ in.   Solve the triangle.

Solution: Sketch the triangle, as in Fig. 9-3.   Then by Eq. 60,

$$B = 90° - A = 90° - 25° = 65°$$

By Eq. 73, $\dfrac{a}{10.0} = \sin A$

$$a = 10.0 \sin 25° = 10.0(0.423) = 4.23 \text{ in.}$$

By Eq. 74, $\dfrac{b}{10.0} = \cos A$

$$b = 10.0 \cos 25° = 10.0(0.906) = 9.06 \text{ in.}$$

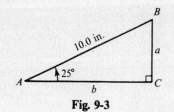

Fig. 9-3

Check: By Eq. 72, $(4.23)^2 + (9.06)^2 \stackrel{?}{=} (10.0)^2$

$$17.9 + 82.1 = 100 \quad \text{checks}$$

As a quick approximate check of any triangle, see if the longest side is opposite the largest angle and if the shortest side is opposite the smallest angle.

Common Error	As far as possible, use only the *given information* for each computation. This will prevent errors from earlier computations from being carried along into later ones.

Example 9.6:   In the previous example, after finding side a, we could have computed side b with the Pythagorean Theorem,

$$b^2 = (10.0)^2 - (4.23)^2$$

$$b = 9.06 \text{ in.}$$

This is not recommended, for a mistake in the computation of side a would have made the computation for side b incorrect as well.

(b) Two Sides Known

1. Draw a diagram.
2. Find the missing angles by relating them to the given sides by means of the trigonometric functions.
3. Find the third side by the Pythagorean Theorem.
4. Check that the computed angles are complementary.
5. Check the side with an appropriate trigonometric function.

Example 9.7:   In right triangle ABC, $a = 12.6$ cm   and $b = 18.2$ cm.   Solve the triangle.

Solution: Sketch the triangle, as in Fig. 9-4.   Then by Eq. 75,

$$\tan A = \frac{a}{b} = \frac{12.6}{18.2} = 0.692$$

$$A = 34.7°$$

By Eq. 76,

$$\cot B = \frac{a}{b} = 0.692$$

$$B = 55.3°$$

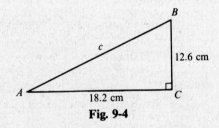

Fig. 9-4

By Eq. 72,

$$c^2 = (12.6)^2 + (18.2)^2 = 490$$

$$c = 22.1 \text{ cm}$$

Check: By Eq. 60,

$$A + B = 34.7° + 55.3° = 90° \qquad \text{checks}$$

By Eq. 73,

$$\sin 34.7° \stackrel{?}{=} \frac{12.6}{22.1}$$

$$0.569 \cong 0.570 \qquad \text{checks}$$

9.5 COFUNCTIONS

The sine of angle A, Fig. 9-5, is

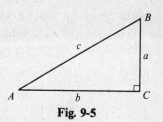

Fig. 9-5

$$\sin A = \frac{a}{c}$$

But a/c is *also* the cosine of angle B. Therefore,

Similarly,

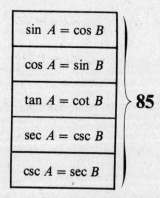

$$\sin A = \cos B$$
$$\cos A = \sin B$$
$$\tan A = \cot B$$
$$\sec A = \csc B$$
$$\csc A = \sec B$$

$\Big\}$ **85**

In general, a trigonometric function of an acute angle is equal to the corresponding cofunction of the complementary angle.

Example 9.8: If the cosine of the acute angle A in a right triangle ABC is equal to 0.586, find the sine of the other acute angle, B.

Solution: By Eq. 85, $\sin B = \cos A = 0.586$.

9.6 RECTANGULAR COMPONENTS OF A VECTOR

(a) Definitions

1. A *scalar quantity* is one that has magnitude only and can be represented by a single number.

   Example 9.9:   Mass, temperature, and time are scalar quantities.

2. A *vector quantity* is one that has direction as well as magnitude. Vector symbols are usually shown in boldface.

   Example 9.10:   Force, velocity, and acceleration are vector quantities.

   A vector is represented graphically, Fig. 9-6, by a line segment which has the same *direction* as the vector quantity. The length of the line is proportional to the *magnitude* of the vector, and an arrowhead gives the *sense* of the vector.

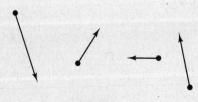

Fig. 9-6

(b) Resolution of Vectors. To *resolve* a vector means to replace it by two other vectors, called *components*.

Example 9.11:   Force **F**, Fig. 9-7, is shown resolved into a *vertical component* $\mathbf{F}_v$ and a *horizontal component* $\mathbf{F}_h$.

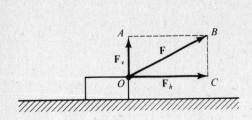

Fig. 9-7

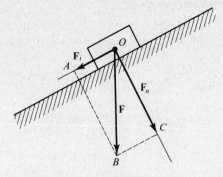

Fig. 9-8

Example 9.12:   The force **F** in Fig. 9-8 is shown resolved into two components.   The component F_t parallel to the inclined surface is called the *tangential* component, and the component F_n perpendicular to the surface is called the *normal* component.

When two components are at right angles to each other they are called *rectangular components*. Note that in Figs. 9-7 and 9-8, for example, the original vector and its rectangular components form right triangles OAB and OBC.   We can thus use right-triangle trigonometry to resolve a vector into rectangular components, as in the following examples.

Example 9.13:   A projectile has a velocity of 225 ft/s, and a direction of 25° from the horizontal, Fig. 9-9. Resolve this velocity into its vertical and horizontal components V_v and V_h.

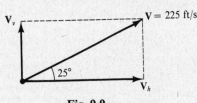

Fig. 9-9

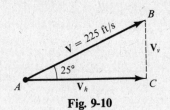

Fig. 9-10

Solution: From Fig. 9-10, in triangle ABC,

$$\frac{V_v}{225} = \sin 25°$$

So, $$V_v = 225 \sin 25° = 95.1 \text{ ft/s}$$

Similarly, $$V_h = 225 \cos 25° = 204 \text{ ft/s}$$

The component of a vector along some axis is also referred to as a *projection* of the vector onto that axis.

Example 9.14:   Find the projection of a vector of magnitude 438 onto an axis which makes an angle of 57° with the vector.

Solution: In Fig. 9-11, we draw the axis OC an angular distance of 57° from the vector OA, and drop a perpendicular from A to axis OC.   Then, by Eq. 74,

$$\frac{OB}{438} = \cos 57°$$

So $$OB = 438 \cos 57° = 438(0.545) = 239 \text{ units}$$

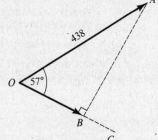

Fig. 9-11

(c)   Resultants.   Two or more vectors may be replaced by a single vector, the *resultant*, or *vector sum*.

When the two vectors we wish to replace by their resultant are at right angles to each other, find the resultant by the Pythagorean Theorem and the angle by means of the trigonometric ratios.

Example 9-15:   Find the resultant of a 158-N vertical force and a 214-N horizontal force, Fig. 9-12.

Solution: By Eq. 72 in triangle *ABC*,

$$R^2 = (158)^2 + (214)^2 = 70\,760$$

$$R = 266 \text{ N}$$

And by Eq. 75,

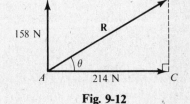

Fig. 9-12

$$\tan \theta = \frac{158}{214} = 0.738$$

$$\theta = 36.4°$$

9.7   IMPEDANCE, REACTANCE, AND PHASE ANGLE

The *reactance X* is a measure of how much the capacitance and inductance retard the flow of current in an ac circuit, Fig. 9-13.   It is the difference between the *inductive reactance* X_L and the *capacitive reactance* X_C.

Reactance	$X = X_L - X_C$	**214**

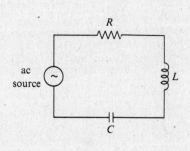

Fig. 9-13

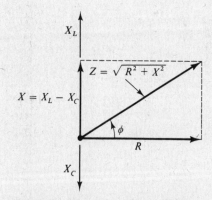

Fig. 9-14   Vector impedance diagram.

The *impedance Z* is a measure of how much the flow of current in an ac circuit is retarded by all circuit elements.   The magnitude of the impedance is related to the total resistance R and reactance X by

Magnitude of Impedance	$|Z| = \sqrt{R^2 + X^2}$	**215**

The impedance, resistance, and reactance form the three sides of a right triangle, the *vector impedance diagram* of Fig. 9-14.   The angle ϕ between Z and R is called the *phase angle*.

Phase Angle	$\phi = \arctan \dfrac{X}{R}$	**216**

9.8   POWER IN AN AC CIRCUIT: THE POWER TRIANGLE

(a)   Apparent Power.   The *apparent power* P_a delivered by an ac source is the product of the effective (rms) voltage V and the effective (rms) current I.

Apparent Power	$P_a = VI$ (volt-amperes)	**217**

(Note that apparent power is *not* given in watts.)

(b)   Average Power.   The *average power* P is that component of the power that gets converted to heat by the resistive elements of the circuit.

Average Power	$P = VI \cos \phi$ (watts)	**218**

where ϕ is the phase angle.

(c)   Reactive Power.   The *reactive power* P_q engaged in shuttling energy to and from the capacitors and inductors remains within the circuit.

Reactive Power	$P_q = VI \sin \phi$ (vars)	**219**

(Note that reactive power is also not given in watts but in volt-amperes-reactive.)

(d)   Geometry.   Geometrically, the apparent, average, and reactive power form the three sides of a right triangle, called the *power triangle*, with ϕ as one of the angles, as in Fig. 9-15.

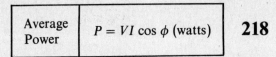

(e)   Power Factor.   The power factor is defined as

Fig. 9-15   The power triangle.

Power Factor	$F_p = \cos \phi$	**220**

Solved Problems

RIGHT TRIANGLES

9.1   Sketch the following right triangles and find all the missing parts:

   (*a*) $B = 35°$, $c = 2.5$ cm   (*b*) $a = 1.0$ in., $b = 2.0$ in.   (*c*) $A = 55°$, $a = 3.50$ cm

Solution

   (*a*)   In Fig. 9-16*a*, first draw an angle of 35° and label it B. Then along one side we measure 2.5 cm. From the other end of this leg, draw line AC perpendicular to BC to complete the triangle.

   To find angle A, we use Eq. 60,

$$A + B = 90°$$

$$A = 90° - 35° = 55°$$

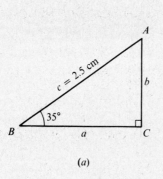 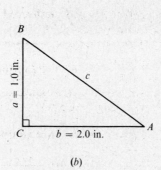 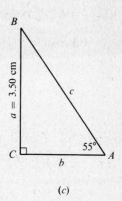

(a)　　　　　　　　　　　　(b)　　　　　　　　　　　　(c)

Fig. 9-16

By Eq. 73,

$$\sin 35° = \frac{b}{2.5}$$

$$b = 2.5 \sin 35° = 2.5(0.574) = 1.4 \text{ cm}$$

By Eq. 74,

$$\cos 35° = \frac{a}{2.5}$$

$$a = 2.5 \cos 35° = 2.5(0.819) = 2.0 \text{ cm}$$

(b)　In Fig. 9-16b, draw right angle C.　Measure 1.0 in. along one of the legs, and label it a.　Along the other leg measure 2.0 in., and label it b.　Draw the hypotenuse to connect the two legs.

　　To find the missing side c, use Eq. 72, the Pythagorean Theorem.

$$c^2 = a^2 + b^2 = 1^2 + 2^2 = 5$$
$$c = 2.2 \text{ in.}$$

By Eq. 75,

$$\tan A = \frac{1}{2}$$
$$A = 27°$$

Similarly,

$$\tan B = \frac{2}{1}$$
$$B = 63°$$

(c)　First draw a right angle, C, as in Fig. 9-16c.　Measure 3.50 cm along one leg, and label this leg a. From Eq. 60,

$$B = 90° - A = 90° - 55° = 35°$$

Angle B is then drawn and the sides extended to close the triangle.　Then by Eq. 73,

$$\sin 55° = \frac{3.50}{c}$$

$$c = \frac{3.50}{\sin 55°} = \frac{3.50}{0.819} = 4.27 \text{ cm}$$

and by Eq. 75,

$$\tan 55° = \frac{3.50}{b}$$

$$b = \frac{3.50}{\tan 55°} = \frac{3.50}{1.43} = 2.45 \text{ cm}$$

9.2　　　In right triangle ABC, Fig. 9-17, $A = 30°$ and $b = 2.50$ in.　Solve the triangle.

Solution

By Eq. 60,

$$B = 90° - A = 90° - 30° = 60°$$

By Eq. 77,

$$\frac{c}{b} = \sec A$$

$$c = 2.50 \sec 30° = 2.50(1.15) = 2.89 \text{ in.}$$

By Eq. 75,

$$\frac{a}{b} = \tan A$$

$$a = 2.50 \tan 30° = 2.5(0.577) = 1.44 \text{ in.}$$

Check:　　　By Eq. 72,

$$(2.89)^2 \overset{?}{=} (1.44)^2 + (2.50)^2$$

$$8.35 \overset{?}{=} 2.07 + 6.25$$

$$8.35 \cong 8.32 \quad \text{checks}$$

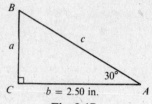

Fig. 9-17

9.3　　　In right triangle ABC, Fig. 9-18, $B = 65°$ and $b = 2.00$ m.　Solve the triangle.

Solution

By Eq. 60,

$$A = 90° - B = 90° - 65° = 25°$$

By Eq. 78,

$$\frac{c}{b} = \csc B$$

$$c = 2.00 \csc 65° = 2.00(1.103) = 2.21 \text{ m}$$

By Eq. 76,

$$\frac{a}{b} = \cot B$$

$$a = 2.00 \cot 65° = 2.00(0.466) = 0.933 \text{ in.}$$

Check:　　　By Eq. 72,

$$(2.21)^2 \overset{?}{=} (0.933)^2 + (2.00)^2$$

$$4.88 \overset{?}{=} 0.870 + 4.00$$

$$4.88 \cong 4.87 \quad \text{checks}$$

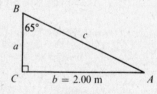

Fig. 9-18

9.4　　　In right triangle ABC, Fig. 9-19, $a = 5.50$ in. and $c = 7.50$ in.　Solve the triangle.

Solution

By Eq. 73,

$$\sin A = \frac{a}{c} = \frac{5.50}{7.50} = 0.733$$

So,

$$A = 47.2°$$

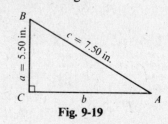

Fig. 9-19

By Eq. 74,

$$\cos B = \frac{a}{c} = 0.733$$

So,

$$B = 42.8°$$

By Eq. 72,

$$b^2 = (7.50)^2 - (5.50)^2 = 26.0$$
$$b = 5.10 \text{ in.}$$

Check: By Eq. 60,

$$A + B = 47.2° + 42.8° = 90° \qquad \text{checks}$$

By Eq. 74,

$$\cos 47.2° \stackrel{?}{=} \frac{5.10}{7.50}$$

$$0.679 \cong 0.680 \qquad \text{checks}$$

9.5   In right triangle ABC, Fig. 9-20, $a = 3.00$ m and $b = 4.00$ m.   Solve the triangle.

Solution

By Eq. 75,

$$\tan A = \frac{a}{b} = \frac{3.00}{4.00} = 0.750$$

$$A = 36.8°$$

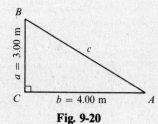

Fig. 9-20

By Eq. 76,

$$\cot B = \frac{a}{b} = 0.750$$

$$B = 53.2°$$

By Eq. 72,

$$c^2 = (3.00)^2 + (4.00)^2 = 25.0$$
$$c = 5.00 \text{ m}$$

Check: By Eq. 60,

$$A + B = 36.8° + 53.2° = 90° \qquad \text{checks}$$

By Eq. 73,

$$\sin 36.8° \stackrel{?}{=} \frac{3.00}{5.00}$$

$$0.599 \cong 0.600 \qquad \text{checks}$$

Note that this is a 3-4-5 right triangle, as described on page 179.

9.6   A guy wire from the top of an antenna is anchored 23.5 m from the base of the antenna and makes an angle of 65° with the ground.   Find   (a) the height h of the antenna and   (b) the length L of the wire.

Solution

By Eq. 75,

$$\frac{h}{23.5} = \tan 65°$$

$$h = 23.5(2.145) = 50.4 \text{ m}$$

By Eq. 74,

$$\cos 65° = \frac{23.5}{L}$$

$$L = \frac{23.5}{0.4226} = 55.6 \text{ m}$$

9.7   Find the angle of elevation of the sun, if a tower 127 ft high casts a shadow 186 ft long.

Solution

From Fig. 9-21,

$$\tan A = \frac{127}{186}$$

$$= 0.683$$

So, $A = 34.3°$

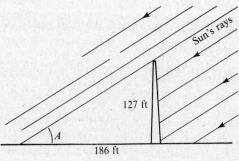

Fig. 9-21

9.8   A roadway is inclined at an angle of 10.5°.   Find the change in elevation of the roadway in a horizontal distance of 255 m.

Solution

Let x = the change in elevation, in meters.   Then

$$\tan 10.5° = \frac{x}{255}$$

$$x = 255 \tan 10.5° = 47.3 \text{ m}$$

9.9   From an airplane flying at 12 600 ft the angle of depression to an airport is 18.3°.   Assuming the terrain to be horizontal between the airport and the spot below the airplane, find the horizontal distance from the airplane to the airport.

Solution

From Statement 56, in Fig. 9-22

$$A = 18.3°$$

So, $\tan 18.3° = \dfrac{12\,600}{x}$

$$x = \frac{12\,600}{\tan 18.3°} = \frac{12\,600}{0.331} = 38\,100 \text{ ft}$$

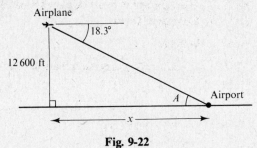

Fig. 9-22

9.10 One end of a prism is a right triangle having one angle equal to 42.5° and a hypotenuse of 115 mm. The length of the prism is 583 mm. Find its volume.

Solution

Let x = the length of the side opposite the given angle and y = the length of the adjacent side. Then,

$$\frac{x}{115} = \sin 42.5°$$

$$x = 115 \sin 42.5° = 77.7 \text{ mm}$$

and

$$\frac{y}{115} = \cos 42.5°$$

$$y = 115 \cos 42.5° = 84.8 \text{ mm}$$

By Eq. 58,

$$\text{Area} = \frac{77.7(84.8)}{2} = 3290 \text{ mm}^2$$

and by Eq. 109,

$$\text{Volume} = 3290(583) = 1\,920\,000 \text{ mm}^3$$

9.11 Find AB, BC, DE, and BE in the truss of Fig. 9-23.

Solution

Since $AC = 51.2$ m,

$$\frac{BC}{51.2} = \tan 33.6°$$

$$BC = 51.2 \tan 33.6° = 34.0 \text{ m}$$

Also,

$$\frac{51.2}{AB} = \cos 33.6°$$

$$AB = \frac{51.2}{\cos 33.6°} = 61.5 \text{ m}$$

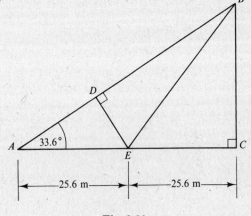

Fig. 9-23

In triangle ADE,

$$\frac{DE}{25.6} = \sin 33.6°$$

$$DE = 25.6 \sin 33.6° = 14.2 \text{ m}$$

In triangle BCE,

$$(BE)^2 = (25.6)^2 + (34.0)^2 = 1811$$

$$BE = 42.6 \text{ m}$$

9.12 How far apart must two stakes on a 5° slope be placed so that the horizontal distance between them is 1000 m?

Solution

Let r = the distance between the stakes, along the slope. Since, by Eq. 74, $x/r = \cos A$,

$$r = \frac{1000}{\cos A}$$

$$= \frac{1000}{\cos 5°} = \frac{1000}{0.9962} = 1004 \text{ m}$$

SHOP TRIGONOMETRY

The trigonometric ratios find many applications in the machine shop and the tool room.

Laying Out Angles

9.13   A machinist making the part shown in Fig. 9-24 wants to know the dimension x. Find this dimension.

    Solution

        From Eq. 75,

$$\frac{x}{3.550} = \tan 18°$$

$$x = 3.550 \tan 18°$$

$$= 3.550(0.3249)$$

$$= 1.153 \text{ in.}$$

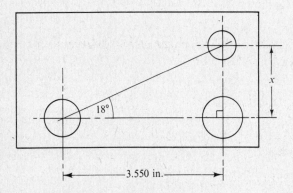

Fig. 9-24   Laying out angles.

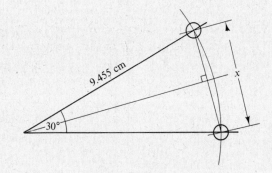

Fig. 9-25   A bolt circle.

Bolt Circles

9.14   A drawing, Fig. 9-25, calls for a bolt circle with a radius of 9.455 cm, containing 12 holes equally spaced. Find the straight-line distance between the holes so that they may be stepped off with dividers.

    Solution

        The angle between adjacent holes is $360/12 = 30°$. By Eq. 73,

$$\frac{\frac{x}{2}}{9.455} = \sin 15°$$

$$\frac{x}{2} = 9.455(0.2588) = 2.447$$

$$x = 4.894 \text{ cm}$$

9.15   The bolt circle of Fig. 9-26 is to be made on a jig borer.   Find the dimensions x_1, y_1, x_2, and y_2.

Solution

From Eq. 73,

$$\frac{y_1}{7.325} = \sin 60°$$

$$y_1 = 7.325(0.8660) = 6.344 \text{ in.}$$

and

$$\frac{y_2}{7.325} = \sin 30°$$

$$y_2 = 7.325(0.500) = 3.663 \text{ in.}$$

By Eq. 74,

$$\frac{x_1}{7.325} = \cos 60°$$

$$x_1 = 7.325(0.500) = 3.663 \text{ in.}$$

and

$$\frac{x_2}{7.325} = \cos 30°$$

$$x_2 = 7.325(0.866) = 6.344 \text{ in.}$$

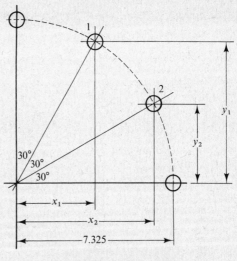

Fig. 9-26   A bolt circle.

Tapers

9.16   An 8.000-in.-long bar, Fig. 9-27, is to taper 2.5°, and be 1.000 in. in diameter at the narrow end.   Find the diameter of the larger end and the taper per foot.

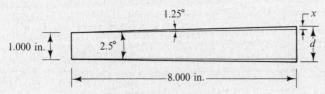

Fig. 9-27   Tapered shaft.

Solution

The angle between the centerline and the surface is half of 2.5°, or 1.25°.   Finding the distance x, Fig. 9-24, by Eq. 75,

$$\frac{x}{8.000} = \tan 1.25°$$

$$x = 8.000 \tan 1.25° = 8.000(0.0218) = 0.1744$$

The diameter of the large end will equal the diameter of the small end plus twice the distance x,

$$d = 1.000 + 2(0.1744) = 1.349 \text{ in.}$$

The *taper per foot* is the change in diameter per foot of length.   The change in diameter is $1.349 - 1.000 = 0.349$.   So,

$$\text{Taper per foot} = \frac{0.349}{8} \times 12 = 0.524 \text{ in./ft}$$

Finding Circle Diameters When the Center Is Inaccessible

Method 1. Lay a scale across the curve to be measured, as in Fig. 9-28, and measure the chord $2c$ and the sagitta h.

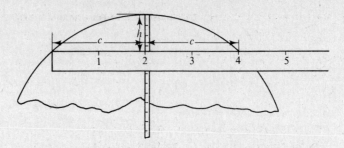

Fig. 9-28   Finding the diameter when the center is inaccessible.

9.17   Write an equation for the diameter D of a circle in terms of the sagitta h and the chord length $2c$.

Solution

In Fig. 9-29, extend line SP to the center O and draw radius OQ. From the Pythagorean Theorem, in triangle OPQ,

$$(r - h)^2 + c^2 = r^2$$

Squaring, $r^2 - 2rh + h^2 + c^2 = r^2$

$$2rh = h^2 + c^2$$

So,

$$D = 2r = \frac{h^2 + c^2}{h}$$

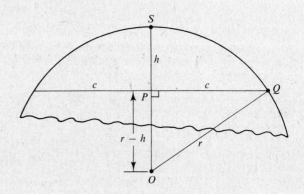

Fig. 9-29   Derivation of equation.

9.18   Find the radius of curvature of the corner of the casting, Fig. 9-30, if the chord length is 6.00 cm and the sagitta is 1.58 cm.

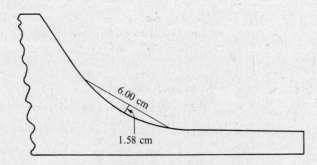

Fig. 9-30   Finding the radius in a casting.

Solution

Using the equation from Problem 9.17,

$$D = \frac{(1.58)^2 + (3.00)^2}{1.58} = 7.28 \text{ cm}$$

So, $r = 3.64 \text{ cm}$

Method 2. Lay a square over the part to be measured, as in Fig. 9-31, and measure the distance h.

9.19   Find the radius of the part if $h = 3.00$ in. and angle A is 60.0°.   See Fig. 9-31.

Solution

By Eq. 101, line OQ bisects angle A.   Triangle OPQ is thus a 30-60-90 right triangle since, by Statement 100, OP and PQ are perpendicular.   So,

$$\frac{r}{h+r} = \sin 30.0° = \frac{1}{2}$$

$$2r = h + r$$

$$r = h = 3.00 \text{ in.}$$

Thus, with a 60° square, the radius of the part will always equal the distance h.

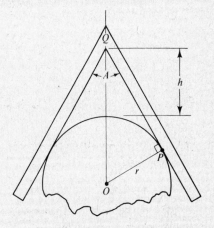

Measuring a diameter with a square.
Fig. 9-31

The Sine Bar.   A convenient tool for setting up or measuring angles in the shop is the *sine bar*, one version of which is shown in Fig. 9-32.

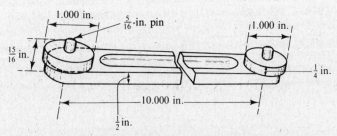

Fig. 9-32   Ten-inch sine bar. (*From Colvin and Stanley,* "*American Machinist's Handbook,*" *McGraw-Hill Book Company, New York, 1945.*)

9.20   A 10-in. sine bar is to be set at an angle of 32.5° to a milling machine table, so that a workpiece can later be clamped at this angle for milling.   Find the height H, Fig. 9-33.

Solution

By Eq. 73,

$$\frac{Y}{10} = \sin 32.5°$$

$$Y = 10 \sin 32.5°$$

$$= 5.373 \text{ in.}$$

Adding the radii of the disks,

$$H = 5.373 + 1$$

$$= 6.373 \text{ in.}$$

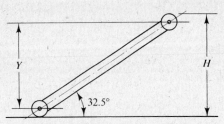

Using the sine bar to lay out an angle.
Fig. 9-33

9.21 A 10-in. sine bar is clamped to a workpiece and the vertical distance between the disks measured at 3.187 in., Fig. 9-34. Find the angle A.

Solution

By Eq. 73,

$$\sin A = \frac{3.187}{10} = 0.3187$$

$$A = 18.58°$$

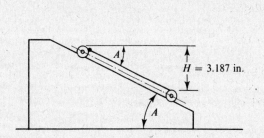

Measuring an angle with the sine bar.
Fig. 9-34

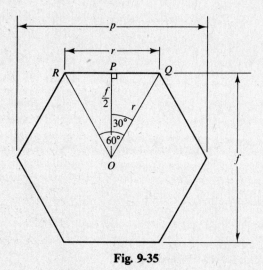

Fig. 9-35

Hexagonal Stock and Bolt Heads

9.22 Find the distance across the flats f, in terms of the distance across the corners p, for the hexagonal stock in Fig. 9-35.

Solution

Connecting two corners to the center produces an equilateral triangle of side r, where $r = p/2$. Line OP will be the perpendicular bisector of QR and will bisect the 60° angle. In triangle OPQ,

$$\frac{\frac{f}{2}}{r} = \cos 30°$$

Solving for f, $f = 2r \cos 30°$

Since $r = p/2$, $f = 2\left(\frac{p}{2}\right) \cos 30° = p \cos 30°$

9.23 A bolt head measures 1.875 cm across the flats. Find the distance across the corners and the width of each flat.

Solution

Using the equation from Problem 9.22 with $f = 1.875$ cm,

$$p = \frac{1.875}{\cos 30°} = \frac{1.875}{0.8660} = 2.165 \text{ cm} = \text{distance across the corners}$$

and $r = \dfrac{p}{2} = 1.083 \text{ cm} \doteq \text{width of each flat}$

VECTORS

9.24   Find the x and y components of the following vectors.   The angles given are those between the vector and the x axis.

    (a)   Magnitude = 212, $\theta = 37.2°$ (c)   Magnitude = 8.36, $\theta = 18.8°$

    (b)   Magnitude = 29.3, $\theta = -62.3°$ (d)   Magnitude = 49.2, $\theta = 155°$

Solution

(a)   In Fig. 9-36a,
$$\frac{V}{212} = \sin 37.2°$$
$$V = 212 \sin 37.2° = 212(0.605) = 128$$

Also,
$$\frac{H}{212} = \cos 37.2°$$
$$H = 212 \cos 37.2° = 212(0.797) = 169$$

(b)   In Fig. 9-36b,
$$\frac{V}{29.3} = \sin(-62.3°)$$
$$V = 29.3 \sin(-62.3°) = 29.3(-0.885) = -25.9$$

and
$$\frac{H}{29.3} = \cos(-62.3°)$$
$$H = 29.3 \cos(-62.3°) = 29.3(0.465) = 13.6$$

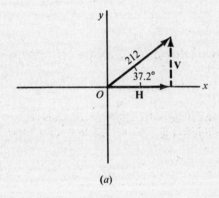

(a)

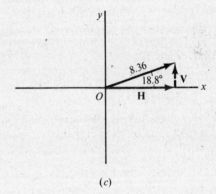

(c)

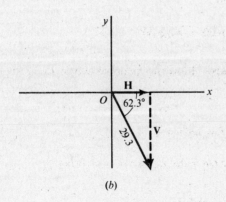

(b)

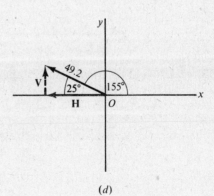

(d)

Fig. 9-36

(c) In Fig. 9-36c,

$$\frac{V}{8.36} = \sin 18.8°$$

$$V = 8.36 \sin 18.8° = 8.36(0.322) = 2.69$$

Also,

$$\frac{H}{8.36} = \cos 18.8°$$

$$H = 8.36 \cos 18.8° = 8.36(0.947) = 7.91$$

(d) In Fig. 9-36d,

$$\frac{V}{49.2} = \sin 155°$$

$$V = 49.2 \sin 155° = 49.2(0.423) = 20.8$$

and

$$\frac{H}{49.2} = \cos 155°$$

$$H = 49.2 \cos 155° = 49.2(-0.906) = -44.6$$

9.25 Find the resultant of the following pairs of vectors, Fig. 9-37, and the angle between resultant and the horizontal.

(a) 115 vertical, 203 horizontal   (c) 83.6 vertical, −63.8 horizontal

(b) −3.85 vertical, 2.99 horizontal   (d) −11.2 vertical, −8.37 horizontal

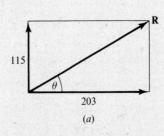

(a)

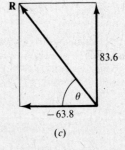

(c)

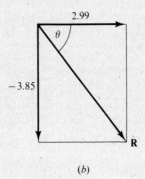

(b)

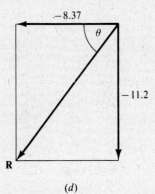

(d)

Fig. 9-37

Solution

(a) By Eq. 72,

$$R = \sqrt{(115)^2 + (203)^2} = 233$$

By Eq. 75,

$$\tan \theta = \frac{115}{203} = 0.567$$

$$\theta = 29.5°$$

(b) By Eq. 72,

$$R = \sqrt{(3.85)^2 + (2.99)^2} = 4.87$$

By Eq. 75,

$$\tan \theta = \frac{3.85}{2.99} = 1.29$$

$$\theta = 52.2°$$

(c) By Eq. 72,

$$R = \sqrt{(83.6)^2 + (63.8)^2} = 105$$

By Eq. 75,

$$\tan \theta = \frac{83.6}{63.8} = 1.31$$

$$\theta = 52.7°$$

(d) By Eq. 72,

$$R = \sqrt{(8.37)^2 + (11.2)^2} = 14.0$$

By Eq. 75,

$$\tan \theta = \frac{11.2}{8.37} = 1.34$$

$$\theta = 53.2°$$

9.26 The stress **S** on the inclined surface of a bar in tension, Fig. 9-38, is 4730 lb/in². Resolve this stress into its normal component **N** and its tangential component (the shearing stress) **T**.

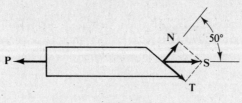

Fig. 9-38

Solution

From Eq. 73,

$$\frac{T}{S} = \sin 50°$$

$$T = S \sin 50° = 4730(0.7660) = 3620 \text{ lb/in}^2$$

From Eq. 74,

$$\frac{N}{S} = \cos 50°$$

$$N = S \cos 50° = 4730(0.6428) = 3040 \text{ lb/in}^2$$

9.27 A projectile is launched at an angle of 37.0° to the horizontal, Fig. 9-39, with a speed of 2740 m/s. Find the vertical and horizontal components of this velocity.

Solution

From Eq. 73,

$$\frac{V}{2740} = \sin 37.0°$$

$$V = 2740 \sin 37.0° = 2740(0.6018) = 1650 \text{ m/s}$$

and from Eq. 74,

$$\frac{H}{2740} = \cos 37.0°$$

$$H = 2740 \cos 37.0° = 2740(0.7986) = 2190 \text{ m/s}$$

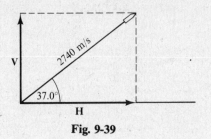

Fig. 9-39

9.28 A point on a rotating wheel has a tangential velocity of 275 cm/s. Find the **x** and **y** components of the velocity when in the position shown in Fig. 9-40.

Solution

Noting that angle A is 32.5°, we get, by Eq. 73,

$$\frac{V_y}{275} = \sin 32.5°$$

$$V_y = 275 \sin 32.5°$$

$$= 275(0.537) = 148 \text{ cm/s}$$

and by Eq. 74,

$$\frac{V_x}{275} = \cos 32.5°$$

$$V_x = 275 \cos 32.5°$$

$$= 275(0.843) = 232 \text{ cm/s}$$

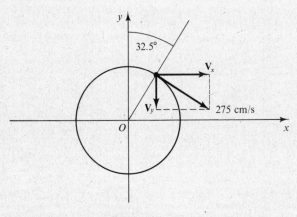

Fig. 9-40

IMPEDANCE, REACTANCE, AND PHASE ANGLE

9.29 The capacitive reactance of a certain circuit is 4590 Ω, the inductive reactance is 2480 Ω, and the resistance is 3270 Ω. (Remember that Ω is the abbreviation for ohms.) Find the reactance and magnitude of the impedance of the circuit.

Solution

By Eq. 214,

$$X = 2480 - 4590 = -2110 \ \Omega$$

and by Eq. 215,

$$|Z| = \sqrt{(3270)^2 + (-2110)^2} = 3892 \ \Omega$$

9.30 Find the phase angle for the circuit of Problem 9.29.

Solution

By Eq. 216,

$$\phi = \arctan \frac{-2110}{3270} = \arctan (-0.6453) = -32.8°$$

9.31   A circuit has a reactance of 1250 Ω and a resistance of 2830 Ω.   Find the magnitude of the impedance and the phase angle.

Solution

By Eq. 215,

$$|Z| = \sqrt{(1250)^2 + (2830)^2} = 3094 \ \Omega$$

By Eq. 216,

$$\tan \phi = \frac{1250}{2830} = 0.4417$$
$$\phi = 23.8°$$

9.32   A circuit has a capacitive reactance of 5430 Ω, an inductive reactance of 8320 Ω, and a total impedance of 4150 Ω.   Find the resistance and the phase angle.

Solution

By Eq. 214,

$$X = 8320 - 5430 = 2890 \ \Omega$$

By Eq. 215,

$$R = \sqrt{(4150)^2 - (2890)^2} = 2978 \ \Omega$$

By Eq. 216,

$$\tan \phi = \frac{2890}{2978} = 0.9704$$
$$\phi = 44.1°$$

9.33   A circuit has a resistance of 8350 Ω, an impedance of 15 200 Ω, an inductive reactance of 2120 Ω, and a positive phase angle.   Find the capacitive reactance and the phase angle.

Solution

By Eq. 215,

$$X = \sqrt{(15\,200)^2 - (8350)^2} = 12\,700 \ \Omega$$

By Eq. 214,

$$X_C = X_L - X = 2120 - 12\,700 = -10\,580 \ \Omega$$

By Eq. 216,

$$\tan \phi = \frac{12\,700}{8350} = 1.521$$
$$\phi = 56.7°$$

9.34   A circuit has a reactance of 55 400 Ω and a phase angle of 38.3°.   Find the resistance and the magnitude of the impedance.

Solution

By Eq. 216,

$$\tan 38.3° = \frac{55\,400}{R}$$
$$R = \frac{55\,400}{0.7898} = 70\,100 \ \Omega$$

By Eq. 215,

$$|Z| = \sqrt{(70\,100)^2 + (55\,400)^2} = 89\,400 \ \Omega$$

9.35 A circuit has a resistance of 12.2 Ω and a phase angle of $-53.2°$. Find the reactance and the magnitude of the impedance.

Solution

By Eq. 216,

$$\frac{X}{12.2} = \tan(-53.2°)$$

By Eq. 215,

$$X = 12.2(-1.337) = -16.3 \ \Omega$$

$$|Z| = \sqrt{(12.2)^2 + (-16.3)^2} = 20.4 \ \Omega$$

9.36 A circuit has an impedance of 923 Ω and a phase angle of 73.9°. Find the resistance and the reactance.

Solution

From the impedance diagram, Fig. 9-39,

$$\frac{X}{Z} = \sin \phi$$

So, $X = Z \sin \phi = 923 \sin 73.9° = 887 \ \Omega$

Similarly, $R = Z \cos \phi = 923 \cos 73.9° = 256 \ \Omega$

POWER TRIANGLE

9.37 A current of 2.35 A rms is delivered to a load at 158 V rms, with a phase angle of 28°. Find the apparent power, the reactive power, the average power, and the power factor.

Solution

By Eq. 217,

$$P_a = VI = 158(2.35) = 371 \text{ VA}$$

By Eq. 218,

$$P_q = 371 \sin 28° = 371(0.4695) = 174 \text{ vars}$$

By Eq. 219,

$$P = 371 \cos 28° = 371(0.8829) = 328 \text{ W}$$

By Eq. 220,

$$F_p = \cos 28° = 0.883$$

9.38 The effective current and voltage in an ac circuit are 0.583 A and 95.4 V respectively. The phase angle is 28.8°. Find (a) the apparent power, (b) the average power, (c) the reactive power, (d) the power factor.

Solution

(a) By Eq. 217,

$$P_a = (0.583)(95.4) = 55.6 \text{ VA}$$

(b) By Eq. 219,

$$P = 55.6 \cos 28.8° = 48.7 \text{ W}$$

(c) By Eq. 218,

$$P_q = 55.6 \sin 28.8° = 26.8 \text{ vars}$$

(d) By Eq. 220,

$$F_p = \cos 28.8° = 0.876$$

9.39 The rms voltage in an ac circuit is 24.7 V, the rms current is 7.92 A, and the power factor is 0.755. Find (a) the apparent power, (b) the phase angle, (c) the average power, (d) the reactive power.

Solution

(a) By Eq. 217,

$$P_a = (24.7)(7.92) = 196 \text{ VA}$$

(b) By Eq. 220,

$$F_p = \cos \phi = 0.755$$
$$\phi = 41.0°$$

(c) By Eq. 218,

$$P = 196(0.755) = 148 \text{ W}$$

(d) By Eq. 219,

$$P_q = 196 \sin 41.0° = 129 \text{ vars}$$

9.40 The impedance in an ac circuit is 5480 Ω, the resistance is 3960 Ω, and the reactive power is 524 vars. Find the (a) reactance, (b) phase angle, (c) power factor, (d) average power.

Solution

(a) By Eq. 215,

$$X = \sqrt{(5480)^2 - (3960)^2} = 3788 \ \Omega$$

(b) By Eq. 216,

$$\tan \phi = \frac{3788}{3960} = 0.9566$$
$$\phi = 43.7°$$

(c) By Eq. 220,

$$F_p = \cos 43.7° = 0.723$$

(d) From the power triangle, Fig. 9-15,

$$\tan \phi = \frac{P_q}{P}$$

$$P = \frac{P_q}{\tan \phi}$$

$$P = \frac{524}{0.9566} = 548 \text{ W}$$

Supplementary Problems

9.41   In right triangle ABC, $B = 27.3°$;   find A.

9.42   Two legs of a right triangle are 36 in. and 2.7 ft.   Find the length of the hypotenuse.

9.43   The hypotenuse of a right triangle is 29 cm and one leg is 13 cm.   Find the other leg.

9.44   In right triangle ABC, $a = 6.70$, $b = 3.20$,   and $c = 7.40$.   Write the six trigonometric functions of angle A.

9.45   If the sine of the acute angle $A = 0.379$, find the cosine of the other acute angle in the same right triangle.

9.46   Find the resultant of the following pairs of vectors and the angle the resultant makes with the horizontal:

   (*a*)   27 units vertical, 39 units horizontal   (*c*)   − 7.1 units vertical, 29 units horizontal

   (*b*)   100 units vertical, 107 units horizontal   (*d*)   3.9 units vertical, 2.6 units horizontal

9.47   In the right triangle ABC, $A = 17°$ and $c = 9.25$ m.   Solve the triangle.

9.48   In the right triangle ABC, $a = 7.11$ cm   and $b = 2.30$ cm.   Solve the triangle.

9.49   A ball is thrown upward with a velocity of 25 ft/s at an angle of 46° with the horizontal.   Resolve this vector into its horizontal and vertical components.

9.50   Two ropes are attached to a tent peg.   One rope, at an angle of 102° to the horizontal, has a tension of 95 lb and the other, at 12° to the ground, has a tension of 120 lb.   Find the resultant force on the tent peg.

9.51   A circular horizontal curve has a chord length of 265 m and a radius of 600 m.   The maximum distance between the chord and the curve (the sagitta) is 20 m.   Find the central angle.

9.52   The distance taped on the horizontal between points A and B is 3025 ft.   If the ground slopes at a 22° angle, find the distance along the slope from A to B.

9.53   The reactance of a circuit is 3030 Ω and the resistance is 5280 Ω.   Find the phase angle.

9.54   For the above circuit, find the impedance.

9.55   A 7-in. long bar is to taper 3° and be 0.500 in. in diameter at the small end.   Find the diameter of the larger end and the taper per foot.

Answers to Supplementary Problems

9.41 62.7°

9.42 48 in.

9.43 26 cm

9.44 $\sin A = 0.905$, $\cos A = 0.432$, $\tan A = 2.09$, $\csc A = 1.10$, $\sec A = 2.31$, $\cot A = 0.478$

9.45 0.379

9.46 (a) $R = 47$, $\theta = 34.7°$ (b) $R = 147$, $\theta = 43.1°$
         (c) $R = 30$, $\theta = -13.8°$ (d) $R = 4.7$, $\theta = 56.3°$

9.47 $c = 90°$, $B = 73°$, $a = 2.70$ m, $b = 8.85$ m

9.48 $C = 90°$, $c = 7.47$ cm, $B = 17.9°$, $A = 72.1°$

9.49 vertical = 18 ft/s, horizontal = 17 ft/s

9.50 $R = 153$ lb at 50.4°

9.51 25.7°

9.52 slope distance = 3263 ft

9.53 $\phi = 30°$

9.54 $z = 6090\ \Omega$

9.55 $d = 0.867$ in., taper/ft = 0.629 in./ft

Chapter 10

Radicals and Complex Numbers

10.1 RADICALS

(a) Terminology. The expression $\sqrt[n]{a}$ is called a *radical*.

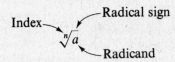

It consists of a *radicand*, a *radical sign*, and an *index*, and it is read "the *n*th root of *a*." It is also referred to as being in *radical form*.

When the index is 2, it is not usually written. Thus $\sqrt{x}$ means the same thing as $\sqrt[2]{x}$.

(b) Principal Roots. To avoid the ambiguity in an expression such as

$$\left.\begin{array}{r} \sqrt{4} = 2 \\ \sqrt{4} = -2 \end{array}\right\}?$$

we give the *principal root* only. The principal root of a positive number is always positive.

Example 10.1

$$\sqrt{4} = +2 \text{ only} \qquad \text{not } \pm 2$$
$$-\sqrt{4} = -(+2) = -2$$
$$\pm\sqrt{4} = \pm(+2) = \pm 2$$
$$\sqrt[3]{8} = +2$$

The principal root of a negative number is negative if the index is *odd*.

Example 10.2

$$\sqrt[3]{-8} = -2$$
$$\sqrt[5]{-32} = -2$$

The principal root of a negative number is imaginary if the index is *even*.

Example 10.3

$$\sqrt{-4} \text{ is imaginary.}$$
$$\sqrt[4]{-16} \text{ is imaginary.}$$

10.2 FRACTIONAL EXPONENTS: RELATION BETWEEN EXPONENTS AND RADICALS

An expression raised to a fractional exponent can be rewritten in *radical form*, because

$$a^{1/n} = \sqrt[n]{a} \qquad \textbf{15}$$

and

$$a^{m/n} = \sqrt[n]{a^m} = (\sqrt[n]{a})^m \qquad \textbf{16}$$

Example 10.4:   Write in radical form:

(a) $y^{1/3}$ (b) $x^{1/2}$ (c) $z^{5/3}$

Solution: By Eqs. 15 and 16,

(a) $y^{1/3} = \sqrt[3]{y}$ (b) $x^{1/2} = \sqrt{x}$ (c) $z^{5/3} = \sqrt[3]{z^5} = (\sqrt[3]{z})^5$

Example 10.5:   Write in exponential form:

(a) $\sqrt{m}$ (b) $\sqrt[5]{w}$ (c) $\sqrt[3]{p^2}$ (d) $(\sqrt[5]{q})^8$

Solution: By Eqs. 15 and 16,

(a) $\sqrt{m} = m^{1/2}$ (b) $\sqrt[5]{w} = w^{1/5}$ (c) $\sqrt[3]{p^2} = p^{2/3}$ (d) $(\sqrt[5]{q})^8 = q^{8/5}$

10.3 RULES OF RADICALS

The rules for multiplying and dividing radicals and for raising radicals to a power are similar to those for exponents.   They are as follows:

(a) Products

Root of a Product	$\sqrt[n]{ab} = \sqrt[n]{a}\,\sqrt[n]{b}$	**17**

Example 10.6:   Write $\sqrt{8}\,\sqrt{5}$ as a single radical.

Solution: By Eq. 17,

$$\sqrt{8}\,\sqrt{5} = \sqrt{(8)(5)} = \sqrt{40}$$

Common Error	There is no similar rule for the root of a sum. $$\sqrt[n]{a + b} \neq \sqrt[n]{a} + \sqrt[n]{b}$$

Example 10.7: $\sqrt{2} + \sqrt{3}$ does not equal $\sqrt{5}$.

Common Error	Equation 17 does not hold when a and b are both negative.

Example 10.8

$$(\sqrt{-4})^2 = \sqrt{-4}\,\sqrt{-4} \neq \sqrt{(-4)(-4)} \neq \sqrt{16} = +4$$

Instead, convert to imaginary numbers, as shown in Section 10.7.

(b) Quotients

| Root of a Quotient | $\sqrt[n]{\dfrac{a}{b}} = \dfrac{\sqrt[n]{a}}{\sqrt[n]{b}}$ | **18** |

Example 10.9: Write $\sqrt{15} \div \sqrt{3}$ as a single radical.

Solution: By Eq. 18,

$$\frac{\sqrt{15}}{\sqrt{3}} = \sqrt{\frac{15}{3}} = \sqrt{5}$$

(c) Powers

| Root of a Power | $\sqrt[n]{a^m} = (\sqrt[n]{a})^m$ | **19** |

Example 10.10:   Evaluate $\sqrt[3]{8^5}$.

Solution: By Eq. 19,

$$\sqrt[3]{8^5} = (\sqrt[3]{8})^5$$

But $\sqrt[3]{8} = 2$; so,

$$(\sqrt[3]{8})^5 = 2^5 = 32$$

10.4 SIMPLIFYING RADICALS

To write a radical in its *simplest form* means to

1. Remove any factors from the radicand.
2. Reduce the index to its lowest possible value.
3. Rewrite any fractions so that the denominator does not contain a radical.

(a) Removing Factors from the Radicand

1. Factor the radicand so that one or more factors has an exponent which is a multiple of the index.
2. Remove any such factor from the radicand and write its root outside the radical sign.

   Example 10.11:   Simplify the radical $\sqrt{18x^5}$.

   Solution: The index is 2, and so we seek factors which have exponents which are a multiple of 2. Thus 18 factors into (3^2) and (2), and x^5 factors into $(x^2)^2$ and (x). So,

   $$\sqrt{18x^5} = \sqrt{(3^2)(2)(x^2)^2(x)}$$

   Removing the perfect squares from the radicand,

   $$\sqrt{18x^5} = 3x^2\sqrt{2x}$$

(b) Reducing the Index

1. Write the radical in exponential form, with each factor of the radicand having a fractional exponent.
2. Reduce each fractional exponent to lowest terms and convert back to radical form.

Example 10.12:   Simplify $\sqrt[6]{16a^4b^{12}}$.

Solution: Writing 16 as 4^2 and switching to exponential form,

$$\sqrt[6]{16a^4b^{12}} = (4^2a^4b^{12})^{1/6}$$

and by Eq. 11,

$$= 4^{2/6}a^{4/6}b^{12/6}$$

Reducing each exponent,

$$4^{2/6}a^{4/6}b^{12/6} = 4^{1/3}a^{2/3}b^2 = b^2(4a^2)^{1/3}$$

Returning to radical form,

$$b^2(4a^2)^{1/3} = b^2\sqrt[3]{4a^2}$$

(c) Rationalizing the Denominator. If the denominator of a fraction is a square root, multiply the denominator (and the numerator too, of course) by the original denominator. This will result in removal of the radical in the denominator.

Example 10.13

$$\sqrt{\frac{2}{3}} = \frac{\sqrt{2}}{\sqrt{3}} = \frac{\sqrt{2}}{\sqrt{3}} \times \frac{\sqrt{3}}{\sqrt{3}} = \frac{\sqrt{2}\sqrt{3}}{\sqrt{3}\sqrt{3}} = \frac{\sqrt{6}}{\sqrt{9}} = \frac{\sqrt{6}}{3}$$

Example 10.14

$$\frac{w}{\sqrt{x}} = \frac{w}{\sqrt{x}} \times \frac{\sqrt{x}}{\sqrt{x}} = \frac{w\sqrt{x}}{\sqrt{x}\sqrt{x}} = \frac{w\sqrt{x}}{x}$$

If the denominator is a cube root, we multiply numerator and denominator of the fraction by that quantity that will make the radicand in the denominator a perfect cube. Thus if the denominator is $\sqrt[3]{a}$, we multiply the fraction by $\sqrt[3]{a^2}/\sqrt[3]{a^2}$.

Example 10.15

$$\frac{7}{\sqrt[3]{2}} = \frac{7}{\sqrt[3]{2}} \times \frac{\sqrt[3]{2^2}}{\sqrt[3]{2^2}} = \frac{7\sqrt[3]{2^2}}{\sqrt[3]{2^3}} = \frac{7\sqrt[3]{4}}{2}$$

If a fraction has as its denominator a radical of index n, rationalize the denominator by multiplying numerator and denominator of the fraction by that quantity which will make the radicand a perfect nth power.

Example 10.16

$$\frac{4ab}{\sqrt[4]{c}} = \frac{4ab}{\sqrt[4]{c}} \times \frac{\sqrt[4]{c^3}}{\sqrt[4]{c^3}} = \frac{4ab\sqrt[4]{c^3}}{\sqrt[4]{c^4}} = \frac{4ab}{c}\sqrt[4]{c^3}$$

10.5 OPERATIONS WITH RADICALS

(a) Addition and Subtraction of Radicals. Two radicals are *similar* if, after simplification, they have the *same index* and the *same radicand*.

Example 10.17: $2\sqrt{5}$, $3\sqrt{5}$, and $-\sqrt{5}$ are similar radicals because the index in each is 2 and the radicand in each is 5.

Example 10.18: $2\sqrt{x}$ and $5\sqrt[4]{16x^2}$ are similar because

$$5\sqrt[4]{16x^2} = 5\sqrt[4]{16}\sqrt[4]{x^2} = 5(2)\sqrt{x} = 10\sqrt{x}$$

Example 10.19: $3\sqrt{y}$ and $3\sqrt[3]{y}$ are not similar because one of the radicals has an index of 2 and the other has an index of 3.

Radicals are added or subtracted by *combining similar radicals*.

Example 10.20:

$$2\sqrt{x} + 3\sqrt{x} - \sqrt{x} = (2 + 3 - 1)\sqrt{x} = 4\sqrt{x}$$

Example 10.21:   Add

$$5\sqrt{3y} - \sqrt{27y} + \sqrt{12y}$$

Solution: Simplifying each radical,

$$5\sqrt{3y} - \sqrt{27y} + \sqrt{12y} = 5\sqrt{3y} - \sqrt{9(3y)} + \sqrt{4(3y)} = 5\sqrt{3y} - 3\sqrt{3y} + 2\sqrt{3y}$$

Adding,
$$= (5 - 3 + 2)\sqrt{3y} = 4\sqrt{3y}$$

(b) Multiplication of Radicals. If the radicals have the *same index*, they may be multiplied by using Eq. 17.

Example 10.22

$$(2\sqrt[3]{x})(4\sqrt[3]{y}) = 8\sqrt[3]{x}\sqrt[3]{y} = 8\sqrt[3]{xy}$$

If the *indices are different*, convert to exponential form, multiply using Eq. 8, and convert back to radical form.

Example 10.23

$$\sqrt{x}\sqrt[3]{y} = x^{1/2} \times y^{1/3} = x^{3/6} \times y^{2/6} = (x^3 y^2)^{1/6} = \sqrt[6]{x^3 y^2}$$

(c) Division of Radicals. If the *indices are the same*, divide by using Eq. 18. Rationalize the denominator where necessary.

Example 10.24

$$\frac{\sqrt[3]{4x^2}}{\sqrt[3]{2x}} = \sqrt[3]{\frac{4x^2}{2x}} = \sqrt[3]{2x}$$

If the *indices are different*, convert to exponential form, divide using Eq. 12, and convert back to radical form.

Example 10.25

$$\frac{\sqrt[3]{x^2}}{\sqrt{x}} = \frac{x^{2/3}}{x^{1/2}} = \frac{x^{4/6}}{x^{3/6}} = x^{4/6 - 3/6} = x^{1/6} = \sqrt[6]{x}$$

When the denominator is a *binomial* containing square roots, multiply numerator and denominator of the fraction by the *conjugate* of that binomial. A conjugate is a binomial having the same two terms as the original binomial, but with the sign of the second term changed.

Example 10.26:   Divide

$$\frac{2}{1 + \sqrt{x}}$$

Solution: The conjugate of the denominator $1 + \sqrt{x}$ is $1 - \sqrt{x}$. Multiplying numerator and denominator by $1 - \sqrt{x}$,

$$\frac{2}{1 + \sqrt{x}} \cdot \frac{1 - \sqrt{x}}{1 - \sqrt{x}} = \frac{2(1 - \sqrt{x})}{1 + \sqrt{x} - \sqrt{x} - x} = \frac{2(1 - \sqrt{x})}{1 - x}$$

This method always results in the removal of the radical from the denominator.

10.6 RADICAL EQUATIONS

A *radical equation* is one in which the unknown is under a radical sign.   To solve a radical equation

1. Isolate the radical on one side of the equals sign.
2. Raise both sides of the equation to a power equal to the index of the radical.
3. If there is more than one radical in the equation, repeat steps 1 and 2 until all radicals are gone.
4. Complete the solution in the usual way.

Example 10.27:   Solve the equation, $\sqrt{x - 2} = 3$.

Solution: Squaring both sides and transposing,

$$x - 2 = 9$$

$$x = 11$$

Check: $\sqrt{11 - 2} = \sqrt{9} = 3$ checks

Common Error	The squaring process often introduces *extraneous roots*. These are discarded because they do not satisfy the original equation.

Example 10.28:   Solve

$$\sqrt{4x^2 + 1} = 2x - 1$$

Solution: Squaring both sides,

$$4x^2 + 1 = (2x - 1)^2 = 4x^2 - 4x + 1$$

Transposing, $4x = 0$

So, $x = 0$

Check: $\sqrt{4(0)^2 + 1} \overset{?}{=} 2(0) - 1$

$$\sqrt{1} \overset{?}{=} -1$$

$$1 \neq -1   \text{doesn't check}$$

10.7 COMPLEX NUMBERS

(a) Imaginary Numbers.   To evaluate the expression

$$\sqrt{-1}$$

we need a number which, when squared, equals -1.   There is no real number whose square equals -1, and so we define a new number, called the *imaginary unit*.

$$i = \sqrt{-1}$$

(The letter j is also used to designate the imaginary unit.)

An *imaginary number* has the form bi, where b is any real number and i is the imaginary unit.

Example 10.29:   Write $\sqrt{-4}$ as an imaginary number.

Solution: By Eq. 17,

$$\sqrt{-4} = \sqrt{4}\sqrt{-1}$$

But since $\sqrt{-1} = i$,

$$\sqrt{4}\sqrt{-1} = \sqrt{4}\,i$$
$$= 2i$$

Common Error	Always convert radicals to imaginary numbers *before* performing other operations, or contradictions may result.

Example 10.30:   Multiply $\sqrt{-4}$ by $\sqrt{-4}$.

Solution: Converting to imaginary numbers,

$$\sqrt{-4}\sqrt{-4} = (2i)(2i) = 4i^2$$

Since $i^2 = -1$,

$$4i^2 = -4$$

It is *incorrect* to write,

$$\sqrt{-4}\sqrt{-4} = \sqrt{(-4)(-4)} = \sqrt{16} = +4$$

(b) Complex Numbers.   A *complex number* has the form $a + bi$, where a is a real number and bi is an imaginary number.   The number a is called the *real part*, and bi is called the *imaginary part*.

$$a + bi$$
Real part———↑   ↑———Imaginary part

Example 10.31:   Write the expression $\sqrt{-9} - 5$ in the form $a + bi$.

Solution: The real part is -5, and the imaginary part is $\sqrt{-9} = \sqrt{9}\sqrt{-1} = 3i$;   so,

$$\sqrt{-9} - 5 = -5 + 3i$$

(c) Powers of i

Since

squaring i, we get

Cubing i,

Raising i to the fourth power,

| $i = \sqrt{-1}$ |
| $i^2 = -1$ |
| $i^3 = i^2(i) = -i$ |
| $i^4 = (i^2)^2 = (-1)^2 = 1$ |

123

For powers higher than four, the values repeat in a cyclic manner.

$$i^5 = i^4(i) = i$$
$$i^6 = i^5(i) = i^2 = -1$$

and so on.

Example 10.32: Evaluate (a) i^8 and (b) i^{15}.

Solution:

(a) By Eq. 10,

$$i^8 = (i^4)^2 = (1)^2 = 1$$

(b) By Eq. 10,

$$i^{15} = (i^3)^5 = (-i)^5 = -i$$

(d) Addition of Complex Numbers. To add complex numbers, add the real parts and the imaginary parts separately, and express the result in the form $a + bi$.

Addition of Complex Numbers	$(a + bi) + (c + di) = (a + c) + (b + d)i$	**124**

Example 10.33: Add $(6 - 3i)$ to $(3 + 5i)$.

Solution: By Eq. 124,

$$(6 - 3i) + (3 + 5i) = (6 + 3) + (-3 + 5)i = 9 + 2i$$

(e) Subtraction of Complex Numbers. To subtract complex numbers, subtract the real and imaginary parts separately, and write the result in the form $a + bi$.

Subtraction of Complex Numbers	$(a + bi) - (c + di) = (a - c) + (b - d)i$	**125**

Example 10.34: Subtract $(2 - 5i)$ from $(4 + 3i)$.

Solution: From Eq. 125,

$$(4 + 3i) - (2 - 5i) = (4 - 2) + [3 - (-5)]i = 2 + 8i$$

(f) Multiplication of Complex Numbers. Multiply complex numbers the same way you would multiply any two binomials (Eq. 36), replace i^2 with -1, and write the result in the form $a + bi$.

Multiplication of Complex Numbers	$(a + bi)(c + di) = (ac - bd) + (ad + bc)i$	**126**

Example 10.35: Multiply $(5 - 2i)$ by $(3 + 4i)$.

Solution: Multiplying as two binomials, using Eq. 36,

$$(5 - 2i)(3 + 4i) = 5(3) + 5(4i) + (-2i)(3) + (-2i)(4i) = 15 + 20i - 6i - 8i^2$$

Combining like terms and replacing i^2 by -1,

$$(5 - 2i)(3 + 4i) = 15 + 14i + 8 = 23 + 14i$$

The same result can be obtained from Eq. 126,

$$(5 - 2i)(3 + 4i) = [(5)(3) - (-2)(4)] + [(5)(4) + (-2)(3)]i = (15 + 8) + (20 - 6)i = 23 + 14i$$

(g) The Conjugate of a Complex Number. The *conjugate* of a complex number is obtained by changing the sign of its imaginary part.

Example 10.36

The conjugate of $3 + 5i$ is $3 - 5i$.

The conjugate of $-2 - 4i$ is $-2 + 4i$.

The conjugate of $a + bi$ is $a - bi$.

(h) Division of Complex Numbers. Write the division as a fraction, and multiply numerator and denominator by the conjugate of the denominator. Simplify, replacing i^2 by -1, and write the result in the form $a + bi$.

Example 10.37:   Divide $(6 + 2i)$ by $(2 - 4i)$.

Solution: Multiplying numerator and denominator by the conjugate of the denominator,

$$\frac{6 + 2i}{2 - 4i} = \frac{6 + 2i}{2 - 4i} \times \frac{2 + 4i}{2 + 4i} = \frac{12 + 24i + 4i + 8i^2}{4 + 8i - 8i - 16i^2}$$

Collecting terms and replacing i^2 by -1,

$$\frac{6 + 2i}{2 - 4i} = \frac{4 + 28i}{20} = \frac{1}{5} + \frac{7}{5}i$$

Example 10.38:   Divide $(a + bi)$ by $(c + di)$.

Solution: Multiplying numerator and denominator by $c - di$,

$$\frac{a + bi}{c + di} = \frac{a + bi}{c + di} \times \frac{c - di}{c - di} = \frac{ac - adi + bci - bdi^2}{c^2 + cdi - cdi - d^2i^2} = \frac{(ac + bd) + (bc - ad)i}{c^2 + d^2}$$

which gives us the general formula,

Division of Complex Numbers	$\dfrac{a + bi}{c + di} = \dfrac{ac + bd}{c^2 + d^2} + \dfrac{bc - ad}{c^2 + d^2}i$

127

Solved Problems

PRINCIPAL ROOTS

10.1   Find the principal root.

(a) $\sqrt[4]{16}$ (b) $\sqrt[9]{-1}$ (c) $\sqrt{25}$ (d) $\sqrt{\dfrac{1}{4}}$ (e) $\sqrt[3]{-64}$ (f) $\sqrt{0.01}$

Solution

(a) $\sqrt[4]{16} = \sqrt[4]{2^4} = 2$ (c) $\sqrt{25} = 5$ (e) $\sqrt[3]{-64} = \sqrt[3]{(-4)^3} = -4$

(b) $\sqrt[9]{-1} = -1$ (d) $\sqrt{\dfrac{1}{4}} = \dfrac{\sqrt{1}}{\sqrt{4}} = \dfrac{1}{2}$ (f) $\sqrt{0.01} = (10^{-2})^{1/2} = 10^{-1} = 0.1$

EXPONENTIAL AND RADICAL FORM

10.2   Express in radical form.

(a) $x^{1/5}$ (c) $x^{1/3}y^{2/3}$ (e) $(w - z)^{1/2}$ (g) $x^0y^{-1/3}$

(b) $w^{5/8}$ (d) $[(a + b)^3]^{1/2}$ (f) $\dfrac{a^{-1/2}}{b^{1/2}}$ (h) $\dfrac{1}{(x + 3)^{-1/5}}$

Solution

(a) $x^{1/5} = \sqrt[5]{x}$

(e) $(w - z)^{1/2} = \sqrt{w - z}$

(b) $w^{5/8} = \sqrt[8]{w^5}$

(f) $\dfrac{a^{-1/2}}{b^{1/2}} = \dfrac{1}{a^{1/2}b^{1/2}} = \dfrac{1}{(ab)^{1/2}} = \dfrac{1}{\sqrt{ab}}$

(c) $x^{1/3}y^{2/3} = (xy^2)^{1/3} = \sqrt[3]{xy^2}$

(g) $x^0 y^{-1/3} = \dfrac{1}{y^{1/3}} = \dfrac{1}{\sqrt[3]{y}}$

(d) $[(a + b)^3]^{1/2} = \sqrt{(a + b)^3}$

(h) $\dfrac{1}{(x + 3)^{-1/5}} = (x + 3)^{1/5} = \sqrt[5]{x + 3}$

10.3 Express in exponential form.

(a) $\sqrt{x}$ (c) $\sqrt[3]{y^2}$ (e) $\sqrt[3]{(x + y)^2}$ (g) $\dfrac{1}{\sqrt[3]{a + b^2}}$

(b) $\sqrt[8]{r}$ (d) $\sqrt[n]{x^3}$ (f) $\sqrt[4]{\dfrac{a^2}{b}}$ (h) $\sqrt[n]{g^{4m}}$

Solution

(a) $\sqrt{x} = x^{1/2}$ (c) $\sqrt[3]{y^2} = y^{2/3}$ (e) $\sqrt[3]{(x + y)^2} = (x + y)^{2/3}$ (g) $\dfrac{1}{\sqrt[3]{a + b^2}} = \dfrac{1}{(a + b^2)^{1/3}}$

(b) $\sqrt[8]{r} = r^{1/8}$ (d) $\sqrt[n]{x^3} = x^{3/n}$ (f) $\sqrt[4]{\dfrac{a^2}{b}} = \left(\dfrac{a^2}{b}\right)^{1/4} = \dfrac{a^{1/2}}{b^{1/4}}$ (h) $\sqrt[n]{g^{4m}} = g^{4m/n}$

SIMPLIFYING RADICALS

10.4 Write in simplest form:

(a) $\sqrt{8x^4y^2}$ (d) $\sqrt[3]{-2^8}$ (g) $\sqrt{50a^5b^3}$ (j) $\sqrt[3]{ay^5b^4}$ (m) $a\sqrt{4ab^3}$

(b) $\sqrt[4]{81x^8y^4}$ (e) $\sqrt[4]{32y^2}$ (h) $\sqrt[3]{a^7}$ (k) $\sqrt[6]{9a^8b^6}$ (n) $\sqrt[3]{256x^4}$

(c) $\sqrt[3]{27a^4y^2}$ (f) $5\sqrt[3]{-24x^5y^2}$ (i) $\sqrt[4]{64a^2b^2}$ (l) $8\sqrt[3]{\dfrac{a^2}{8}}$

Solution

(a) $\sqrt{8x^4y^2} = \sqrt{2(2)^2(x^2)^2y^2} = 2x^2y\sqrt{2}$

(b) $\sqrt[4]{81x^8y^4} = \sqrt[4]{(3)^4(x^2)^4y^4} = 3x^2y$

(c) $\sqrt[3]{27a^4y^2} = \sqrt[3]{(3)^3a(a^3)y^2} = 3a\sqrt[3]{ay^2}$

(d) $\sqrt[3]{-2^8} = \sqrt[3]{(-1)(2)^6(2)^2} = \sqrt[3]{(-1)(2^2)^3(2^2)} = -2^2\sqrt[3]{2^2} = -4\sqrt[3]{4}$

(e) $\sqrt[4]{32y^2} = \sqrt[4]{16(2)y^2} = \sqrt[4]{(2)^4(2)y^2} = 2\sqrt[4]{2y^2}$

(f) $5\sqrt[3]{-24x^5y^2} = 5\sqrt[3]{(-8)(3)x^3x^2y^2} = 5\sqrt[3]{(-2)^3(3)x^3x^2y^2} = -10x\sqrt[3]{3x^2y^2}$

(g) $\sqrt{50a^5b^3} = \sqrt{(25)(2)(a^4)(a)(b^2)b} = \sqrt{(5)^2(2)(a^2)^2(a)(b^2)b} = 5a^2b\sqrt{2ab}$

(h) $\sqrt[3]{a^7} = \sqrt[3]{(a^6)a} = \sqrt[3]{(a^2)^3a} = a^2\sqrt[3]{a}$

(i) $\sqrt[4]{64a^2b^2} = (8^2a^2b^2)^{1/4} = (8ab)^{2/4} = (8ab)^{1/2} = \sqrt{8ab} = 2\sqrt{2ab}$

(j) $\sqrt[3]{ay^5b^4} = \sqrt[3]{ay^3y^2b^3b} = by\sqrt[3]{aby^2}$

(k) $\sqrt[6]{9a^8b^6} = \sqrt[6]{3^2a^8b^6} = (3^2a^8b^6)^{1/6} = 3^{2/6}a^{8/6}b^{6/6} = 3^{1/3}a^{4/3}b = b(3a^4)^{1/3} = b\sqrt[3]{3a^4} = ab\sqrt[3]{3a}$

(l) $8\sqrt[3]{\dfrac{a^2}{8}} = 8\dfrac{\sqrt[3]{a^2}}{\sqrt[3]{8}} = 8\dfrac{\sqrt[3]{a^2}}{2} = 4\sqrt[3]{a^2}$

(m) $a\sqrt{4ab^3} = a\sqrt{2^2ab^2b} = 2ab\sqrt{ab}$

(n) $\sqrt[3]{256x^4} = \sqrt[3]{64(4)x^3x} = \sqrt[3]{4^3(4)x^3x} = 4x\sqrt[3]{4x}$

10.5   Simplify

(a) $\dfrac{10}{\sqrt{5}}$ (e) $\sqrt[3]{\dfrac{-8}{125}}$ (i) $\sqrt{8x^{-1}}$ (m) $\dfrac{3xy}{\sqrt[4]{x}}$

(b) $\dfrac{5}{\sqrt[3]{4}}$ (f) $\sqrt[3]{\dfrac{-27}{64}}$ (j) $\dfrac{2x}{\sqrt[3]{x^2}}$ (n) $\dfrac{3a}{2b}\sqrt{\dfrac{18ab^2}{2a^3}}$

(c) $\dfrac{8}{\sqrt{2}}$ (g) $\sqrt{\dfrac{x^6}{y^8}}$ (k) $\dfrac{4x^2y}{\sqrt[5]{-x}}$ (o) $\dfrac{3x}{\sqrt{3x}}$

(d) $\dfrac{8x}{\sqrt[3]{3}}$ (h) $\sqrt[3]{\dfrac{a^4}{-7}}$ (l) $\sqrt{\dfrac{x-y}{x+y}}$ (p) $\sqrt{\dfrac{7x^9}{2x^5y^3}}$

Solution

(a) $\dfrac{10}{\sqrt{5}} = \dfrac{10}{\sqrt{5}} \times \dfrac{\sqrt{5}}{\sqrt{5}} = \dfrac{10\sqrt{5}}{5} = 2\sqrt{5}$

(b) $\dfrac{5}{\sqrt[3]{4}} = \dfrac{5}{\sqrt[3]{2^2}} \times \dfrac{\sqrt[3]{2}}{\sqrt[3]{2}} = \dfrac{5\sqrt[3]{2}}{\sqrt[3]{2^3}} = \dfrac{5\sqrt[3]{2}}{2}$

(c) $\dfrac{8}{\sqrt{2}} = \dfrac{8}{\sqrt{2}} \times \dfrac{\sqrt{2}}{\sqrt{2}} = \dfrac{8\sqrt{2}}{2} = 4\sqrt{2}$

(d) $\dfrac{8x}{\sqrt[3]{3}} = \dfrac{8x}{\sqrt[3]{3}} \times \dfrac{\sqrt[3]{3^2}}{\sqrt[3]{3^2}} = \dfrac{8x\sqrt[3]{3^2}}{\sqrt[3]{3^3}} = \dfrac{8x\sqrt[3]{9}}{3}$

(e) $\sqrt[3]{\dfrac{-8}{125}} = \sqrt[3]{\dfrac{(-2)^3}{5^3}} = -\dfrac{2}{5}$

(f) $\sqrt[3]{\dfrac{-27}{64}} = \sqrt[3]{\dfrac{(-3)^3}{4^3}} = -\dfrac{3}{4}$

(g) $\sqrt{\dfrac{x^6}{y^8}} = \sqrt{\dfrac{(x^3)^2}{(y^4)^2}} = \dfrac{x^3}{y^4}$

(h) $\sqrt[3]{\dfrac{a^4}{-7}} = \sqrt[3]{\dfrac{(-1)a^4}{7}} = -\sqrt[3]{\dfrac{a^4}{7} \times \dfrac{7^2}{7^2}} = -\dfrac{\sqrt[3]{49a^4}}{\sqrt[3]{7^3}} = -\dfrac{a}{7}\sqrt[3]{49a}$

(i) $\sqrt{8x^{-1}} = \sqrt{\dfrac{8}{x}} = \sqrt{\dfrac{8}{x} \cdot \dfrac{x}{x}} = \sqrt{\dfrac{8x}{x^2}} = \dfrac{\sqrt{2^2(2)x}}{x} = \dfrac{2\sqrt{2x}}{x}$

(j) $\dfrac{2x}{\sqrt[3]{x^2}} = \dfrac{2x}{\sqrt[3]{x^2}} \cdot \dfrac{\sqrt[3]{x}}{\sqrt[3]{x}} = \dfrac{2x\sqrt[3]{x}}{\sqrt[3]{x^3}} = \dfrac{2x\sqrt[3]{x}}{x} = 2\sqrt[3]{x}$

(k) $\dfrac{4x^2y}{\sqrt[5]{-x}} = \dfrac{4x^2y}{-\sqrt[5]{x}} = -\dfrac{4x^2y}{\sqrt[5]{x}} \cdot \dfrac{\sqrt[5]{x^4}}{\sqrt[5]{x^4}} = -\dfrac{4x^2y\sqrt[5]{x^4}}{\sqrt[5]{x^5}} = -\dfrac{4x^2y\sqrt[5]{x^4}}{x} = -4xy\sqrt[5]{x^4}$

(l) $\sqrt{\dfrac{x-y}{x+y}} = \sqrt{\dfrac{x-y}{x+y} \cdot \dfrac{x+y}{x+y}} = \dfrac{\sqrt{(x-y)(x+y)}}{\sqrt{(x+y)^2}} = \dfrac{\sqrt{x^2-y^2}}{x+y}$

(m) $\dfrac{3xy}{\sqrt[4]{x}} = \dfrac{3xy}{\sqrt[4]{x}} \times \dfrac{\sqrt[4]{x^3}}{\sqrt[4]{x^3}} = \dfrac{3xy\sqrt[4]{x^3}}{\sqrt[4]{x^4}} = \dfrac{3xy\sqrt[4]{x^3}}{x} = 3y\sqrt[4]{x^3}$

(n) $\dfrac{3a}{2b}\sqrt{\dfrac{18ab^2}{2a^3}} = \dfrac{3a}{2b}\sqrt{\dfrac{9b^2}{a^2}} = \dfrac{3a}{2b} \times \dfrac{3b}{a} = \dfrac{9ab}{2ab} = \dfrac{9}{2}$

(o) $\dfrac{3x}{\sqrt{3x}} = \dfrac{3x}{\sqrt{3x}} \times \dfrac{\sqrt{3x}}{\sqrt{3x}} = \sqrt{3x}$

(p) $\sqrt{\dfrac{7x^9}{2x^5y^3}} = \sqrt{\dfrac{7x^4}{2y^3}} = \dfrac{x^2}{y}\sqrt{\dfrac{7}{2y}} = \dfrac{x^2}{y}\sqrt{\dfrac{7}{2y} \times \dfrac{2y}{2y}} = \dfrac{x^2}{y}\sqrt{\dfrac{14y}{4y^2}} = \dfrac{x^2\sqrt{14y}}{2y^2}$

ADDITION AND SUBTRACTION OF RADICALS

10.6   Combine and simplify.

(a) $\sqrt{3} - 7\sqrt{2} + \sqrt{3} - \sqrt{2}$ (d) $\sqrt{\dfrac{9}{8}} + \sqrt{\dfrac{25}{18}}$ (g) $\sqrt{8a} - \sqrt[4]{\dfrac{a^2}{4}} + \sqrt[6]{8a^3}$

(b) $\sqrt[3]{16} + \sqrt[3]{-54} - \sqrt[3]{2}$ (e) $2\sqrt{50} + \sqrt{32}$ (h) $\dfrac{4\sqrt[5]{x}}{3} - \dfrac{5\sqrt[5]{x}}{2} + \dfrac{1}{4}\sqrt[5]{x}$

(c) $\sqrt{54} - 3\sqrt{24}$ (f) $\sqrt{45} - \sqrt{\dfrac{5}{4}}$ (i) $\sqrt{\dfrac{2x}{y}} - 4\sqrt{\dfrac{x^3y}{8}} + 3\sqrt{\dfrac{y}{2x^3}}$

Solution

(a) $\sqrt{3} - 7\sqrt{2} + \sqrt{3} - \sqrt{2} = 2\sqrt{3} - 8\sqrt{2}$

(b) $\sqrt[3]{16} + \sqrt[3]{-54} - \sqrt[3]{2} = 2\sqrt[3]{2} - 3\sqrt[3]{2} - \sqrt[3]{2} = -2\sqrt[3]{2}$

(c) $\sqrt{54} - 3\sqrt{24} = 3\sqrt{6} - 6\sqrt{6} = -3\sqrt{6}$

(d) $\sqrt{\dfrac{9}{8}} + \sqrt{\dfrac{25}{18}} = \dfrac{3}{2\sqrt{2}} + \dfrac{5}{3\sqrt{2}} = \dfrac{3}{2\sqrt{2}} \times \dfrac{\sqrt{2}}{\sqrt{2}} + \dfrac{5}{3\sqrt{2}} \times \dfrac{\sqrt{2}}{\sqrt{2}} = \dfrac{3\sqrt{2}}{4} + \dfrac{5\sqrt{2}}{6} = \dfrac{9\sqrt{2}}{12} + \dfrac{10\sqrt{2}}{12} = \dfrac{19\sqrt{2}}{12}$

(e) $2\sqrt{50} + \sqrt{32} = 2\sqrt{5^2(2)} + \sqrt{4^2(2)} = 10\sqrt{2} + 4\sqrt{2} = 14\sqrt{2}$

(f) $\sqrt{45} - \sqrt{\dfrac{5}{4}} = 3\sqrt{5} - \dfrac{\sqrt{5}}{2} = \dfrac{5}{2}\sqrt{5}$

(g) $\sqrt{8a} - \sqrt[4]{\dfrac{a^2}{4}} + \sqrt[6]{8a^3} = 2\sqrt{2a} - \sqrt{\dfrac{a}{2}} + \sqrt{2a} = 2\sqrt{2a} - \dfrac{1}{2}\sqrt{2a} + \sqrt{2a} = \dfrac{5}{2}\sqrt{2a}$

(h) $\dfrac{4\sqrt[5]{x}}{3} - \dfrac{5\sqrt[5]{x}}{2} + \dfrac{1}{4}\sqrt[5]{x} = \left(\dfrac{4}{3} - \dfrac{5}{2} + \dfrac{1}{4}\right)\sqrt[5]{x} = -\dfrac{11\sqrt[5]{x}}{12}$

(i) $\sqrt{\dfrac{2x}{y}} - 4\sqrt{\dfrac{x^3y}{8}} + 3\sqrt{\dfrac{y}{2x^3}} = \dfrac{1}{y}\sqrt{2xy} - x\sqrt{2xy} + \dfrac{3}{2x^2}\sqrt{2xy} = \left(\dfrac{1}{y} - x + \dfrac{3}{2x^2}\right)\sqrt{2xy}$

$$= \dfrac{2x^2 - 2x^3y + 3y}{2x^2y}\sqrt{2xy}$$

MULTIPLICATION OF RADICALS

10.7   Multiply and simplify.

(a) $\sqrt{3}\sqrt{12}$ (c) $\sqrt{2x^3}\sqrt{4x^3}$ (e) $(\sqrt{3a} + 2)(\sqrt{3a} - 2)$

(b) $3\sqrt{3}(\sqrt{3} - \sqrt{6})$ (d) $(\sqrt{x} - \sqrt{y})^2$ (f) $\sqrt[5]{4ab^2}\sqrt[5]{8a^2b}$

(g) $\sqrt[4]{2}\,\sqrt[4]{8}$ (i) $\sqrt{2a^2b}\,\sqrt{8ab}$ (k) $(5\sqrt{ax^3}\,\sqrt{bx})(4\sqrt{cx})$

(h) $\sqrt[3]{3}\,\sqrt[3]{18}$ (j) $\sqrt{x}\,\sqrt[4]{y}$ (l) $(\sqrt{a}-\sqrt{2})(2\sqrt{a}+\sqrt{2})$

Solution

(a) $\sqrt{3}\sqrt{12} = \sqrt{3(12)} = \sqrt{36} = 6$

(b) $3\sqrt{3}(\sqrt{3}-\sqrt{6}) = 3\sqrt{3}\sqrt{3} - 3\sqrt{3}\sqrt{6} = 3(3) - 3\sqrt{18} = 9 - 3\sqrt{9(2)} = 9 - 9\sqrt{2}$

(c) $\sqrt{2x^3}\sqrt{4x^3} = \sqrt{2x^3(4x^3)} = 2\sqrt{2(x^3)^2} = 2x^3\sqrt{2}$

(d) $(\sqrt{x}-\sqrt{y})^2 = \sqrt{x}\sqrt{x} - \sqrt{x}\sqrt{y} - \sqrt{x}\sqrt{y} + \sqrt{y}\sqrt{y} = x - 2\sqrt{xy} + y$

(e) $(\sqrt{3a}+2)(\sqrt{3a}-2) = \sqrt{3a}\sqrt{3a} + 2\sqrt{3a} - 2\sqrt{3a} - 4 = 3a - 4$

(f) $\sqrt[5]{4ab^2}\,\sqrt[5]{8a^2b} = \sqrt[5]{32a^3b^3} = 2\sqrt[5]{a^3b^3}$

(g) $\sqrt[4]{2}\,\sqrt[4]{8} = \sqrt[4]{16} = 2$

(h) $\sqrt[3]{3}\,\sqrt[3]{18} = \sqrt[3]{54} = \sqrt[3]{27(2)} = 3\sqrt[3]{2}$

(i) $\sqrt{2a^2b}\,\sqrt{8ab} = \sqrt{16a^3b^2} = 4ab\sqrt{a}$

(j) $\sqrt{x}\,\sqrt[4]{y} = x^{1/2}y^{1/4} = x^{2/4}y^{1/4} = (x^2y)^{1/4} = \sqrt[4]{x^2y}$

(k) $(5\sqrt{ax^3}\,\sqrt{bx})(4\sqrt{cx}) = 20\sqrt{(ax^3)(bx)(cx)} = 20\sqrt{abcx^5} = 20\sqrt{abc(x^2)^2x} = 20x^2\sqrt{abcx}$

(l) $(\sqrt{a}-\sqrt{2})(2\sqrt{a}+\sqrt{2}) = 2\sqrt{a}\sqrt{a} + \sqrt{2}\cdot2\sqrt{a} - 2\sqrt{2}\sqrt{a} - 2 = 2a - \sqrt{2a} - 2$

DIVISION OF RADICALS

10.8 Divide and simplify.

(a) $16 \div 2\sqrt{2}$ (d) $(2\sqrt[3]{4} + 3\sqrt[3]{3} + 4\sqrt[3]{6}) \div \sqrt[3]{6}$ (g) $\dfrac{8\sqrt[4]{32x^2}}{2\sqrt[4]{2x^3}}$

(b) $(4\sqrt{2} - 2\sqrt{3} + 2\sqrt{6}) \div 2\sqrt{6}$ (e) $\dfrac{\sqrt[6]{16a}}{\sqrt[3]{4a^2}}$ (h) $\dfrac{\sqrt{2xy}}{\sqrt{4xy^2}}$

(c) $(3\sqrt[3]{3} - 2\sqrt[3]{2}) \div \sqrt[3]{2}$ (f) $\dfrac{5\sqrt[3]{2x^3}}{25\sqrt[3]{4x^5}}$

Solution

(a) $16 \div 2\sqrt{2} = \dfrac{16}{2\sqrt{2}} = \dfrac{8}{\sqrt{2}} \times \dfrac{\sqrt{2}}{\sqrt{2}} = \dfrac{8\sqrt{2}}{2} = 4\sqrt{2}$

(b) $\dfrac{4\sqrt{2} - 2\sqrt{3} + 2\sqrt{6}}{2\sqrt{6}} = 2\dfrac{\sqrt{2}}{\sqrt{6}} - \dfrac{\sqrt{3}}{\sqrt{6}} + \dfrac{\sqrt{6}}{\sqrt{6}} = 2\sqrt{\dfrac{2}{6}} - \sqrt{\dfrac{3}{6}} + 1 = 2\sqrt{\dfrac{1}{3}} - \sqrt{\dfrac{1}{2}} + 1 = \dfrac{2\sqrt{3}}{3} - \dfrac{\sqrt{2}}{2} + 1$

(c) $\dfrac{3\sqrt[3]{3} - 2\sqrt[3]{2}}{\sqrt[3]{2}} = 3\sqrt[3]{\dfrac{3}{2}} - 2 = 3\sqrt[3]{\dfrac{3}{2} \times \dfrac{4}{4}} - 2 = 3\dfrac{\sqrt[3]{12}}{\sqrt[3]{8}} - 2 = \dfrac{3\sqrt[3]{12}}{2} - 2$

(d) $\dfrac{2\sqrt[3]{4} + 3\sqrt[3]{3} + 4\sqrt[3]{6}}{\sqrt[3]{6}} = \dfrac{2\sqrt[3]{4}}{\sqrt[3]{6}} + \dfrac{3\sqrt[3]{3}}{\sqrt[3]{6}} + \dfrac{4\sqrt[3]{6}}{\sqrt[3]{6}} = 2\sqrt[3]{\dfrac{4}{6}} + 3\sqrt[3]{\dfrac{3}{6}} + 4 = 2\sqrt[3]{\dfrac{2}{3}} + 3\sqrt[3]{\dfrac{1}{2}} + 4$

$= 2\sqrt[3]{\dfrac{2}{3} \times \dfrac{3^2}{3^2}} + 3\sqrt[3]{\dfrac{1}{2} \times \dfrac{4}{4}} + 4 = \dfrac{2}{3}\sqrt[3]{18} + \dfrac{3}{2}\sqrt[3]{4} + 4$

$(e)\ \dfrac{\sqrt[6]{16a}}{\sqrt[3]{4a^2}} = \dfrac{\sqrt[6]{16a}}{\sqrt[3]{4a^2}} \times \dfrac{\sqrt[3]{2a}}{\sqrt[3]{2a}} = \dfrac{(16a)^{1/6}(2a)^{1/3}}{\sqrt[3]{8a^3}} = \dfrac{(16a)^{1/6}(2a)^{2/6}}{2a} = \dfrac{[16a(2a)^2]^{1/6}}{2a} = \dfrac{(64a^3)^{1/6}}{2a} = \dfrac{2a^{3/6}}{2a} = \dfrac{\sqrt{a}}{a}$

$(f)\ \dfrac{5\sqrt[3]{2x^3}}{25\sqrt[3]{4x^5}} = \dfrac{x\sqrt[3]{2}}{5x\sqrt[3]{4x^2}} = \dfrac{1}{5}\sqrt[3]{\dfrac{2}{4x^2}} = \dfrac{1}{5}\sqrt[3]{\dfrac{1}{2x^2}\cdot\dfrac{4x}{4x}} = \dfrac{1}{5}\sqrt[3]{\dfrac{4x}{8x^3}} = \dfrac{\sqrt[3]{4x}}{10x}$

$(g)\ \dfrac{8\sqrt[4]{32x^2}}{2\sqrt[4]{2x^3}} = 4\sqrt[4]{\dfrac{32x^2}{2x^3}} = 4\sqrt[4]{\dfrac{16x^2}{x^3}\times\dfrac{x}{x}} = 4\sqrt[4]{\dfrac{16x^3}{x^4}} = \dfrac{8}{x}\sqrt[4]{x^3}$

$(h)\ \dfrac{\sqrt{2xy}}{\sqrt{4xy^2}} = \sqrt{\dfrac{2xy}{4xy^2}} = \sqrt{\dfrac{1}{2y}} = \sqrt{\dfrac{1}{2y}\times\dfrac{2y}{2y}} = \dfrac{\sqrt{2y}}{2y}$

10.9   Divide and simplify.

$(a)\ \dfrac{2}{\sqrt{3}+1}$

$(b)\ \dfrac{\sqrt{3}}{\sqrt{3}-\sqrt{2}}$

$(c)\ \dfrac{\sqrt{2}}{3\sqrt{2}-2\sqrt{3}}$

$(d)\ \dfrac{1}{1+\sqrt{2}}$

$(e)\ \dfrac{3\sqrt{6}-\sqrt{18}}{2\sqrt{6}}$

$(f)\ \dfrac{3\sqrt{3}-\sqrt{2}}{2\sqrt{3}+\sqrt{2}}$

$(g)\ x\div(1+\sqrt{x})$

$(h)\ \dfrac{2\sqrt{a}+3\sqrt{b}}{\sqrt{a}-\sqrt{b}}$

$(i)\ (\sqrt{a}-b)\div(a-\sqrt{b})$

$(j)\ \dfrac{\sqrt{x}-\sqrt{y}}{\sqrt{x}+\sqrt{y}}$

Solution

$(a)\ \dfrac{2}{\sqrt{3}+1} = \dfrac{2}{\sqrt{3}+1} \times \dfrac{\sqrt{3}-1}{\sqrt{3}-1} = \dfrac{2\sqrt{3}-2}{3+\sqrt{3}-\sqrt{3}-1} = \dfrac{2\sqrt{3}-2}{2} = \sqrt{3}-1$

$(b)\ \dfrac{\sqrt{3}}{\sqrt{3}-\sqrt{2}} = \dfrac{\sqrt{3}}{\sqrt{3}-\sqrt{2}} \times \dfrac{\sqrt{3}+\sqrt{2}}{\sqrt{3}+\sqrt{2}} = \dfrac{3+\sqrt{6}}{3+\sqrt{6}-\sqrt{6}-2} = \dfrac{3+\sqrt{6}}{1} = 3+\sqrt{6}$

$(c)\ \dfrac{\sqrt{2}}{3\sqrt{2}-2\sqrt{3}} = \dfrac{\sqrt{2}}{3\sqrt{2}-2\sqrt{3}} \times \dfrac{3\sqrt{2}+2\sqrt{3}}{3\sqrt{2}+2\sqrt{3}} = \dfrac{3(2)+2\sqrt{6}}{9(2)+6\sqrt{6}-6\sqrt{6}-4(3)} = \dfrac{6+2\sqrt{6}}{18-12} = 1+\dfrac{\sqrt{6}}{3}$

$(d)\ \dfrac{1}{1+\sqrt{2}} = \dfrac{1}{1+\sqrt{2}} \times \dfrac{1-\sqrt{2}}{1-\sqrt{2}} = \dfrac{1-\sqrt{2}}{1+\sqrt{2}-\sqrt{2}-2} = \dfrac{1-\sqrt{2}}{-1} = -1+\sqrt{2} = \sqrt{2}-1$

$(e)\ \dfrac{3\sqrt{6}-\sqrt{18}}{2\sqrt{6}} = \dfrac{3\sqrt{6}-\sqrt{18}}{2\sqrt{6}} \times \dfrac{\sqrt{6}}{\sqrt{6}} = \dfrac{3(6)-\sqrt{18}\sqrt{6}}{2(6)} = \dfrac{18-\sqrt{108}}{12} = \dfrac{18-6\sqrt{3}}{12} = \dfrac{3}{2}-\dfrac{\sqrt{3}}{2}$

$(f)\ \dfrac{3\sqrt{3}-\sqrt{2}}{2\sqrt{3}+\sqrt{2}} = \dfrac{3\sqrt{3}-\sqrt{2}}{2\sqrt{3}+\sqrt{2}} \times \dfrac{2\sqrt{3}-\sqrt{2}}{2\sqrt{3}-\sqrt{2}} = \dfrac{6(3)-2\sqrt{6}-3\sqrt{6}+2}{4(3)+2\sqrt{6}-2\sqrt{6}-2} = \dfrac{20-5\sqrt{6}}{10} = 2-\dfrac{\sqrt{6}}{2}$

$(g)\ \dfrac{x}{1+\sqrt{x}} = \dfrac{x}{1+\sqrt{x}} \times \dfrac{1-\sqrt{x}}{1-\sqrt{x}} = \dfrac{x-x\sqrt{x}}{1+\sqrt{x}-\sqrt{x}-x} = \dfrac{x-x\sqrt{x}}{1-x}$

$(h)\ \dfrac{2\sqrt{a}+3\sqrt{b}}{\sqrt{a}-\sqrt{b}} = \dfrac{2\sqrt{a}+3\sqrt{b}}{\sqrt{a}-\sqrt{b}} \times \dfrac{\sqrt{a}+\sqrt{b}}{\sqrt{a}+\sqrt{b}} = \dfrac{2a+2\sqrt{ab}+3\sqrt{ab}+3b}{a+\sqrt{ab}-\sqrt{ab}-b} = \dfrac{2a+5\sqrt{ab}+3b}{a-b}$

$(i)\ \dfrac{\sqrt{a}-b}{a-\sqrt{b}} = \dfrac{\sqrt{a}-b}{a-\sqrt{b}} \times \dfrac{a+\sqrt{b}}{a+\sqrt{b}} = \dfrac{a\sqrt{a}+\sqrt{ab}-ab-b\sqrt{b}}{a^2+a\sqrt{b}-a\sqrt{b}-b} = \dfrac{a\sqrt{a}+\sqrt{ab}-ab-b\sqrt{b}}{a^2-b}$

$(j)\ \dfrac{\sqrt{x}-\sqrt{y}}{\sqrt{x}+\sqrt{y}} = \dfrac{\sqrt{x}-\sqrt{y}}{\sqrt{x}+\sqrt{y}} \times \dfrac{\sqrt{x}-\sqrt{y}}{\sqrt{x}-\sqrt{y}} = \dfrac{x-\sqrt{xy}-\sqrt{xy}+y}{x+\sqrt{xy}-\sqrt{xy}-y} = \dfrac{x-2\sqrt{xy}+y}{x-y}$

RADICAL EQUATIONS

10.10   Solve and check.

(a) $\sqrt{x} = 6$ (d) $\sqrt{1 + 3b} = 5$ (g) $\sqrt{1 + a} = \sqrt{2a - 7}$

(b) $\sqrt{8 + 7a} = 6$ (e) $\sqrt{3a - 2} - 5 = 0$ (h) $3 - \sqrt{2y} + \sqrt{2y - 9} = 0$

(c) $\sqrt[5]{x^2} = 9$ (f) $7 - \sqrt{5w - 6} = 0$ (i) $\sqrt[4]{3z + 1} = 2$

Solution

(a) $\sqrt{x} = 6$

Squaring both sides,
$$x = 36$$

Check: $\sqrt{36} = 6$ checks

(b) $\sqrt{8 + 7a} = 6$

Squaring both sides,
$$8 + 7a = 36$$

Transposing, $$7a = 28$$
$$a = 4$$

Check: $\sqrt{8 + 7(4)} = \sqrt{36} = 6$ checks

(c) $\sqrt[5]{x^2} = 9$

By Eq. 19,
$$\sqrt[5]{x^2} = (\sqrt[5]{x})^2 = 9$$

Taking the square root of both sides,
$$\sqrt[5]{x} = \pm 3$$

Raising to the fifth power,
$$x = 3^5 = 243$$

and $$(-3)^5 = -243$$

Check: $\sqrt[5]{(243)^2} = \sqrt[5]{59\,049} = 9$ checks

$\sqrt[5]{(-243)^2} = \sqrt[5]{59\,049} = 9$ checks

(d) $\sqrt{1 + 3b} = 5$

Squaring both sides,
$$1 + 3b = 25$$

Transposing, $$3b = 24$$
$$b = 8$$

Check: $\sqrt{1 + 3(8)} = \sqrt{25} = 5$ checks

(e) $\sqrt{3a - 2} - 5 = 0$

Transposing and squaring both sides,
$$3a - 2 = 25$$
$$3a = 27$$
$$a = 9$$

Check: $\sqrt{3(9) - 2} - 5 = \sqrt{25} - 5 = 0$ checks

(f) $7 - \sqrt{5w - 6} = 0$

Transposing and squaring,

$$5w - 6 = 49$$
$$5w = 55$$
$$w = 11$$

Check: $7 - \sqrt{5(11) - 6} = 7 - \sqrt{49} = 0$ checks

(g) $\sqrt{1 + a} = \sqrt{2a - 7}$

Squaring both sides,

$$1 + a = 2a - 7$$
$$a = 8$$

Check:

$$\sqrt{1 + 8} \overset{?}{=} \sqrt{2(8) - 7}$$
$$\sqrt{9} \overset{?}{=} \sqrt{16 - 7}$$
$$3 = 3 \quad \text{checks}$$

(h) $3 - \sqrt{2y} + \sqrt{2y - 9} = 0$

Transposing, $\sqrt{2y - 9} = \sqrt{2y} - 3$

Squaring both sides,

$$2y - 9 = 2y - 6\sqrt{2y} + 9$$

Collecting terms, $6\sqrt{2y} = 18$

$$\sqrt{2y} = 3$$

Squaring both sides, $2y = 9$

$$y = \frac{9}{2}$$

Check:

$$3 - \sqrt{2\left(\frac{9}{2}\right)} + \sqrt{2\left(\frac{9}{2}\right) - 9} = 3 - \sqrt{9} + \sqrt{0} = 3 - 3 = 0 \quad \text{checks}$$

(i) $\sqrt[4]{3z + 1} = 2$

Raising both sides to the fourth power,

$$3z + 1 = 2^4 = 16$$
$$3z = 15$$
$$z = 5$$

Check: $\sqrt[4]{3(5) + 1} = \sqrt[4]{16} = 2$ checks

IMAGINARY NUMBERS

10.11 Write the following as imaginary numbers:

(a) $\sqrt{-9}$ (b) $2\sqrt{-25}$ (c) $\sqrt{\dfrac{-25}{16}}$ (d) $\sqrt{-\dfrac{1}{3}}$

Solution

(a) $\sqrt{-9} = \sqrt{9(-1)} = \sqrt{9}\sqrt{-1} = 3i$

(b) $2\sqrt{-25} = 2\sqrt{25(-1)} = 2\sqrt{25}\sqrt{-1} = 2(5)i = 10i$

(c) $\sqrt{\dfrac{-25}{16}} = \sqrt{\dfrac{25(-1)}{16}} = \dfrac{\sqrt{25}\sqrt{-1}}{\sqrt{16}} = \dfrac{5}{4}i$

(d) $\sqrt{-\dfrac{1}{3}} = \sqrt{\dfrac{1}{3}(-1)} = \sqrt{\dfrac{1}{3}}\sqrt{-1} = \sqrt{\dfrac{1}{3}}i = \dfrac{\sqrt{3}}{3}i$

10.12 Evaluate

(a) i^7 (b) i^{10} (c) i^{16} (d) i^{25}

Solution

(a) $i^7 = i^3 i^4 = (-i)(1) = -i$ (c) $i^{16} = (i^4)^4 = 1^4 = 1$

(b) $i^{10} = (i^5)^2 = (i^4 i)^2 = [(1)(i)]^2 = i^2 = -1$ (d) $i^{25} = (i^5)^5 = i^5 = i$

ADDITION AND SUBTRACTION OF COMPLEX NUMBERS

10.13 Combine and simplify.

(a) $(3 - 2i) + (5 + 3i)$ (c) $\left(\dfrac{1}{3} - \dfrac{1}{6}i\right) - \left(\dfrac{5}{6} + \dfrac{2}{3}i\right)$

(b) $(6 - 3i) - (7 + i)$ (d) $(m + ni) + (n + mi)$

Solution

(a) $(3 - 2i) + (5 + 3i) = 3 - 2i + 5 + 3i = 8 + i$

(b) $(6 - 3i) - (7 + i) = 6 - 3i - 7 - i = -1 - 4i$

(c) $\left(\dfrac{1}{3} - \dfrac{1}{6}i\right) - \left(\dfrac{5}{6} + \dfrac{2}{3}i\right) = \dfrac{1}{3} - \dfrac{1}{6}i - \dfrac{5}{6} - \dfrac{2}{3}i = \dfrac{2}{6} - \dfrac{5}{6} - \dfrac{1}{6}i - \dfrac{4}{6}i = -\dfrac{3}{6} - \dfrac{5}{6}i = -\dfrac{1}{2} - \dfrac{5}{6}i$

(d) $(m + ni) + (n + mi) = m + ni + n + mi = (m + n) + (n + m)i$

MULTIPLICATION OF COMPLEX NUMBERS

10.14 Multiply and simplify.

(a) $\sqrt{-5}\sqrt{-20}$ (c) $(3i)^2$ (e) $(6 - 3i)(3 + i)$

(b) $(2i)(7i)$ (d) $4i(i - 3)$ (f) $(2 - 3i)^2$

Solution

(a) $\sqrt{-5}\sqrt{-20} = (\sqrt{5}i)(\sqrt{20}i) = \sqrt{5}\sqrt{20}i^2 = \sqrt{100}(-1) = -10$

(b) $(2i)(7i) = 14i^2 = 14(-1) = -14$

(c) $(3i)^2 = 3^2 i^2 = 9i^2 = 9(-1) = -9$

(d) $4i(i - 3) = 4i^2 - 12i = 4(-1) - 12i = -4 - 12i$

(e) $(6 - 3i)(3 + i) = 18 + 6i - 9i - 3i^2 = 18 - 3i - 3(-1) = 21 - 3i$

(f) $(2 - 3i)^2 = 4 - 6i - 6i + 9i^2 = 4 - 12i + 9(-1) = -5 - 12i$

DIVISION OF COMPLEX NUMBERS

10.15   Divide.

$$(a)\ \frac{2-3i}{1+2i} \qquad (b)\ \frac{1+i}{2-i} \qquad (c)\ \frac{-5-4i}{-1+i}$$

Solution

$(a)\ \dfrac{2-3i}{1+2i} = \dfrac{2-3i}{1+2i} \times \dfrac{1-2i}{1-2i} = \dfrac{2-4i-3i+6i^2}{1+2i-2i-4i^2} = \dfrac{2-7i+6(-1)}{1-4(-1)} = \dfrac{-4-7i}{5} = -\dfrac{4}{5}-\dfrac{7}{5}i$

$(b)\ \dfrac{1+i}{2-i} = \dfrac{1+i}{2-i} \times \dfrac{2+i}{2+i} = \dfrac{2+i+2i+i^2}{4+2i-2i-i^2} = \dfrac{2+3i+(-1)}{4-(-1)} = \dfrac{1+3i}{5} = \dfrac{1}{5}+\dfrac{3}{5}i$

$(c)\ \dfrac{-5-4i}{-1+i} = \dfrac{-5-4i}{-1+i} \times \dfrac{-1-i}{-1-i} = \dfrac{5+5i+4i+4i^2}{1-i+i-i^2} = \dfrac{5+9i+4(-1)}{1-(-1)} = \dfrac{1+9i}{2} = \dfrac{1}{2}+\dfrac{9}{2}i$

Supplementary Problems

PRINCIPAL ROOTS

10.16   Find the principal root.

$$(a)\ \sqrt{49} \qquad (b)\ \sqrt[3]{125} \qquad (c)\ \sqrt[12]{1} \qquad (d)\ \sqrt{\frac{1}{16}}$$

EXPONENTIAL AND RADICAL FORM

10.17   Express in radical form.

$$(a)\ y^{1/3} \qquad (b)\ a^{1/3}c^{4/3} \qquad (c)\ \frac{x^{1/2}}{y^{-1/2}} \qquad (d)\ \frac{1}{(y-2)^{-1/2}}$$

10.18   Express in exponential form.

$$(a)\ \sqrt[3]{x^2 y^3} \qquad (b)\ \sqrt[4]{(x-y)^2} \qquad (c)\ \sqrt[a]{b^4} \qquad (d)\ \sqrt[5]{\frac{x^3}{b}}$$

SIMPLIFYING RADICALS

10.19   Write in simplest form.

$(a)\ \sqrt{7x^2y^4} \qquad (b)\ \sqrt[3]{27y^8} \qquad (c)\ \sqrt[4]{ab^5x^{16}} \qquad (d)\ \sqrt{90ac^4} \qquad (e)\ \sqrt[3]{216x^9} \qquad (f)\ \sqrt[6]{9a^6x^{36}}$

$(g)\ b\sqrt[3]{-27x^3}$

10.20   Simplify.

$(a)\ \dfrac{12}{\sqrt{2}} \qquad (b)\ \dfrac{8}{x^2\sqrt{9x^4}} \qquad (c)\ \sqrt{\dfrac{x^4}{y^6}} \qquad (d)\ \dfrac{5a}{\sqrt{5a}} \qquad (e)\ \dfrac{2x}{y}\sqrt{\dfrac{9Nm^2}{2}} \qquad (f)\ \sqrt[3]{\dfrac{-8}{27}}$

$(g)\ \sqrt{\dfrac{a-b}{x+y}} \qquad (h)\ \sqrt{x^2y+ax^2}$

ADDITION AND SUBTRACTION OF RADICALS

10.21   Combine and simplify.

(a) $2\sqrt{12} + \sqrt{27}$ (b) $\sqrt{32} + \sqrt{18} - \sqrt{50}$ (c) $3\sqrt{\dfrac{9}{2}} - \sqrt{\dfrac{49}{2}} + \sqrt{\dfrac{25}{2}}$

(d) $\sqrt[3]{81} - 2\sqrt[3]{24} + \sqrt[3]{192}$

MULTIPLICATION OF RADICALS

10.22   Multiply and simplify.

(a) $\sqrt{2}\sqrt{6}$ (b) $2\sqrt{3}(\sqrt{2} - \sqrt{7})$ (c) $(\sqrt{8} - 2)^2$ (d) $\sqrt{x}\sqrt[3]{y}$ (e) $5\sqrt{xy^3}\sqrt{bx}$

(f) $(\sqrt{a} - 1)\sqrt[4]{9}$

DIVISION OF RADICALS

10.23   Divide and simplify.

(a) $4 \div 3\sqrt{5}$ (b) $(2\sqrt{2} + \sqrt{3}) \div \sqrt{6}$ (c) $\dfrac{3\sqrt{4a^3}}{7\sqrt{9a^3y^4}}$

10.24   Divide and simplify.

(a) $\dfrac{6}{1 + \sqrt{3}}$ (b) $\dfrac{\sqrt{5}}{\sqrt{6} + \sqrt{2}}$ (c) $\dfrac{\sqrt{x} + \sqrt{y}}{\sqrt{x}}$ (d) $2a \div (2 - \sqrt{ax^2y})$

RADICAL EQUATIONS

10.25   Solve and check.

(a) $\sqrt[3]{y} = 7$ (b) $\sqrt[3]{x - 1} = 9$ (c) $\sqrt{7x - 3} = 10$ (d) $\sqrt{9x} + \sqrt{x} = 3$

(e) $2 + \sqrt{6y} + \sqrt{6y - 9} = 11$

COMPLEX NUMBERS

10.26   Write the following as imaginary numbers.

(a) $3\sqrt{-36}$ (b) $\sqrt{-24}$ (c) $3\sqrt{-\dfrac{1}{2}}$

10.27   Evaluate.

(a) i^9 (b) i^{17} (c) i^{11}

10.28   Combine and simplify.

(a) $(3 + i) + (-2 - 3i)$ (b) $(1 - i) - (6 + 2i)$ (c) $\left(\dfrac{1}{2} - \dfrac{i}{3}\right) + \left(\dfrac{1}{3} + \dfrac{i}{2}\right)$

10.29   Multiply and simplify.

(a) $\sqrt{-3}\sqrt{-12}$ (b) $(3i)(5i)$ (c) $(3 - 2i)(4 + 3i)$ (d) $(2 - i)^2$

10.30   Divide and simplify.

(a) $\dfrac{3 - 4i}{5 - 2i}$ (b) $\dfrac{-3 + 2i}{6 - 3i}$

Answers to Supplementary Problems

10.16 (a) 7   (b) 5   (c) 1   (d) $\dfrac{1}{4}$

10.17 (a) $\sqrt[3]{y}$ (b) $\sqrt[3]{ac^4}$ (c) $\sqrt{xy}$ (d) $\sqrt{y-2}$

10.18 (a) $x^{2/3}y$ (b) $(x-y)^{1/2}$ (c) $b^{4/a}$ (a) $\left(\dfrac{x^3}{b}\right)^{1/5}$

10.19 (a) $xy^2\sqrt{7}$ (b) $3y^2\sqrt[3]{y^2}$ (c) $bx^4\sqrt[4]{ab}$ (d) $3c^2\sqrt{10a}$ (e) $6x^3$ (f) $ax^6\sqrt[3]{3}$
    (g) $-3bx$

10.20 (a) $6\sqrt{2}$ (b) $\dfrac{8}{3x^4}$ (c) $\dfrac{x^2}{y^3}$ (d) $\sqrt{5a}$ (e) $\dfrac{3mx}{y}\sqrt{2N}$ (f) $\dfrac{-2}{3}$

    (g) $\dfrac{\sqrt{ax-bx+ay-by}}{x+y}$ (h) $x\sqrt{a+y}$

10.21 (a) $7\sqrt{3}$ (b) $2\sqrt{2}$ (c) $\dfrac{7\sqrt{2}}{2}$ (d) $3\sqrt[3]{3}$

10.22 (a) $2\sqrt{3}$ (b) $2\sqrt{6}-2\sqrt{21}$ (c) $12-8\sqrt{2}$ (d) $\sqrt[6]{x^3y^2}$ (e) $5xy\sqrt{by}$
    (f) $\sqrt{3a}-\sqrt{3}$

10.23 (a) $\dfrac{4\sqrt{5}}{15}$ (b) $\dfrac{2\sqrt{3}}{3}+\dfrac{\sqrt{2}}{2}$ (c) $\dfrac{2}{7y^2}$

10.24 (a) $-3+3\sqrt{3}$ (b) $\dfrac{\sqrt{30}-\sqrt{10}}{4}$ (c) $\dfrac{x+\sqrt{xy}}{x}$ (d) $\dfrac{4a+2ax\sqrt{ay}}{4-ax^2y}$

10.25 (a) 343   (b) 730   (c) $\dfrac{103}{7}$ (d) $\dfrac{9}{16}$ (e) $\dfrac{25}{6}$

10.26 (a) $18i$ (b) $2\sqrt{6}\,i$ (c) $\dfrac{3}{2}\sqrt{2}i$

10.27 (a) i (b) i (c) $-i$

10.28 (a) $1-2i$ (b) $-5-3i$ (c) $\dfrac{5}{6}+\dfrac{1}{6}i$

10.29 (a) -6 (b) -15 (c) $18+i$ (d) $3-4i$

10.30 (a) $\dfrac{23}{29}-\dfrac{14}{29}i$ (b) $-\dfrac{8}{15}+\dfrac{1}{15}i$

Chapter 11

Systems of Equations

11.1 DEFINITIONS

(a) Systems of Equations

Example 11.1

$$\begin{vmatrix} 3x + 2y = & 5 \\ x - 3y = & -2 \end{vmatrix}$$

is a *system* of two equations in two *variables*.

Example 11.2

$$\begin{cases} x - 2y + z = 1 \\ 2x + y - z = 2 \\ x - 3y + 2z = 5 \end{cases}$$

is a system of three equations in three variables.

(b) Solutions. The *solution* to a system of two equations in two variables is a pair of numbers, x and y, which will satisfy *each* of the equations.

Example 11.3: Do the values $x = 1$ and $y = -2$ satisfy the system

$$\begin{vmatrix} 2x - y = & 4 \\ x + 2y = & -3 \end{vmatrix}$$

Solution: Substituting $(1, -2)$ into the first equation,

$$2(1) - (-2) \overset{2}{=} 4$$
$$2 + 2 = 4 \quad \text{checks}$$

and into the second equation,

$$1 + 2(-2) \overset{2}{=} -3$$
$$1 - 4 = -3 \quad \text{checks}$$

(c) Linear Systems

Example 11.4: The system

$$\begin{vmatrix} 2x - 3y = 5 \\ x + y = 2 \end{vmatrix}$$

is linear because all the variables are of *first degree*.

Example 11.5: The system

$$\begin{vmatrix} x^2 + y^2 = 9 \\ x + 2y = 3 \end{vmatrix}$$

is *not* linear, because it contains some second degree terms.

11.2 SOLVING LINEAR SYSTEMS GRAPHICALLY

To solve a pair of linear equations graphically, plot the two equations on the same coordinate axes. The coordinates of their point of intersection is the approximate solution to the pair of equations.

Example 11.6:   Solve the following system graphically:

$$\begin{vmatrix} x + y = 5 \\ x - y = 3 \end{vmatrix}$$

Solution: We graph each line as was shown in Section 6.11. Two points are needed to graph each line, although a third point will serve as a check. For the first line, we obtain two points by first letting y equal zero and solving for x, and then letting x equal zero and solving for y.

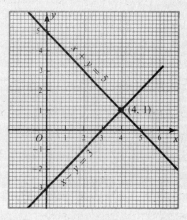

$$x = 5 \quad \text{when} \quad y = 0 \qquad \text{and} \qquad y = 5 \quad \text{when} \quad x = 0$$

For the second line,

$$x = 3 \quad \text{when} \quad y = 0 \qquad \text{and} \qquad y = -3 \quad \text{when} \quad x = 0$$

Their point of intersection is found graphically, Fig. 11-1, and is the solution to the pair of equations. It is $x = 4$, $y = 1$.

Check: Substituting (4, 1) into the first equation,

$$4 + 1 \overset{?}{=} 5 \qquad \text{checks}$$

Substituting the same values into the second equation,

$$4 - 1 \overset{?}{=} 3 \qquad \text{checks}$$

Fig. 11-1

11.3 ADDITION-SUBTRACTION METHOD

(a) Object of Method. The object of this method and of those that follow is to *eliminate* one unknown from the pair of equations. The resulting single equation with one unknown is then solved by the methods of Chapter 4.

To eliminate an unknown using the addition-subtraction method, first multiply one or both of the given equations by some factor that will make the coefficients of one unknown the same in both equations. Then add or subtract the two modified equations so that the unknown drops out.

Example 11.7:   Solve the following system by the addition-subtraction method:

$$\begin{vmatrix} x + 2y = -4 \\ 2x - y = -3 \end{vmatrix}$$

Solution: Multiply the second equation by 2 and add the two equations vertically.

$$\begin{array}{r} x + 2y = -4 \\ 4x - 2y = -6 \\ \hline 5x \qquad = -10 \end{array}$$

Dividing by 5,

$$x = -2$$

Substitute $x = -2$ into either of the equations to find y. Substituting into the first of the original equations,

$$-2 + 2y = -4$$
$$2y = -2$$

Solving for y,

$$y = -1$$

Check: Substituting $x = -2$ and $y = -1$ into the first original equation,

$$-2 + 2(-1) \overset{?}{=} -4$$

$$-2 - 2 = -4 \qquad \text{checks}$$

Substituting into the second original equation,

$$2(-2) - (-1) \overset{?}{=} -3$$

$$-4 + 1 = -3 \qquad \text{checks}$$

(b) Fractional Equations.   When the unknowns have *fractional coefficients*, multiply each equation by its LCD.   This will eliminate the denominators and the system may then be solved as in the previous example.

Example 11.8:   Solve the system of equations

$$\frac{2x}{3} + \frac{y}{5} = \frac{1}{15}$$

$$\frac{x}{2} - \frac{2y}{5} = \frac{2}{3}$$

Solution: Multiplying the first equation by 15 and the second equation by 30, we eliminate the denominators.

$$10x + 3y = 1$$

$$15x - 12y = 20$$

The coefficients of the y terms in both equations can now be made equal by multiplying the first equation by 4.

$$\begin{array}{r} 40x + 12y = 4 \\ 15x - 12y = 20 \\ \hline 55x \qquad\quad = 24 \end{array}$$

Adding.

$$x = \frac{24}{55}$$

Substituting back into the first original equation,

$$\frac{2}{3}\left(\frac{24}{55}\right) + \frac{y}{5} = \frac{1}{15}$$

Transposing,

$$\frac{y}{5} = \frac{1}{15} - \frac{2}{3}\left(\frac{24}{55}\right)$$

Multiplying by 5,

$$y = \frac{1}{3} - \frac{2}{3}\left(\frac{24}{11}\right) = \frac{11}{33} - \frac{48}{33} = -\frac{37}{33}$$

Check: Substituting into the first original equation,

$$\frac{2}{3}\left(\frac{24}{55}\right) + \frac{1}{5}\left(-\frac{37}{33}\right) \overset{?}{=} \frac{1}{15}$$

$$\frac{48}{165} - \frac{37}{165} \overset{?}{=} \frac{1}{15}$$

$$\frac{11}{165} = \frac{1}{15} \qquad \text{checks}$$

Substituting into the second original equation,

$$\frac{1}{2}\left(\frac{24}{55}\right) - \frac{2}{5}\left(-\frac{37}{33}\right) \stackrel{?}{=} \frac{2}{3}$$

$$\frac{12}{55} + \frac{74}{165} \stackrel{?}{=} \frac{2}{3}$$

$$\frac{36}{165} + \frac{74}{165} \stackrel{?}{=} \frac{2}{3}$$

$$\frac{110}{165} = \frac{2}{3} \quad \text{checks}$$

The same method—multiplying by the LCD—may be used when the *unknowns are in the denominators.*

Example 11.9:　　Solve for x and y:

$$\frac{3}{2x} - \frac{1}{2y} = -3$$

$$\frac{2}{x} + \frac{3}{y} = 7$$

Solution:　　Multiplying the first equation by $2xy$ and the second equation by xy,

$$3y - x = -6xy$$

$$2y + 3x = 7xy$$

We now make the coefficients of the x terms equal by multiplying the first equation by 3.

$$9y - 3x = -18xy$$
$$2y + 3x = 7xy$$

Adding,
$$11y = -11xy$$

Dividing by $-11y$,

$$-1 = x$$

Substituting into the first equation,

$$\frac{3}{2(-1)} - \frac{1}{2y} = -3$$

Transposing,
$$-\frac{1}{2y} = -3 + \frac{3}{2} = -\frac{3}{2}$$

Cross-multiplying,
$$-6y = -2$$

$$y = \frac{1}{3}$$

Equations having *unknowns in the denominator* may also be solved by introducing new variables which are the reciprocals of the original variables, as in the following example.

Example 11.10:　　Solve the pair of equations in the previous example by a different method.

Solution:　　We introduce two new variables. Let $w = 1/x$ and $z = 1/y$. Substituting into the original equations,

$$\frac{3w}{2} - \frac{z}{2} = -3$$

$$2w + 3z = 7$$

We now solve for w and z.   Multiplying the first equation by 6,

$$9w - 3z = -18$$

$$2w + 3z = 7$$

Adding,

$$11w = -11$$

$$w = -1$$

Since $w = 1/x$,

$$w = \frac{1}{x} = -1$$

$$x = -1$$

Substituting this value back into one of the original equations will give us y, as in the previous example.

(c)   Literal Equations.   If the coefficients of the unknowns are letters, the system of equations may still be solved by the addition-subtraction method.   Simply treat the literal coefficients as if they were numbers, and obtain solutions for the unknowns in terms of the literal coefficients.

Example 11.11: .   Solve for x and y.

$$3x + ay = 5$$

$$bx - 2y = c$$

Solution: Multiplying the first equation by 2 and the second equation by a,

$$6x + 2ay = 10$$

$$abx - 2ay = ac$$

Adding,

$$6x + abx = 10 + ac$$

$$(6 + ab)x = 10 + ac$$

$$x = \frac{10 + ac}{6 + ab}$$

Substituting this expression into the first equation

$$3\left(\frac{10 + ac}{6 + ab} \right) + ay = 5$$

$$ay = 5 - \frac{30 + 3ac}{6 + ab}$$

Using the LCD, $6 + ab$, and combining terms on the right-hand side,

$$ay = \frac{30 + 5ab - 30 - 3ac}{6 + ab} = \frac{5ab - 3ac}{6 + ab}$$

$$y = \frac{5b - 3c}{6 + ab}$$

Example 11.12:   Solve by the addition-subtraction method.

$$a_1 x + b_1 y = c_1$$

$$a_2 x + b_2 y = c_2$$

Solution: Multiplying the first equation by b_2 and the second equation by b_1,

$$a_1 b_2 x + b_1 b_2 y = c_1 b_2$$

$$a_2 b_1 x + b_1 b_2 y = c_2 b_1$$

Subtracting,

$$(a_1 b_2 - a_2 b_1)x = c_1 b_2 - c_2 b_1$$

Dividing by $a_1 b_2 - a_2 b_1$,

$$x = \frac{c_1 b_2 - c_2 b_1}{a_1 b_2 - a_2 b_1}$$

Now solving for y, we multiply the first equation by a_2 and the second equation by a_1,

$$a_1 a_2 x + a_2 b_1 y = a_2 c_1$$

$$a_1 a_2 x + a_1 b_2 y = a_1 c_2$$

Subtracting the upper equation from the lower,

$$(a_1 b_2 - a_2 b_1)y = a_1 c_2 - a_2 c_1$$

Dividing by $\quad a_1 b_2 - a_2 b_1,$ $$y = \frac{a_1 c_2 - a_2 c_1}{a_1 b_2 - a_2 b_1}$$

This result may be summarized as follows:

> The solution to the set of equations
> $$a_1 x + b_1 y = c_1$$
> $$a_2 x + b_2 y = c_2$$
> is $\quad x = \dfrac{c_1 b_2 - c_2 b_1}{a_1 b_2 - a_2 b_1}$ and $y = \dfrac{a_1 c_2 - a_2 c_1}{a_1 b_2 - a_2 b_1}$

122

11.4 SUBSTITUTION METHOD

To solve a pair of equations by the substitution method, first solve one of the equations for either of the unknowns.   Then substitute this expression into the other equation, and solve.

Example 11.13:   Solve the set of equations of Example 11.7 by substitution.

Solution: First solve either equation for one of the variables.   From the first equation,

$$x = -4 - 2y$$

This expression is now substituted for x in the second equation.

$$2(-4 - 2y) - y = -3$$

Solving for y, we remove the parentheses.

$$-8 - 4y - y = -3$$

Transposing and combining like terms,

$$-5y = 5$$

$$y = -1$$

Substituting back into the first equation,

$$x + 2(-1) = -4$$

$$x = -4 + 2 = -2$$

as in Example 11.7.

11.5 SYSTEMS HAVING NO UNIQUE SOLUTION

If both variables drop out when you try to solve a pair of equations, it means that the equations have no single (unique) solution.

(a) Dependency.   A *dependent system* is one where the two equations plot as a single line, so that any solution to one equation is also a solution to the other equation. If both variables drop out and an equality results, the system is a dependent one.

Example 11.14:   Solve the system

$$2x - 3y = 5$$
$$4x - 6y = 10$$

Solution: Multiplying the first equation by 2,

$$4x - 6y = 10$$
$$\underline{4x - 6y = 10}$$

Subtracting, $$0 = 0$$

Both variables have dropped out and we are left with an equality, so the system is dependent.

(b)   Inconsistency.   An *inconsistent system* is one which plots as two parallel lines. If both variables drop out and an inequality results, the system is inconsistent.

Example 11.15:   Solve the system

$$x - y = 3$$
$$2x - 2y = 1$$

Solution: Multiplying the first by -2 and adding,

$$-2x + 2y = -6$$
$$\underline{2x - 2y = 1}$$
$$0 = -5$$

Both variables have dropped out and an inequality results, showing that the system is inconsistent.

11.6   DETERMINANTS

(a)   Definitions.   A determinant is a *square array* of numbers written between vertical bars, such as,

$$\begin{vmatrix} a_1 & b_1 \\ a_2 & b_1 \end{vmatrix}$$

The determinant shown has four *elements*, a_1, b_1, a_2, and b_2. It is of *second order*, because it has two rows and two columns.

The *value* of a second-order determinant is the product of the elements on its *principal diagonal* minus the product of the elements on the *secondary diagonal*.

$$\boxed{\begin{vmatrix} a_1 & b_1 \\ a_2 & b_2 \end{vmatrix} = a_1 b_2 - a_2 b_1} \qquad \textbf{120}$$

Example 11.16:   Find the value of the determinant

$$\begin{vmatrix} 2 & 5 \\ -1 & 3 \end{vmatrix}$$

Solution: By Eq. 120,

$$\begin{vmatrix} 2 & 5 \\ -1 & 3 \end{vmatrix} = (2)(3) - (5)(-1) = 6 + 5 = 11$$

Example 11.17:   Find the value of the determinant

$$\begin{vmatrix} 3 & 2 \\ 0 & 5 \end{vmatrix}$$

Solution: By Eq. 120,

$$\begin{vmatrix} 3 & 2 \\ 0 & 5 \end{vmatrix} = (3)(5) - (2)(0) = 15$$

Example 11.18:   Solve for x.

$$\begin{vmatrix} 2 & 3 \\ 4 & x \end{vmatrix} = 6$$

Solution: By Eq. 120,

$$\begin{vmatrix} 2 & 3 \\ 4 & x \end{vmatrix} = 2x - 12 = 6$$

Transposing, $2x = 18$

$$x = 9$$

(b)   Solving a System of Two Equations by Determinants.   In Example 11.12, we saw that the solution to the set of equations

$$a_1 x + b_1 y = c_1$$

$$a_2 x + b_2 y = c_2$$

was

$$x = \frac{c_1 b_2 - c_2 b_1}{a_1 b_2 - a_2 b_1} \qquad y = \frac{a_1 c_2 - a_2 c_1}{a_1 b_2 - a_2 b_1}$$

Note that the denominator $a_1 b_2 - a_2 b_1$ is the same as the value of the determinant in Eq. 120,

$$\begin{vmatrix} a_1 & b_1 \\ a_2 & b_2 \end{vmatrix} = a_1 b_2 - a_2 b_1$$

The numerator $c_1 b_2 - c_2 b_1$ can also be written as the determinant,

$$\begin{vmatrix} c_1 & b_1 \\ c_2 & b_2 \end{vmatrix} = c_1 b_2 - c_2 b_1$$

Thus the solution for x can be written as the ratio of two determinants

$$x = \frac{c_1 b_2 - c_2 b_1}{a_1 b_2 - a_2 b_1} = \frac{\begin{vmatrix} c_1 & b_1 \\ c_2 & b_2 \end{vmatrix}}{\begin{vmatrix} a_1 & b_1 \\ a_2 & b_2 \end{vmatrix}}$$

Similarly, the solution for y is

$$y = \frac{a_1 c_2 - a_2 c_1}{a_1 b_2 - a_2 b_1} = \frac{\begin{vmatrix} a_1 & c_1 \\ a_2 & c_2 \end{vmatrix}}{\begin{vmatrix} a_1 & b_1 \\ a_2 & b_2 \end{vmatrix}}$$

Example 11.19:   Solve by determinants,

$$4x + 2y = 5$$
$$3x - 4y = 1$$

Solution

1. Set up and evaluate the determinant whose elements are the coefficients of the x and y terms.

$$D = \begin{vmatrix} 4 & 2 \\ 3 & -4 \end{vmatrix} = (4)(-4) - (2)(3) = -22$$

This is called the *determinant of the system*, or *the determinant of the coefficients*.

2. Form another determinant in which the column of x coefficients $\begin{vmatrix} 4 \\ 3 \end{vmatrix}$ is replaced by the column of constants $\begin{vmatrix} 5 \\ 1 \end{vmatrix}$.

$$\begin{vmatrix} 5 & 2 \\ 1 & -4 \end{vmatrix}$$

The value of this determinant divided by D will give x.

$$x = \frac{\begin{vmatrix} 5 & 2 \\ 1 & -4 \end{vmatrix}}{D} = \frac{(5)(-4) - (2)(1)}{-22} = \frac{-22}{-22} = 1$$

3. Replace the y coefficients in the determinant of the system by the column of constants,

$$\begin{vmatrix} 4 & 5 \\ 3 & 1 \end{vmatrix}$$

and divide by D to obtain y.

$$y = \frac{\begin{vmatrix} 4 & 5 \\ 3 & 1 \end{vmatrix}}{D} = \frac{(4)(1) - (5)(3)}{-22} = \frac{-11}{-22} = \frac{1}{2}$$

This method of solving systems of equations is called *Cramer's Rule* and is summarized as follows:

Cramer's Rule	The solution to the set of equations $$a_1 x + b_1 y = c_1$$ $$a_2 x + b_2 y = c_2$$ is $$x = \frac{c_1 b_2 - c_2 b_1}{a_1 b_2 - a_2 b_1} = \frac{\begin{vmatrix} c_1 & b_1 \\ c_2 & b_2 \end{vmatrix}}{\begin{vmatrix} a_1 & b_1 \\ a_2 & b_2 \end{vmatrix}} \qquad y = \frac{a_1 c_2 - a_2 c_1}{a_1 b_2 - a_2 b_1} = \frac{\begin{vmatrix} a_1 & c_1 \\ a_2 & c_2 \end{vmatrix}}{\begin{vmatrix} a_1 & b_1 \\ a_2 & b_2 \end{vmatrix}}$$

122

11.7 THREE EQUATIONS IN THREE UNKNOWNS

To solve a system of three equations, start by eliminating one unknown from any two of the three equations. This may be done either by the addition-subtraction method, or by substitution. Next, eliminate the *same* unknown from a different pair of equations. This will result in a pair of equations containing only two unknowns, which may then be solved by any of the previous methods.

Example 11.20:   Solve the following system of three equations for x, y, and z:

$$2x + y + z = 7$$
$$x + y + 2z = 18$$
$$x + 2y + z = 15$$

Solution: Subtracting the first equation from the second to eliminate y,

$$-x + z = 11$$

Multiplying the second equation by -2 and adding it to the third to eliminate y,

$$
\begin{aligned}
x + 2y + z &= 15 \\
-2x - 2y - 4z &= -36 \\
\hline
-x \qquad\ - 3z &= -21
\end{aligned}
$$

we now have two equations in the two unknowns, x and z.

$$
\begin{aligned}
-x + z &= 11 \\
-x - 3z &= -21 \\
\hline
\end{aligned}
$$

Subtracting,
$$4z = 32$$
$$z = 8$$

Substituting back,

$$x = z - 11 = 8 - 11 = -3$$

and, from the first original equation,

$$y = 7 - 2x - z = 7 - (-6) - 8 = 5$$

So our solution is

$$x = -3 \qquad y = 5 \qquad z = 8$$

To check, substitute the three values obtained into each of the three original equations.

Check: Substituting into the first equation

$$2(-3) + 5 + 8 \overset{?}{=} 7$$
$$-6 + 5 + 8 \overset{?}{=} 7$$
$$-6 + 13 = 7 \qquad \text{checks}$$

In the second equation,

$$-3 + 5 + 2(8) \overset{?}{=} 18$$
$$-3 + 21 = 18 \qquad \text{checks}$$

and in the third equation

$$-3 + 2(5) + 8 \overset{?}{=} 15$$
$$-3 + 18 = 15 \qquad \text{checks}$$

11.8 SOLVING A SYSTEM OF THREE EQUATIONS BY DETERMINANTS

A *third-order determinant* is one having three rows and three columns.   Its *value* is

$$\begin{vmatrix} a_1 & b_1 & c_1 \\ a_2 & b_2 & c_2 \\ a_3 & b_3 & c_3 \end{vmatrix} = a_1 b_2 c_3 + a_3 b_1 c_2 + a_2 b_3 c_1 - (a_3 b_2 c_1 + a_1 b_3 c_2 + a_2 b_1 c_3)$$ **121**

A convenient way to evaluate a third-order determinant is to, first, recopy the first two columns to the right of the original three, then multiply the elements along each diagonal of the extended determinant and combine them as in Eq. 121.

Example 11.21:   Evaluate the determinant

$$\begin{vmatrix} 1 & 4 & 3 \\ 5 & 1 & 2 \\ 3 & 0 & 1 \end{vmatrix}$$

Solution: Repeating the first two columns,

$$\begin{vmatrix} 1 & 4 & 3 & | & 1 & 4 \\ 5 & 1 & 2 & | & 5 & 1 \\ 3 & 0 & 1 & | & 3 & 0 \end{vmatrix}$$

Multiplying along the diagonals,

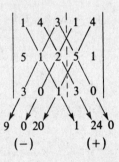

$$\begin{array}{ccccc} 9 & 0 & 20 & \quad 1 & 24 & 0 \\ & (-) & & & (+) \end{array}$$

Those products from diagonals sloping downward to the right are taken as positive, and those sloping downward to the left are taken as negative.

So, the value of this determinant is

$$1 + 24 + 0 - (9 + 0 + 20) = 25 - 29 = -4$$

The procedure for solving a system of three equations by determinants is similar to that used for solving a set of two equations, and it will be explained by means of an example.

Example 11.22:   Solve the following system of equations by determinants.

$$\begin{aligned} 2x + \ y + \ z &= \ 7 \\ x + \ y + 2z &= 18 \\ x + 2y + \ z &= 15 \end{aligned}$$

Solution

1. Set up and evaluate the determinant whose elements are the coefficients of the x, y, and z terms (the determinant of the system).

$$D = \begin{vmatrix} 2 & 1 & 1 & | & 2 & 1 \\ 1 & 1 & 2 & | & 1 & 1 \\ 1 & 2 & 1 & | & 1 & 2 \end{vmatrix} = 2 + 2 + 2 - (1 + 8 + 1) = 6 - 10 = -4$$

2. Form a new determinant in which the column of coefficients of the x terms

$$\begin{vmatrix} 2 \\ 1 \\ 1 \end{vmatrix}$$

is replaced by the column of constants

$$\begin{vmatrix} 7 \\ 18 \\ 15 \end{vmatrix}$$

The value of this determinant, divided by D, will give x.

$$\begin{vmatrix} 7 & 1 & 1 \\ 18 & 1 & 2 \\ 15 & 2 & 1 \end{vmatrix} \begin{matrix} 7 & 1 \\ 18 & 1 \\ 15 & 2 \end{matrix} = 7 + 30 + 36 - (15 + 28 + 18) = 73 - 61 = 12$$

$$x = \frac{12}{D} = \frac{12}{-4} = -3$$

3. Replace the y coefficients in the determinant of the system with the column of constants. Evaluate this new determinant and divide it by D to obtain y.

$$\begin{vmatrix} 2 & 7 & 1 \\ 1 & 18 & 2 \\ 1 & 15 & 1 \end{vmatrix} \begin{matrix} 2 & 7 \\ 1 & 18 \\ 1 & 15 \end{matrix} = 36 + 14 + 15 - (18 + 60 + 7) = 65 - 85 = -20$$

So,

$$y = \frac{-20}{-4} = 5$$

4. We find z by substituting x and y back into one of the original equations. From the first equation,

$$z = 7 - 2x - y = 7 - (-6) - 5 = 8$$

These are the same values that were obtained and checked in Example 20.

11.9 WORD PROBLEMS HAVING MORE THAN ONE UNKNOWN

In Chapter 5 we solved word problems which had only one unknown. We set up problems having more than one unknown in the same way we did then, except that we now need a different letter for each unknown, and must write as many equations as there are unknowns.

Example 11.23: An airplane flies 1500 miles in 4.0 hours against the wind, and takes only 3.5 hours to return with the wind. Find the wind speed and the speed of the plane in still air.

Solution: Let x = speed of plane in still air (mi/h) and w = wind speed (mi/h). Then, by Eq. 177,

$$(x - w)4.0 = 1500$$

and

$$(x + w)3.5 = 1500$$

Rewriting,

$$4.0x - 4.0w = 1500$$
$$3.5x + 3.5w = 1500$$

Multiplying the first equation by 3.5 and the second by 4,

$$14x - 14w = 5250$$

Adding,

$$\underline{14x + 14w = 6000}$$
$$28x \qquad = 11\,250$$
$$x = 400 \text{ mi/h}$$

and

$$3.5w = 1500 - 3.5x = 100$$
$$w = 29 \text{ mi/h}$$

Solved Problems

GRAPHICAL METHOD

11.1 Solve graphically for x and y.

(a) $\quad x - 3y = 0$ $\qquad(b)$ $\quad 3x - y + 6 = 0$ $\qquad(c)$ $\quad x = 8 - 2y$

$\qquad 2x + y = 0$ $\qquad\qquad 2x + 3y - 7 = 0$ $\qquad\qquad y = 7 - 2x$

Solution

(a) We obtain two points on each line by arbitrarily assigning values to one variable and solving for the other variable.   For the first equation,

$$x = 3y$$

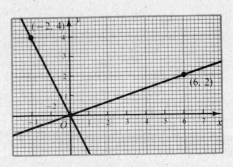

When $y = 0$, $x = 0$ and when $y = 2$, $x = 3(2) = 6$. We plot the points $(0,0)$ and $(6,2)$ and connect them to obtain the first line.   Similarly for the second line, when $y = 0$, $x = 0$ and when $y = 4$, $x = -2$. Plotting the lines, Fig. 11-2, their point of intersection—the solution to the set of equations—is

$$x = 0 \qquad y = 0$$

Fig. 11-2

(b) First equation: When $y = 0$, $x = -2$; when $x = 0$, $y = 6$.
Second equation: When $y = 0$, $x = 7/2$; when $x = 0$, $y = 7/3$.
Plotting the two lines, their point of intersection is, Fig. 11-3,

$$x = -1 \qquad \text{and} \qquad y = 3$$

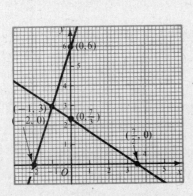

Fig. 11-3

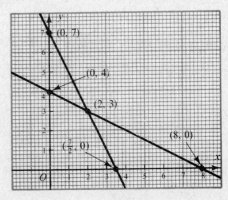

Fig. 11-4

(c) First equation: When $y = 0$, $x = 8$; when $x = 0$, $y = 4$.
Second equation: When $y = 0$, $x = 7/2$; when $x = 0$, $y = 7$.
The lines are plotted in Fig. 11-4, and their point of intersection is

$$x = 2 \qquad \text{and} \qquad y = 3$$

ADDITION-SUBTRACTION METHOD

11.2 Solve for x and y by addition and subtraction.

(a) $\quad x + y = 2$ $\qquad(b)$ $\quad x - 2y = 8$ $\qquad(c)$ $\quad x + 4y + 5 = 0$ $\qquad(d)$ $\quad 4x = 7 - 3y$

$\qquad 4x - 3y = 8$ $\qquad\qquad x + y = 6$ $\qquad\qquad 7y = 1 - 5x$ $\qquad\qquad 2x = 10 + 5y$

Solution

(a)
$$x + y = 2$$
$$4x - 3y = 8$$

To eliminate the y term, multiply the first equation by 3 and add.

$$3x + 3y = 6$$
$$\underline{4x - 3y = 8}$$
$$7x \quad\;\; = 14$$
$$x = 2$$

Substituting into the first equation,

$$y = 2 - x = 2 - 2 = 0$$

(b)
$$x - 2y = 8$$
$$x + y = 6$$

Subtracting the second equation from the first,

$$-3y = 2$$
$$y = \frac{2}{3}$$

Substituting into the second equation,

$$x = 6 - y = 6 - \left(-\frac{2}{3}\right) = 6\frac{2}{3}$$

(c)
$$x + 4y + 5 = 0$$
$$7y = 1 - 5x$$

Rearranging the equations so that like terms are above each other,

$$x + 4y = -5$$
$$5x + 7y = \quad 1$$

Multiplying the first equation by -5 and adding,

$$-5x - 20y = 25$$
$$\underline{5x + \;\; 7y = \;\; 1}$$
$$-13y = 26$$
$$y = -2$$

Substituting into the first equation,

$$x = -5 - 4(-2) = -5 + 8 = 3$$

(d)
$$4x = 7 \;\; - 3y$$
$$2x = 10 + 5y$$

Rearranging the equations,

$$4x + 3y = 7$$
$$2x - 5y = 10$$

Multiplying the second by -2 and adding,

$$4x + \;\; 3y = \quad\;\; 7$$
$$\underline{-4x + 10y = -20}$$
$$13y = -13$$
$$y = -1$$

Substituting back,

$$x = \frac{1}{4}(7 - 3y) = \frac{1}{4}(7 + 3) = \frac{5}{2}$$

FRACTIONAL EQUATIONS

11.3   Solve by any method.

(a) $\dfrac{2x}{5} + \dfrac{y}{2} = 4$ (c) $\dfrac{x+1}{3} + \dfrac{y-2}{2} = 6$ (e) $\dfrac{3}{2x} - \dfrac{2}{3y} = 4$

$\dfrac{x}{3} + \dfrac{3y}{4} = -1$ $\dfrac{2x+2}{3} - \dfrac{y-1}{4} = 2$ $\dfrac{1}{x} + \dfrac{2}{y} = 3$

(b) $\dfrac{x}{2} - \dfrac{2y}{3} = -2$ (d) $\dfrac{2}{x} + \dfrac{1}{y} = 1$

$\dfrac{3x}{5} + \dfrac{y}{2} = 5$ $\dfrac{1}{x} - \dfrac{3}{2y} = 2$

Solution

(a)

$$\frac{2x}{5} + \frac{y}{2} = 4$$

$$\frac{x}{3} + \frac{3y}{4} = -1$$

Multiplying by the LCDs to clear fractions,

$$\begin{aligned} 4x + 5y &= 40 \\ 4x + 9y &= -12 \end{aligned}$$

Subtracting,

$$\begin{aligned} -4y &= 52 \\ y &= -13 \end{aligned}$$

Substituting back,

$$4x = 40 - 5y = 40 + 65 = 105$$

$$x = \frac{105}{4}$$

(b)

$$\frac{x}{2} - \frac{2y}{3} = -2$$

$$\frac{3x}{5} + \frac{y}{2} = 5$$

Multiplying by the LCDs to clear fractions,

$$\begin{aligned} 3x - 4y &= -12 \\ 6x + 5y &= 50 \end{aligned}$$

Multiplying the first equation by -2 and adding,

$$\begin{aligned} -6x + 8y &= 24 \\ 6x + 5y &= 50 \\ \hline 13y &= 74 \\ y &= \frac{74}{13} \end{aligned}$$

Substituting back,

$$3x = 4y - 12 = \frac{296}{13} - \frac{156}{13} = \frac{140}{13}$$

So,
$$x = \frac{140}{39}$$

(c)

$$\frac{x + 1}{3} + \frac{y - 2}{2} = 6$$

$$\frac{2x + 2}{3} - \frac{y - 1}{4} = 2$$

Multiplying by the LCDs to clear fractions,

$$2(x + 1) + 3(y - 2) = 36$$
$$2x + 2 + 3y - 6 = 36$$
$$2x + 3y = 40$$

and
$$4(2x + 2) - 3(y - 1) = 24$$
$$8x + 8 - 3y + 3 = 24$$
$$8x - 3y = 13$$

Adding our two simplified equations,

$$2x + 3y = 40$$
$$\underline{8x - 3y = 13}$$
$$10x \qquad = 53$$
$$x = \frac{53}{10}$$

Substituting back,

$$3y = 40 - 2x = 40 = 2\left(\frac{53}{10}\right) = \frac{400}{10} - \frac{106}{10} = \frac{294}{10}$$

$$y = \frac{294}{30} = \frac{147}{15}$$

(d)

$$\frac{2}{x} + \frac{1}{y} = 1$$

$$\frac{1}{x} - \frac{3}{2y} = 2$$

Multiplying the first equation by the LCD xy, and the second equation by $2xy$,

$$2y + \ x = \ xy$$
$$\underline{2y - 3x = \ 4xy}$$

Subtracting,
$$4x = -3xy$$

Dividing by $-3x$,

$$y = -\frac{4}{3}$$

Substituting back,

$$\frac{2}{x} = 1 - \frac{1}{y} = 1 + \frac{3}{4} = \frac{7}{4}$$

Cross-multiplying, $7x = 8$

$$x = \frac{8}{7}$$

(e)

$$\frac{3}{2x} - \frac{2}{3y} = 4$$

$$\frac{1}{x} + \frac{2}{y} = 3$$

Multiplying the first equation by $6xy$ and the second equation by xy,

$$9y - 4x = 24xy$$
$$y + 2x = 3xy$$

Multiplying the second equation by 2,

$$9y - 4x = 24xy$$
$$2y + 4x = 6xy$$

Adding, $\overline{11y \qquad = 30xy}$

$$x = \frac{11}{30}$$

Substituting back,

$$\frac{2}{y} = 3 - \frac{1}{x} = 3 - \frac{30}{11} = \frac{3}{11}$$

$$\frac{y}{2} = \frac{11}{3}$$

$$y = \frac{22}{3}$$

LITERAL EQUATIONS

11.4   Solve for x and y in terms of a and b.

(a) $2x - 3y = a$ (b) $\dfrac{x}{a} - \dfrac{y}{b} = 1$
        $x + 2y = b$

$$\frac{2x}{a} + \frac{y}{3b} = 2$$

Solution

(a)

$$2x - 3y = a$$
$$x + 2y = b$$

Multiplying the second equation by -2,

$$2x - 3y = a$$
$$-2x - 4y = -2b$$

Adding, $\overline{-7y = a - 2b}$

$$y = \frac{2b - a}{7}$$

Substituting back,

$$x = b - 2y = b - \frac{2(2b - a)}{7} = b - \frac{4b}{7} + \frac{2a}{7} = \frac{3b + 2a}{7}$$

(b)

$$\frac{x}{a} - \frac{y}{b} = 1$$

$$\frac{2x}{a} + \frac{y}{3b} = 2$$

Multiplying the first equation by -2,

$$-\frac{2x}{a} + \frac{2y}{b} = -2$$

$$\frac{2x}{a} + \frac{y}{3b} = 2$$

Adding,

$$\frac{2y}{b} + \frac{y}{3b} = 0$$

$$\frac{7y}{3b} = 0$$

$$y = 0$$

Substituting back,

$$\frac{x}{a} = 1 + \frac{y}{b} = 1 + 0$$

$$x = a$$

SUBSTITUTION METHOD

11.5 Solve by substitution.

(a) $3x + 8y - 7 = 0$ (b) $6x = 3y + 4$ (c) $4x - 2y = 5$ (d) $y = 8x - 3$
       $x - 4y - 9 = 0$ $y = 2x - 2$ $3y + 2x = 7$ $y = 4x + 2$

Solution

(a)

$$3x + 8y - 7 = 0$$
$$x - 4y - 9 = 0$$

From the second equation,

$$x = 4y + 9$$

Substituting this into the first equation,

$$3(4y + 9) + 8y - 7 = 0$$
$$12y + 27 + 8y - 7 = 0$$
$$20y = -20$$
$$y = -1$$

Substituting back into the second equation,

$$x = 4(-1) + 9 = 5$$

(b)

$$y = 2x - 2$$
$$6x = 3y + 4$$

Substituting $(2x - 2)$ for y in the second equation,

$$6x = 3(2x - 2) + 4$$
$$6x = 6x - 6 + 4$$
$$0 = -2$$

The variables drop out, indicating that there is no solution to this set of equations.

(c)

$$4x - 2y = 5$$
$$3y + 2x = 7$$

From the first equation,

$$x = \frac{2y + 5}{4}$$

Substituting into the second equation,

$$3y + \frac{2(2y + 5)}{4} = 7$$

Solving for y,

$$6y + 2y + 5 = 14$$
$$8y = 9$$
$$y = \frac{9}{8}$$

Substituting back,

$$x = \frac{1}{4}\left(\frac{9}{4} + 5\right) = \frac{1}{4}\left(\frac{29}{4}\right) = \frac{29}{16}$$

(d)

$$y = 8x - 3$$
$$y = 4x + 2$$

Substituting,

$$8x - 3 = 4x + 2$$
$$4x = 5$$
$$x = \frac{5}{4}$$

Substituting back,

$$y = 8\left(\frac{5}{4}\right) - 3 = \frac{40}{4} - 3 = 7$$

DETERMINANTS

11.6   Evaluate the following second-order determinants:

(a) $\begin{vmatrix} 1 & 3 \\ 2 & 4 \end{vmatrix}$ (b) $\begin{vmatrix} 8 & 1 \\ 2 & 3 \end{vmatrix}$ (c) $\begin{vmatrix} -2 & 6 \\ -3 & 4 \end{vmatrix}$ (d) $\begin{vmatrix} 9 & -1 \\ 2 & 0 \end{vmatrix}$

Solution

(a) $\begin{vmatrix} 1 & 3 \\ 2 & 4 \end{vmatrix} = 4 - 6 = -2$ (c) $\begin{vmatrix} -2 & 6 \\ -3 & 4 \end{vmatrix} = -8 - (-18) = 10$

(b) $\begin{vmatrix} 8 & 1 \\ 2 & 3 \end{vmatrix} = 24 - 2 = 22$ (d) $\begin{vmatrix} 9 & -1 \\ 2 & 0 \end{vmatrix} = 0 - (-2) = 2$

11.7   Solve for x.

(a) $\begin{vmatrix} x & 1 \\ 3 & 2 \end{vmatrix} = 9$ (b) $\begin{vmatrix} 3 & x \\ 5 & 2 \end{vmatrix} = -8$ (c) $\begin{vmatrix} 1 & -3 \\ x & -4 \end{vmatrix} = 21$

Solution

(a)

$$\begin{vmatrix} x & 1 \\ 3 & 2 \end{vmatrix} = 9$$

Evaluating the determinant,

$$2x - 3 = 9$$

Solving for x,

$$2x = 12$$
$$x = 6$$

(b)

$$\begin{vmatrix} 3 & x \\ 5 & 2 \end{vmatrix} = -8$$
$$3(2) - 5x = -8$$
$$-5x = -14$$
$$x = \frac{14}{5}$$

(c)

$$\begin{vmatrix} 1 & -3 \\ x & -4 \end{vmatrix} = 21$$
$$-4 - (-3x) = 21$$
$$3x = 21 + 4 = 25$$
$$x = \frac{25}{3}$$

11.8   Solve by determinants.

(a) $3x - y = -6$ (b) $x + y = 2$ (c) $x + 4y = -5$ (d) $4x + 3y = 7$
    $2x + 3y = 7$ $4x - 3y = 8$ $5x + 7y = 1$ $2x - 5y = 10$

Solution

(a)

$$3x - y = -6$$
$$2x + 3y = 7$$

The determinant of the system is

$$D = \begin{vmatrix} 3 & -1 \\ 2 & 3 \end{vmatrix} = 9 - (-2) = 11$$

By Eq. 122,

$$x = \frac{\begin{vmatrix} -6 & -1 \\ 7 & 3 \end{vmatrix}}{11} = \frac{-18 - (-7)}{11} = \frac{-11}{11} = -1$$

Similarly,

$$y = \frac{\begin{vmatrix} 3 & -6 \\ 2 & 7 \end{vmatrix}}{11} = \frac{21 - (-12)}{11} = \frac{33}{11} = 3$$

The value of y could also have been obtained by substituting the value $x = -1$ into either of the given equations. Substituting into the first equation,

$$3(-1) - y = -6$$

and transposing,

$$-3 + 6 = y$$
$$y = 3$$

(b)

$$x + y = 2$$
$$4x - 3y = 8$$

By Eq. 122,

$$D = \begin{vmatrix} 1 & 1 \\ 4 & -3 \end{vmatrix} = -3 - 4 = -7$$

So,

$$x = \frac{\begin{vmatrix} 2 & 1 \\ 8 & -3 \end{vmatrix}}{-7} = \frac{-6 - 8}{-7} = 2$$

Substituting into the first equation,

$$y = 2 - x = 2 - 2 = 0$$

(c)

$$x + 4y = -5$$
$$5x + 7y = 1$$

By Eq. 122,

$$D = \begin{vmatrix} 1 & 4 \\ 5 & 7 \end{vmatrix} = 7 - 20 = -13$$

So,

$$x = \frac{\begin{vmatrix} -5 & 4 \\ 1 & 7 \end{vmatrix}}{-13} = \frac{-35 - 4}{-13} = 3$$

Substituting into the first equation,

$$4y = -5 - x = -5 - 3 = -8$$
$$y = -2$$

(d)

$$4x + 3y = 7$$
$$2x - 5y = 10$$

By Eq. 122,

$$D = \begin{vmatrix} 4 & 3 \\ 2 & -5 \end{vmatrix} = -20 - 6 = -26$$

So,

$$x = \frac{\begin{vmatrix} 7 & 3 \\ 10 & -5 \end{vmatrix}}{-26} = \frac{-35 - 30}{-26} = \frac{-65}{-26} = \frac{5}{2}$$

Substituting into the first equation,

$$3y = 7 - 4\left(\frac{5}{2}\right) = \frac{14}{2} - \frac{20}{2} = -\frac{6}{2} = -3$$
$$y = -1$$

THREE EQUATIONS IN THREE UNKNOWNS

11.9 Solve for x, y, and z.

(a) $3x + y = 5$
$2y - 3z = -5$
$x + 2z = 7$

(b) $x - y = 5$
$y - z = -6$
$2x - z = 2$

(c) $\dfrac{x}{10} + \dfrac{y}{5} + \dfrac{z}{20} = \dfrac{1}{4}$
$x + y + z = 6$
$\dfrac{x}{3} + \dfrac{y}{2} + \dfrac{z}{6} = 1$

(d) $\dfrac{1}{x} + \dfrac{2}{y} - \dfrac{1}{z} = -3$
$\dfrac{3}{x} + \dfrac{1}{y} + \dfrac{1}{z} = 4$
$\dfrac{1}{x} - \dfrac{1}{y} + \dfrac{2}{z} = 6$

Solution

(a) From the first equation, $y = 5 - 3x$. Substituting this into the second equation,

$$2(5 - 3x) - 3z = -5$$

Simplifying,

$$10 - 6x - 3z = -5$$
$$-6x = -15 + 3z$$

Or

$$x = \frac{5}{2} - \frac{z}{2}$$

Substituting this into the third equation,

$$\frac{5}{2} - \frac{z}{2} + 2z = 7$$

Solving for z,

$$\frac{3z}{2} = 7 - \frac{5}{2} = \frac{9}{2}$$
$$3z = 9$$
$$z = 3$$

Substituting back,

$$x = \frac{5}{2} - \frac{3}{2} = 1$$

and

$$y = 5 - 3(1) = 2$$

(b) From the first equation, $y = x - 5$. Substituting this into the second equation,

$$x - 5 - z = -6$$

or

$$z = x - 5 + 6 = x + 1$$

Substituting $x + 1$ for z in the third equation,

$$2x - (x + 1) = 2$$
$$2x - x - 1 = 2$$
$$x = 3$$

Substituting back,

$$y = x - 5 = 3 - 5 = -2$$

and

$$z = x + 1 = 4$$

(c) We first clear fractions by multiplying each equation by its LCD.

$$2x + 4y + z = 5$$
$$x + y + z = 6$$
$$2x + 3y + z = 6$$

Subtracting the third equation from the first eliminates x and z, leaving,

$$y = -1$$

Subtracting the second equation from the first,

$$x + 3y = -1$$

or

$$x = -1 - 3y$$

Since $y = -1$,

$$x = -1 - 3(-1) = 2$$

From the second equation,

$$z = 6 - x - y = 6 - 2 + 1 = 5$$

(d) Make the substitution,

$$a = \frac{1}{x} \qquad b = \frac{1}{y} \qquad \text{and} \qquad c = \frac{1}{z}$$

Our equations then become

$$a + 2b - c = -3$$
$$3a + b + c = 4$$
$$a - b + 2c = 6$$

Adding the first and second equations, we get

$$4a + 3b = 1$$

Multiplying the second equation by -2 and adding it to the third,

$$\begin{array}{rcr} -6a - 2b - 2c &=& -8 \\ a - b + 2c &=& 6 \\ \hline -5a - 3b &=& -2 \end{array}$$

or
$$b = \frac{2 - 5a}{3}$$

Substituting back,

$$4a + \frac{3(2 - 5a)}{3} = 1$$

Solving for a,

$$4a + 2 - 5a = 1$$
$$4a - 5a = 1 - 2$$
$$a = 1$$

Substituting back,

$$b = \frac{1}{3}(2 - 5a) = \frac{1}{3}(2 - 5) = -\frac{3}{3} = -1$$

and
$$c = a + 2b + 3 = 1 + 2(-1) + 3 = 1 - 2 + 3 = 2$$

So,
$$x = \frac{1}{a} = 1 \qquad y = \frac{1}{b} = -1 \qquad \text{and} \qquad z = \frac{1}{c} = \frac{1}{2}$$

THIRD-ORDER DETERMINANTS

11.10 Evaluate the following third-order determinants.

(a) $\begin{vmatrix} 1 & 5 & 0 \\ 3 & 2 & -1 \\ 1 & -2 & 3 \end{vmatrix}$
    (b) $\begin{vmatrix} 0 & 1 & -2 \\ 3 & 1 & 5 \\ 0 & -2 & 3 \end{vmatrix}$
    (c) $\begin{vmatrix} 2 & 4 & -1 \\ 3 & 2 & 0 \\ 1 & -1 & 0 \end{vmatrix}$

Solution

Recopy the first two columns and multiply along the diagonals. From Eq. 121,

(a) $\begin{vmatrix} 1 & 5 & 0 \\ 3 & 2 & -1 \\ 1 & -2 & 3 \end{vmatrix} \begin{matrix} 1 & 5 \\ 3 & 2 \\ 1 & -2 \end{matrix} = 6 - 5 + 0 - (0 + 2 + 45) = -46$

(b) $\begin{vmatrix} 0 & 1 & -2 \\ 3 & 1 & 5 \\ 0 & -2 & 3 \end{vmatrix} \begin{matrix} 0 & 1 \\ 3 & 1 \\ 0 & -2 \end{matrix} = 0 + 0 + 12 - (0 + 0 + 9) = 3$

(c) $\begin{vmatrix} 2 & 4 & -1 \\ 3 & 2 & 0 \\ 1 & -1 & 0 \end{vmatrix} \begin{matrix} 2 & 4 \\ 3 & 2 \\ 1 & -1 \end{matrix} = 0 + 0 + 3 - (-2 + 0 + 0) = 5$

11.11 Solve for x.

(a) $\begin{vmatrix} 1 & x & 0 \\ 8 & 2 & 1 \\ 3 & -1 & 2 \end{vmatrix} = 18$
    (b) $\begin{vmatrix} 3 & 0 & 1 \\ 2 & x & -1 \\ -2 & 4 & 2 \end{vmatrix} = -9$

Solution

(a) Evaluating the determinant by Eq. 121,

$$4 + 3x + 0 - (0 - 1 + 16x) = 18$$

Solving for x,

$$3x - 16x = 18 - 4 - 1 = 13$$
$$-13x = 13$$
$$x = -1$$

(b) By Eq. 121,

$$6x + 0 + 8 - (-2x - 12 + 0) = -9$$

Solving for x,

$$6x + 2x = -9 - 8 - 12$$
$$8x = -29$$
$$x = -\frac{29}{8}$$

11.12   Solve for x, y, and z, using determinants.

(a) $x + 2y - z = -3$ (b) $2x + 4y + z = 5$ (c) $5x - 2y - z = 0$
$3x + y + z = 4$ $x + y + z = 6$ $7x + 4y + 2z = 0$
$x - y + z = 6$ $2x + 3y + z = 6$ $6x + 6y - 2z = 5$

Solution

(a) The determinant of the system is

$$D = \begin{vmatrix} 1 & 2 & -1 \\ 3 & 1 & 1 \\ 1 & -1 & 1 \end{vmatrix} \begin{matrix} 1 & 2 \\ 3 & 1 \\ 1 & -1 \end{matrix} = 1 + 2 + 3 - (-1 - 1 + 6) = 2$$

Replacing the x coefficients by the column of constants,

$$\begin{vmatrix} -3 & 2 & -1 \\ 4 & 1 & 1 \\ 6 & -1 & 1 \end{vmatrix} \begin{matrix} -3 & 2 \\ 4 & 1 \\ 6 & -1 \end{matrix} = -3 + 12 + 4 - (-6 + 3 + 8) = 8$$

So,

$$x = \frac{8}{D} = \frac{8}{2} = 4$$

Replacing the y coefficients by the column of constants,

$$\begin{vmatrix} 1 & -3 & -1 \\ 3 & 4 & 1 \\ 1 & 6 & 1 \end{vmatrix} \begin{matrix} 1 & -3 \\ 3 & 4 \\ 1 & 6 \end{matrix} = 4 - 3 - 18 - (-4 + 6 - 9) = -10$$

So,

$$y = \frac{-10}{D} = \frac{-10}{2} = -5$$

Substituting back into the first equation,

$$z = x + 2y + 3 = 4 + 2(-5) + 3 = -3$$

(b) The determinant of the system is

$$D = \begin{vmatrix} 2 & 4 & 1 \\ 1 & 1 & 1 \\ 2 & 3 & 1 \end{vmatrix} \begin{matrix} 2 & 4 \\ 1 & 1 \\ 2 & 3 \end{matrix} = 2 + 8 + 3 - (2 + 6 + 4) = 1$$

Replacing the x coefficients with the column of constants,

$$\begin{vmatrix} 5 & 4 & 1 \\ 6 & 1 & 1 \\ 6 & 3 & 1 \end{vmatrix} \begin{matrix} 5 & 4 \\ 6 & 1 \\ 6 & 3 \end{matrix} = 5 + 24 + 18 - (6 + 15 + 24) = 2$$

So,
$$x = \frac{2}{D} = \frac{2}{1} = 2$$

Replacing the y coefficients with the column of constants,

$$\begin{vmatrix} 2 & 5 & 1 \\ 1 & 6 & 1 \\ 2 & 6 & 1 \end{vmatrix} \begin{matrix} 2 & 5 \\ 1 & 6 \\ 2 & 6 \end{matrix} = 12 + 10 + 6 - (12 + 12 + 5) = -1$$

So,
$$y = \frac{-1}{D} = \frac{-1}{1} = -1$$

Substituting back into the second equation,

$$z = 6 - x - y = 6 - 2 + 1 = 5$$

(c) The determinant of the system is

$$\begin{vmatrix} 5 & -2 & -1 \\ 7 & 4 & 2 \\ 6 & 6 & -2 \end{vmatrix} \begin{matrix} 5 & -2 \\ 7 & 4 \\ 6 & 6 \end{matrix} = -40 - 24 - 42 - (-24 + 60 + 28) = -170$$

Replacing the x coefficients by the column of constants,

$$\begin{vmatrix} 0 & -2 & -1 \\ 0 & 4 & 2 \\ 5 & 6 & -2 \end{vmatrix} \begin{matrix} 0 & -2 \\ 0 & 4 \\ 5 & 6 \end{matrix} = -20 - (-20) = 0$$

So,
$$x = 0$$

Replacing the y coefficients by the column of constants,

$$\begin{vmatrix} 5 & 0 & -1 \\ 7 & 0 & 2 \\ 6 & 5 & -2 \end{vmatrix} \begin{matrix} 5 & 0 \\ 7 & 0 \\ 6 & 5 \end{matrix} = -35 - 50 = -85$$

So,
$$y = \frac{-85}{-170} = \frac{1}{2}$$

Substituting back into the first equation,

$$z = 5x - 2y = 0 - 2\left(\frac{1}{2}\right) = -1$$

NUMBER PROBLEMS

11.13   Find two numbers whose sum is 60 and whose difference is 28.

Solution

Let .
$$x = \text{the larger number}$$
$$y = \text{the smaller number}$$

Then,
$$x + y = 60$$
$$x - y = 28$$

Adding these two equations,
$$2x \qquad = 88$$
$$x = 44$$

Substituting back,
$$y = 60 - 44 = 16$$

11.14 Find the two acute angles in a right triangle if their difference is 40°.

Solution

Let

$$x = \text{the larger acute angle}$$
$$y = \text{the smaller acute angle}$$

So,

$$x - y = 40°$$

and by Eq. 60,

$$x + y = 90°$$

Adding these equations,

$$2x = 130°$$
$$x = 65°$$

Substituting back,

$$y = 90° - 65° = 25°$$

11.15 A certain fraction has a value of 1/3. When the numerator is increased by 3 and the denominator decreased by the same amount, its value becomes 7/9. Find the original fraction.

Solution

Let

$$x = \text{original numerator}$$
$$y = \text{original denominator}$$

Then,

$$\frac{x}{y} = \frac{1}{3}$$

and

$$\frac{x + 3}{y - 3} = \frac{7}{9}$$

Simplifying the first equation,

$$3x = y$$

and the second,

$$9x + 27 = 7y - 21$$
$$9x = 7y - 48$$

Substituting $3x$ for y,

$$9x = 7(3x) - 48$$
$$12x = 48$$

and

$$x = 4$$
$$y = 3(4) = 12$$

So, the original fraction is 4/12.

MIXTURE PROBLEMS

11.16 A dry concrete mixture is available containing 6% cement and 15% sand. How many pounds of cement and how many pounds of sand must be added to 500 lb of this mixture to make a new batch containing 10% cement and 20% sand?

Solution

Let

$$x = \text{pounds of cement to be added}$$
$$y = \text{pounds of sand to be added}$$

There are 0.06(500) lb of cement in the original mixture and 0.10(500 + x + y) lb in the final mixture; so by Eq. 158,

$$0.10(500 + x + y) = 0.06(500) + x$$

Similarly, for the sand,

$$0.20(500 + x + y) = 0.15(500) + y$$

Solving the first equation for x,

$$50 + 0.10x + 0.10y = 30 + x$$
$$0.90x = 20 + 0.10y$$
$$x = 22.2 + 0.11y$$

Substituting into the second equation,

$$100 + 0.20(22.2 + 0.11y) + 0.20y = 75 + y$$
$$100 + 4.44 + 0.022y + 0.20y = 75 + y$$
$$0.778y = 29.4$$
$$y = 37.8 \text{ lb of sand}$$

and $x = 22.2 + 0.11(37.8) = 26.4 \text{ lb of cement}$

11.17   Two solders are available; one contains 60% lead and 40% tin, and a second contains 35% lead and 65% tin.   How much of each must be mixed to produce 200 kg of solder which is 45% lead and 55% tin?

Solution

Let $x =$ kilograms of 60/40 solder needed
                                 $y =$ kilograms of 35/65 solder needed

From Eq. 157,

$$x + y = 200$$

and from Eq. 159,

$$0.60x + 0.35y = 0.45(200)$$
or $60x + 35y = 9000$

Multiplying the first equation by -35 and adding,

$$-35x - 35y = -7000$$
$$\underline{60x + 35y = 9000}$$
$$25x = 2000$$
$$x = 80.0 \text{ kg of 60/40 solder}$$
$$y = 200 - 80 = 12\bar{0} \text{ kg of 35/65 solder}$$

11.18   Two thousand (20$\bar{0}$0) pounds of steel containing 8% nickel is to be made by mixing a steel containing 15% nickel with another containing 6% nickel.   How much of each is needed?

Solution

Let $x =$ pounds of 15% Ni steel needed
                                 $y =$ pounds of 6% Ni steel needed

By Eq. 157,

$$x + y = 20\bar{0}0 \text{ lb}$$
So, $y = 2000 - x$

and by Eq. 159,

$$0.15x + 0.06y = 0.08(2000)$$

or

$$15x + 6y = 16\,000$$

Substituting $(2000 - x)$ for y,

$$15x + 6(2000 - x) = 16\,000$$
$$15x + 12\,000 - 6x = 16\,000$$
$$9x = 4000$$
$$x = 444 \text{ lb}$$

Substituting back,

$$y = 2000 - 444 = 1556 \text{ lb}$$

STATICS PROBLEMS

11.19 A horizontal rod of negligible weight has concentrated loads of 520 lb and 830 lb at its ends. When the smaller of these loads is increased by 150 lb, the center of gravity shifts 1.20 in. toward that end. Find the length of the rod.

Solution

Drawing a diagram, Fig. 11-5, we let

L = length of rod, in.
x = original distance of C.G. from the 830-lb load

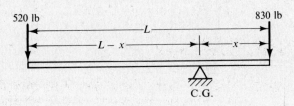

We sum moments about the C.G. By Eq. 175,

Fig. 11-5

$$520(L - x) = 830x$$
$$520L - 520x = 830x$$
$$520L = 1350x$$

or

$$x = 0.385L$$

When the smaller load is increased by 150 lb,

$$(520 + 150)(L - x - 1.20) = 830(x + 1.20)$$
$$670L - 670x - 804 = 830x + 996$$
$$670L = 1500x + 1800$$
$$L = 2.24x + 2.69$$

Substituting $0.385L$ for x,

$$L = 2.24(0.385L) + 2.69 = 0.862L + 2.69$$
$$0.138L = 2.69$$
$$L = 19.5 \text{ in.}$$

11.20 A horizontal bar of negligible weight has concentrated loads of 155 g and 264 g at either end, and it is balanced on a fulcrum, Fig. 11-6. When the smaller load is doubled, the fulcrum must be moved 7.00 cm to keep the bar in balance. How long is the bar?

Solution

Let

L = the length of the bar, cm

x = original distance from fulcrum to 264-g load, cm

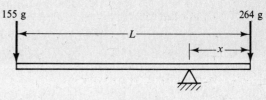

Fig. 11-6

By Eq. 175,

$$264x = 155(L - x) = 155L - 155x$$

or

$$419x = 155L$$

$$x = 0.370L$$

After the smaller load is doubled, we get by Eq. 175,

$$264(x + 7.00) = 2(155)(L - x - 7.00)$$

$$264x + 1850 = 310L - 310x - 2170$$

$$310L = 574x + 4020$$

Substituting $(0.370L)$ for x,

$$310L = 574(0.370L) + 4020$$

$$97.6L = 4020$$

$$L = 41.2 \text{ cm}$$

WORK PROBLEMS

11.21　　On one job, two power shovels excavate $20\,\overline{0}00$ cubic meters (m^3) of earth; the larger shovel working for 41.0 h and the smaller for 35.0 h.

On another job, they remove $42\,\overline{0}00$ m^3, with the larger shovel working 75.0 h and the smaller working 95.0 h.　How much earth can each move in 1 h, working alone?

Solution

Let　　　　　　　x = work rate of larger shovel (m^3/h) working alone

y = work rate of smaller shovel (m^3/h) working alone

By Eq. 163,

$$41x + 35y = 20\,000$$

and

$$75x + 95y = 42\,000$$

From the second equation,

$$x = \frac{42\,000 - 95y}{75} = \frac{8400 - 19y}{15}$$

Substituting into the first equation,

$$41\left(\frac{8400 - 19y}{15}\right) + 35y = 20\,000$$

$$22\,960 - 51.9y + 35y = 20\,000$$

$$16.9y = 2960$$

$$y = 175 \text{ m}^3/\text{h}$$

Substituting back,

$$x = \frac{8400 - 19(175)}{15} = 338 \text{ m}^3/\text{h}$$

11.22   A team of four technicians and three helpers can wire five consoles in 6 days; three technicians and five helpers can wire seven consoles in 8 days. How much can a technician and a helper each do in a day, working alone?

Solution

Let $x =$ technician's work rate, consoles per day, working alone

                             $y =$ helper's work rate, consoles per day, working alone

In one day, 4 technicians and 3 helpers produce

$$4x + 3y \text{ consoles}$$

and in 6 days,

$$6(4x + 3y) \text{ consoles}$$

Since they wire 5 consoles in 6 days,

$$6(4x + 3y) = 5$$

Similarly, $$8(3x + 5y) = 7$$

Removing parentheses,

$$24x + 18y = 5$$
$$\underline{24x + 40y = 7}$$
and subtracting, $-22y = -2$

So,   Helper's rate $= y = \dfrac{1}{11} \cong 0.091$ console per day

Substituting back,

$$6\left[4x + 3\left(\frac{1}{11}\right)\right] = 5$$

$$4x + \frac{3}{11} = \frac{5}{6}$$

$$4x = \frac{5}{6} - \frac{3}{11} = \frac{37}{66}$$

So,   Technician's rate $= x = \dfrac{37}{4(66)} \cong 0.140$ console per day

11.23   During a certain 40-h week, two screw machines produce 85 000 parts, the faster of the two machines working the full time, but with the slower machine down for repairs for 6.0 h. The following week they produce 91 000 parts, with the faster machine down for 3.0 h for lubrication, and the slower machine working 9.0 h overtime. How much can each produce in 1 h, working alone?

Solution

Let $x =$ rate of faster machine, parts per hour

                             $y =$ rate of slower machine, parts per hour

For the first week, by Eqs. 162 and 163,

$$40x + 34y = 85\,000$$

or

$$y = \frac{85\,000 - 40x}{34} = 2500 = 1.18x$$

For the second week,

$$37x + 49y = 91\,000$$

$$x = \frac{91\,000 - 49y}{37} = 2459 - 1.32y$$

Substituting $(2500 - 1.18x)$ for y,

$$x = 2459 - 1.32(2500 - 1.18x) = 2459 - 3300 + 1.56x$$

$$0.56x = 841$$

$$x = 1500 \text{ parts per hour}$$

Substituting back,

$$y = 2500 - 1.18(1500) = 730 \text{ parts per hour}$$

GEOMETRY PROBLEMS

11.24 Cables of two different lengths run from the top of a certain tower to the ground, as in Fig. 11-7. The shorter cables, when drawn taut, reach the ground 12.0 ft from the base of the tower, while the longer cables reach 20.0 ft from the base.   The longer cables are 2.0 ft longer than the short cables.   Find the height of the tower, and the length of the cables.

Solution

Let $x =$ height of tower, ft

   $y =$ length of short cables, ft

Writing the Pythagorean Theorem for each of the right triangles in Fig. 11-7, we have

$$y^2 = x^2 + 144$$

and $(y + 2)^2 = x^2 + 400$

or $y^2 + 4y + 4 = x^2 + 400$

Solving simultaneously, from the first equation,

$$x^2 = y^2 - 144$$

Substituting $y^2 - 144$ for x in the second equation,

$$y^2 + 4y + 4 = y^2 - 144 + 400$$

$$4y = 252$$

$$y = 63.0 \text{ ft} = \text{length of short cables}$$

$$x^2 = (63.0)^2 - 144 = 3825$$

$$x = 61.9 \text{ ft} = \text{height of antenna}$$

$$y + 2 = 65.0 \text{ ft} = \text{length of long cables}$$

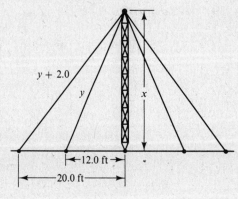

Fig. 11-7

WORD PROBLEMS WITH THREE UNKNOWNS

11.25 Three cylindrical rods are soldered together, as in Fig. 11-8. Find the diameter of each rod if $A = 1.250$ cm, $B = 1.625$ cm, and $C = 1.875$ cm.

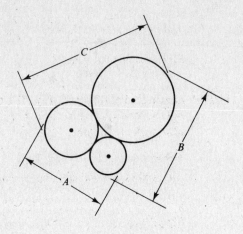

Solution

Let x = diameter of smallest rod, cm

     y = diameter of middle rod, cm

     z = diameter of largest rod, cm

Then, $A = 1.250 = x + y$

             $B = 1.625 = x + z$

             $C = 1.875 = y + z$

From the first of these equations,

$$x = 1.250 - y$$

Substituting into the second equation,

$$1.625 = (1.250 - y) + z$$

or $z = 0.375 + y$

Fig. 11-8

Substituting this into the third equation,

$$1.875 = y + (0.375 + y)$$

$$2y = 1.500$$

$$y = 0.750 \text{ cm}$$

Substituting back,

$$x = 1.250 - 0.750 = 0.500 \text{ cm}$$

and $z = 0.375 + 0.750 = 1.125 \text{ cm}$

DC NETWORK PROBLEMS

11.26 The equations for the three loop currents i_1, i_2, and i_3 in amperes in a certain network are

$$2i_1 + i_2 - i_3 = 5$$
$$3i_1 - 2i_2 + 2i_3 = -3$$
$$i_1 - 3i_2 - 3i_3 = -2$$

Solve this set for the three currents.

Solution

The determinant of the system is, by Eq. 121,

$$D = \begin{vmatrix} 2 & 1 & -1 \\ 3 & -2 & 2 \\ 1 & -3 & -3 \end{vmatrix} = 12 + 2 + 9 - (2 - 12 - 9) = 23 - (-19) = 42$$

Solving for i_1,

$$i_1 = \frac{\begin{vmatrix} 5 & 1 & -1 \\ -3 & -2 & 2 \\ -2 & -3 & -3 \end{vmatrix}}{42} = \frac{30 - 4 - 9 - (-4 - 30 + 9)}{42} = \frac{42}{42} = 1 \text{ A}$$

and for i_2,

$$i_2 = \frac{\begin{vmatrix} 2 & 5 & -1 \\ 3 & -3 & 2 \\ 1 & -2 & -3 \end{vmatrix}}{42} = \frac{18 + 10 + 6 - (3 - 8 - 45)}{42} = \frac{84}{42} = 2 \text{ A}$$

From the first equation, $i_3 = 2i_1 + i_2 - 5 = 2 + 2 - 5 = -1 \text{ A}$

11.27 Apply Kirchhoff's law, Eq. 206, to the circuit of Fig. 11-9 and obtain a set of three equations. Solve these equations simultaneously by any of the methods in this chapter to obtain the values of the three currents.

Solution

Adding the voltage drops around the first loop, starting at A,

$$6.2 - 110i_1 + 12.4 - 150i_1 + 150i_2 - 75.3i_1 = 0$$

Simplifying, $-335i_1 + 150i_2 = -18.6$

$$18.0i_1 - 8.06i_2 = 1$$

In the middle loop, starting at B,

$$-150i_2 + 150i_1 - 12.4 - 572i_2 - 88.5i_2 + 88.5i_3 - 151i_2 = 0$$

Simplifying, $150i_1 - 962i_2 + 88.5i_3 = 12.4$

or $12.1i_1 - 77.6i_2 + 7.14i_3 = 1$

And for the end loop, starting at C,

$$-88.5i_3 + 88.5i_2 - 3.1 - 10.5i_3 = 0$$

Simplifying, $88.5i_2 - 99i_3 = 3.1$

or $28.6i_2 - 31.9i_3 = 1$

Our three equations are thus

$$18.0i_1 - 8.06i_2 \qquad\quad = 1$$
$$12.1i_1 - 77.6i_2 + 7.14i_3 = 1$$
$$28.6i_2 - 31.9i_3 = 1$$

The determinant of the system is

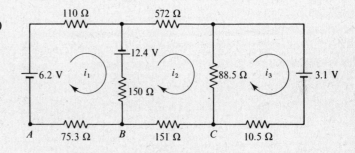

Fig. 11-9

$$D = \begin{vmatrix} 18.0 & -8.06 & 0 \\ 12.1 & -77.6 & 7.14 \\ 0 & 28.6 & -31.9 \end{vmatrix} \begin{matrix} 18.0 & -8.06 \\ 12.1 & -77.6 \\ 0 & 28.6 \end{matrix} = 44\,558 + 0 + 0 - (0 + 3676 + 3111) = 37\,771$$

Solving for i_1,

$$i_1 = \frac{\begin{vmatrix} 1 & -8.06 & 0 \\ 1 & -77.6 & 7.14 \\ 1 & 28.6 & -31.9 \end{vmatrix}}{37\,771} = \frac{2475 - 57.6 + 0 - (0 + 204 + 257)}{37\,771} = \frac{1956}{37\,771} = 51.8 \times 10^{-3} \text{ A} = 51.8 \text{ mA}$$

Solving for i_2,

$$i_2 = \frac{\begin{vmatrix} 18.0 & 1 & 0 \\ 12.1 & 1 & 7.14 \\ 0 & 1 & -31.9 \end{vmatrix}}{37\,771} = \frac{-574 + 0 + 0 - (0 + 129 - 386)}{37\,771} = -\frac{317}{37\,771} = -8.39 \times 10^{-3} \text{ A} = -8.39 \text{ mA}$$

Substituting back,

$$i_3 = \frac{28.6i_2 - 1}{31.9} = \frac{28.6(-8.39 \times 10^{-3}) - 1}{31.9} = -38.9 \times 10^{-3} \text{ A} = -38.9 \text{ mA}$$

Supplementary Problems

11.28 Solve graphically for x and y.

$$2x = 6y = 4$$
$$5x = 2 - y$$

11.29 Solve for x and y by addition or subtraction.

(a) $2x - y = 3$ (b) $x - 4y - 4 = 0$
     $x + 3y = 6$ $2y + 6x = 0$

11.30 Solve for x and y by substitution.

(a) $2x - 6y = 4$ (b) $x = 2y + 3$
     $5x + y - 2 = 0$ $2 - 6x + y = 0$

11.31 Solve for x and y by any method.

(a) $\dfrac{x}{3} - 2y + \dfrac{3}{4} = 0$ (b) $\dfrac{2x-1}{2} + \dfrac{3y+2}{3} = 2$

   $x - \dfrac{2}{3}y = 0$ $\dfrac{x+6}{4} - \dfrac{y-1}{2} = 8$

11.32 Solve for x and y in terms of a and b.

(a) $x + 5y = a$ (b) $\dfrac{2x}{a} - \dfrac{y}{b} = 3$
     $2x - 3y = b$
                              $\dfrac{-x}{a} + \dfrac{3y}{b} = 1$

11.33 Solve the following systems of equations for x, y, and z:

(a) $2x - y = 3$ (b) $x + y = 2$
     $3x + 2z = 2$ $2x - 6z = 5$
      $y - z = -5$ $y + z = -3$

11.34 Evaluate the following second-order determinants:

(a) $\begin{vmatrix} 4 & 2 \\ 3 & 1 \end{vmatrix}$ (b) $\begin{vmatrix} -1 & 6 \\ 3 & -5 \end{vmatrix}$

11.35 Solve by determinants.

(a) $4x - y = 4$ (b) $5x + 7y = -10$
     $x + 3y = -2$ $-3x - 6y = 4$

11.36 Evaluate the following third-order determinants:

(a) $\begin{vmatrix} 1 & 3 & 2 \\ 2 & 0 & 5 \\ 4 & 1 & 6 \end{vmatrix}$ (b) $\begin{vmatrix} 3 & 4 & 7 \\ 2 & 1 & 0 \\ 1 & 0 & 3 \end{vmatrix}$

11.37 Solve for x, y, and z using determinants.

(a) $x - y + z = 2$ (b) $x + y = 2$
     $2x + 3y - 2z = -4$ $2x - 6z = 5$
     $3x + y + 3z = 5$ $y + z = -3$

11.38   Find two numbers whose sum is 60 and whose difference is 14.

11.39   A fraction has a reduced value of 3/4.   When the numerator is decreased by 2 and the denominator is increased by 4, its value becomes 1/2.   Find the original fraction.

11.40   Mixture 1 contains 10% sand and 90% stone, and mixture 2 contains 25% sand and 75% stone.   How many pounds of each mixture must be combined to make 2000 lb of a new mixture containing 15% sand and 85% stone?

11.41   Six slower stamping machines working together with two faster stamping machines can produce 8000 parts in 4 days.   Also, five slower machines working together with four faster machines can produce 15 000 parts in 5 days.   How many parts per day can be produced by a slower machine and by a faster machine, working alone?

11.42   Three amplifiers and five speakers cost $700.   Five amplifiers and eight speakers cost $1150.   Find the cost of each speaker and each amplifier.

Answers to Supplementary Problems

11.28   From the graph we find the coordinates of the point of intersection: $x = 1/2$, $y = -1/2$.

11.29 (a) $x = \dfrac{15}{7}, y = \dfrac{9}{7}$ (b) $x = \dfrac{4}{13}, y = -\dfrac{12}{13}$

11.30 (a) $x = \dfrac{1}{2}, y = -\dfrac{1}{2}$ (b) $x = \dfrac{1}{11}, y = -1\dfrac{5}{11}$

11.31 (a) $x = \dfrac{9}{32}, y = \dfrac{27}{64}$ (b) $x = \dfrac{83}{9}, y = -\dfrac{133}{18}$

11.32 (a) $x = \dfrac{3a + 5b}{13}, y = \dfrac{2a - b}{13}$ (b) $x = 2a, y = b$

11.33 (a) $a = -\dfrac{2}{7}, y = -3\dfrac{4}{7}, z = 1\dfrac{3}{7}$ (b) $x = 6\dfrac{1}{4}, y = -4\dfrac{1}{4}, z = 1\dfrac{1}{4}$

11.34 (a) -2 (b) -13

11.35 (a) $x = \dfrac{10}{13}, y = -\dfrac{12}{13}$ (b) $x = -3\dfrac{5}{9}, y = 1\dfrac{1}{9}$

11.36 (a) 23 (b) -22

11.37 (a) $x = \dfrac{1}{16}, y = -\dfrac{1}{4}, z = 1\dfrac{11}{16}$ (b) $x = 6\dfrac{1}{4}, y = -4\dfrac{1}{4}, z = 1\dfrac{1}{4}$

11.38 $x = 37, y = 23$

11.39 $\dfrac{12}{16}$

11.40   1330 lb of mixture 1 and 667 lb of mixture 2.

11.41   Slower machine makes 143 parts per day.   Faster machine makes 571 parts per day.

11.42 $150 for an amplifier and $50 for a speaker

Chapter 12

The Straight Line

12.1 DISTANCE BETWEEN TWO POINTS

(a) Points on a Coordinate Axis. To find the distance between two points lying on the x (or y) axis, merely take the absolute value of the differences in the abscissas if on the x axis or the ordinates if on the y axis.

Example 12.1: Find the distance between points A and B, and between C and D, in Fig. 12-1.

Solution: The distance between A and B is

$$7 - 3 = 4 \text{ units}$$

and between C and D is

$$4 - (-2) = 6 \text{ units}$$

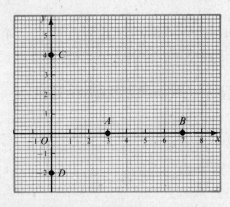

Fig. 12-1

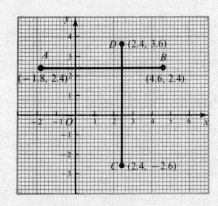

Fig. 12-2

(b) Lines Parallel to a Coordinate Axis. If the line segment formed by joining two points is parallel to the x (or y) axis, its length is found by taking the absolute value of the difference in the abscissas (or ordinates) of the two points.

Example 12.2: Find the length of the line segments AB and CD in Fig. 12-2.

Solution: The length AB is

$$4.6 - (-1.8) = 6.4 \text{ units} \qquad \text{(the difference in the abscissas)}$$

and the length CD is

$$3.6 - (-2.6) = 6.2 \text{ units} \qquad \text{(the difference in the ordinates)}$$

(c) Distance between Any Two Points. Let d be the distance between any two points P_1 and P_2, as in Fig. 12-3. We drop perpendiculars to the x axis and draw a horizontal line through P_1, forming a right triangle whose sides are d, $x_2 - x_1$, and $y_2 - y_1$. Then by the Pythagorean Theorem,

324

$$\boxed{\begin{array}{c|c|c} \text{Distance} & & \\ \text{Formula} & d = \sqrt{(x_2 - x_1)^2 + (y_2 - y_1)^2} & \mathbf{128} \end{array}}$$

Example 12.3:   Find the distance between the points $(3, -2)$ and $(-4, 5)$.

Solution: It does not matter which of the points we call P_1 or P_2, so let us choose $(3, -2)$ as P_1. Then,

$$x_1 = 3 \qquad x_2 = -4$$
$$y_1 = -2 \qquad y_2 = 5$$

So, by Eq. 128,

$$d^2 = (-4 - 3)^2 + [5 - (-2)]^2 = 49 + 49 = 98$$
$$d = \sqrt{98} \cong 9.9 \text{ units}$$

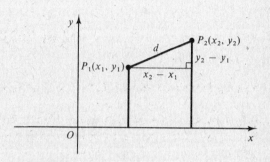

Fig. 12-3   Length of a line segment.

(d)   Directed Distance. We have thus far considered only the *magnitude* of the distance between two points. In some applications, it is necessary to consider *direction* as well. The *directed* distance AB is defined as the distance *from* point A *to* point B.

Example 12.4:   Given two points, $A(4,0)$ and $B(8,0)$, find the directed distances AB and BA.

Solution

$$AB = 8 - 4 = 4$$
$$BA = 4 - 8 = -4$$

12.2   SLOPE

(a)   Rise and Run. The difference in the ordinates of any two points on a line is called the *rise*. The difference between the abscissas of the same points is called the *run*. The *slope m* of the line is the ratio of the rise to the run.

$$\text{Slope } m = \frac{\text{rise}}{\text{run}}$$

If the two points are $P_1(x_1, y_1)$ and $P_2(x_2, y_2)$, the rise will be

$$\text{Rise} = y_2 - y_1$$

and the run,

$$\text{Run} = x_2 - x_1$$

as in Fig. 12-4.

So,

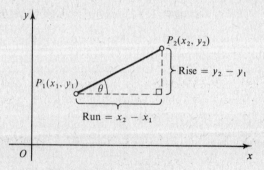

Fig. 12-4   Rise and run.

$$\boxed{\begin{array}{c|c} \text{Slope } m = \dfrac{\text{rise}}{\text{run}} = \dfrac{y_2 - y_1}{x_2 - x_1} & \mathbf{129} \end{array}}$$

Example 12.5:   Find the slope of the line connecting the points $(2, 4)$ and $(1, 1)$.

Solution: Let us call the first of the two points P_1 and the second P_2.   The rise is

$$\text{Rise} = y_2 - y_1 = 1 - 4 = -3$$

and the run is

$$\text{Run} = x_2 - x_1 = 1 - 2 = -1$$

so the slope is

$$m = \frac{-3}{-1} = 3$$

If we had chosen to call the first point P_2 and the second point P_1, we would have gotten

$$m = \frac{-4+1}{-2+1} = \frac{-3}{-1} = 3$$

the same as before.   In general it doesn't matter which of the points is chosen for P_1 or P_2.

Example 12.6:   Find the slope of the line passing through the points $(-3, 5)$ and $(2, 1)$.

Solution: From Eq. 129, taking the point $(-3, 5)$ as P_1 and $(2, 1)$ as P_2,

$$m = \frac{1-5}{2-(-3)} = \frac{-4}{5} = -0.8$$

(b)   Some Special Slopes

1. The slope of a line parallel to the x axis is zero.
2. As a line approaches the vertical, its slope approaches infinity, although the slope is not defined for a line parallel to the y axis.
3. Lines parallel to each other have equal slopes.
4. The slope m_1 of any line is the negative reciprocal of the slope m_2 of a line perpendicular to it.

Slopes of Two Perpendicular Lines	$m_1 = -\dfrac{1}{m_2}$	**136**

(Derivation in Problem 12.10)

Example 12.7:   Find the slope of any line perpendicular to a line which has a slope of 2.

Solution: By Eq. 136,

$$m = -\frac{1}{2}$$

Common Error	The minus sign in Eq. 136 is often forgotten. $$m_1 \neq \frac{1}{m_2}$$

(c)   Pitch.   In some applications, the term *pitch* is used instead of *slope*.

Example 12.8:   Find the pitch of a roof that rises 5 m in a run of 10 m.

Solution: By Eq. 129,

$$\text{Slope} = \frac{5}{10} = \frac{1}{2}$$

So,

$$\text{Pitch} = \text{slope} = \frac{1}{2}$$

12.3 ANGLE OF INCLINATION

(a) Definition. The angle that a line makes with the x axis (or with any line parallel to the x axis) is called the *angle of inclination*, θ. It is always measured counterclockwise (CCW) from the positive x direction as shown in Fig. 12-5.

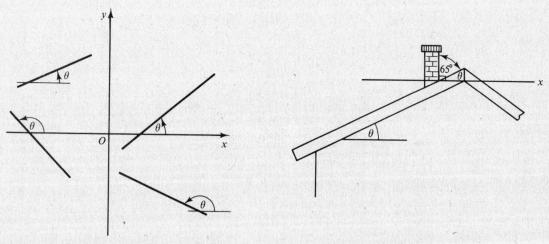

Angles of inclination of several line segments.
Fig. 12-5

Angle of inclination of a roof.
Fig. 12-6

Example 12.9:   Find the angle of inclination of the roof in Fig. 12-6, if the x axis is taken in the horizontal direction.

Solution: If we assume the chimney to be vertical, it makes an angle of 90° with the horizontal. Therefore the angle the roof makes with the chimney is complementary to the angle of inclination. So

$$\theta = 90° - 65° = 25°$$

(b) Relation between Slope and Angle of Inclination. Since the tangent of an angle in a right triangle is equal to the side opposite the angle divided by the side adjacent to it, Eq. 75, it is clear from Fig. 12-4 that

$$\tan \theta = \frac{y_2 - y_1}{x_2 - x_1}$$

but, by Eq. 129,

$$\frac{y_2 - y_1}{x_2 - x_1} = m$$

So,
$$\boxed{m = \tan \theta} \qquad \textbf{130}$$

Example 12.10:   Find the slope of a line having an angle of inclination of 30°.

Solution: By Eq. 130,

$$m = \tan 30° = 0.577$$

Example 12.11:   Find the angle of inclination of a line having a slope of 2.

Solution: By Eq. 130,

$$\tan \theta = 2$$

$$\theta = \arctan 2 = 63.4°$$

Common Error	When the slope is negative, the angle of inclination will be between 90° and 180°. Find the reference angle (see Section 8.6) and subtract it from 180°.

Example 12.12:   Find the angle of inclination of the line connecting the points $(1, 1)$ and $(-5, 3)$.

Solution: By Eq. 129,

$$m = \frac{3-1}{-5-1} = -\frac{1}{3}$$

$$\text{Reference angle} = \arctan\left(\frac{1}{3}\right) = 18.4°$$

Subtracting from 180°,

$$\theta = 180° - 18.4° = 161.6°$$

(c) Angle between Two Lines.   In Fig. 12-7, line L_1 has an angle of inclination θ_1 and line L_2 has an angle of inclination θ_2.   The angle A between the two lines can be found using Eq. 60,

$$A + \theta_1 + (180 - \theta_2) = 180$$

Transposing, $\boxed{A = \theta_2 - \theta_1}$ **137**

Example 12.13:   Find the angle A between line 1 which has an angle of inclination of 25° and line 2 which has an angle of inclination of 65°.

Solution: By Eq. 137, $A = 65° - 25° = 40°$.

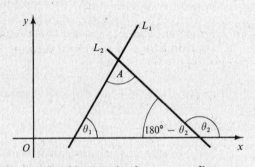

Angle of intersection between two lines.
Fig. 12-7

12.4 EQUATION OF A LINE

(a) General Form. The general linear equation can have at most an x term, a y term, and a constant term.   It does not contain any term greater than first degree.

General Form	$Ax + By + C = 0$	**131**

where A, B, and C are constants.

Example 12.14:   Which of the following are linear (first-degree) equations?

(a) $y = x^2 - 5$ (b) $x - xy = 2$ (c) $y = 2x - 4$

Solution

(a) $y = x^2 - 5$ is of second degree.

(b) $x - xy = 2$ is of second degree.

(c) $y = 2x - 4$ is linear.

Example 12.15:   Write the equation $y = 5 - 3x$ in general form.

Solution: Rearranging, $3x + y - 5 = 0$.

(b) Point-Slope Form. Let x_1 and y_1 be the coordinates of a known point P_1 on a line (Fig. 12-8), and let $P(x, y)$ be any other point on the line. From Eq. 129, the slope is

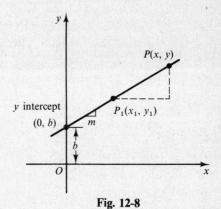

Fig. 12-8

Point-Slope Form	$m = \dfrac{y - y_1}{x - x_1}$	**134**

Example 12.16: Write the equation in general form of the line having a slope of 3 and passing through the point $(2, 5)$.

Solution: Substituting $m = 3$, $x_1 = 2$, $y_1 = 5$ into Eq. 134.

$$3 = \frac{y - 5}{x - 2}$$

Multiplying by $x - 2$,

$$3x - 6 = y - 5$$

or, in general form,

$$3x - y - 1 = 0$$

(c) Slope-Intercept Form. The y intercept is the point where the line cuts the y axis, as in Fig. 12-8. Its coordinates are $(0, b)$, where b is the distance from the point to the origin. Substituting these coordinates into Eq. 134,

$$m = \frac{y - b}{x - 0}$$

Multiplying by x,

$$y - b = mx$$

and transposing, we get,

Slope-Intercept Form	$y = mx + b$	**132**

Example 12.17: Write the equation of the line having a slope of 3 and a y intercept of 2, in both slope-intercept and general form.

Solution: Substituting $m = 3$ and $b = 2$ into Eq. 132,

$$y = 3x + 2 \qquad \text{(in slope-intercept form)}$$

or

$$3x - y + 2 = 0 \qquad \text{(in general form)}$$

Example 12.18: Find the slope and y intercept of the line

$$2x - 6y - 5 = 0$$

Solution: Solving for y,

$$6y = 2x - 5$$

$$y = \frac{1}{3}x - \frac{5}{6}$$

So,

$$m = \frac{1}{3} \qquad \text{and} \qquad b = -\frac{5}{6}$$

(d) Two-Point Form. If two points P_1 and P_2 on a line are known, Fig. 12-9, we can write the slope of the line,

$$m = \frac{y_2 - y_1}{x_2 - x_1}$$

The slope of the line segment connecting P_1 with any other point P on the same line is

$$m = \frac{y - y_1}{x - x_1}$$

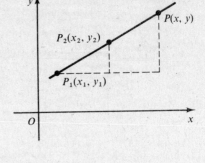

Since these slopes must be equal, we get

Fig. 12-9

Two-Point Form	$\dfrac{y - y_1}{x - x_1} = \dfrac{y_2 - y_1}{x_2 - x_1}$	**133**

Example 12.19:   Write the equation in general form of the line passing through the points $(5, 6)$ and $(1, 3)$.

Solution: Calling the first given point P_2 and the second P_1 and substituting in Eq. 133,

$$\frac{y - 3}{x - 1} = \frac{6 - 3}{5 - 1} = \frac{3}{4}$$

Cross-multiplying, $3x - 3 = 4y - 12$

and transposing, $3x - 4y + 9 = 0$

(e) Intercept Form. For a line that crosses the x axis at $x = a$ and crosses the y axis at $y = b$, Fig. 12-10, we get, by substituting the points $(a, 0)$ and $(0, b)$ into Eq. 133,

$$\frac{y - 0}{x - a} = \frac{b - 0}{0 - a}$$

Cross-multiplying, $bx - ab = -ay$

$$bx + ay = ab$$

Dividing by ab,

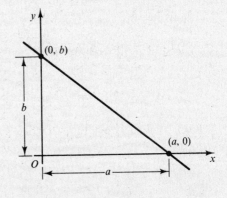

$\dfrac{x}{a} + \dfrac{y}{b} = 1$	**135**

Fig. 12-10

Example 12.20:   Write the equation of a line, in general form, having an x intercept of 5 and a y intercept of 2.

Solution: Substituting into Eq. 135, with $a = 5$ and $b = 2$,

$$\frac{x}{5} + \frac{y}{2} = 1$$

Multiplying by the LCD 10,

$$2x + 5y = 10$$

Transposing, $2x + 5y - 10 = 0$

Solved Problems

DISTANCE BETWEEN TWO POINTS

12.1   Find the lengths of the line segments connecting the following pairs of points:

(a) (0, 9) and (0, −3)   (b) (0, 7) and (0, 2)   (c) (0, −2) and (0, −7)

Solution

Since these points all lie on the y axis, we find the distance between each pair simply by subtracting the ordinates.

(a) $d = 9 − (−3) = 9 + 3 = 12$ (b) $d = 7 − 2 = 5$ (c) $d = −2 − (−7) = −2 + 7 = 5$

12.2   Find the lengths of the line segments connecting the following pairs of points:

(a) (9, 0) and (3, 0)   (b) (6, 0) and (−3, 0)   (c) (7, 0) and (2, 0)

Solution

Since these points all lie on the x axis, we find the distance between each pair by subtracting the abscissas.

(a) $d = 9 − 3 = 6$ (b) $d = 6 − (−3) = 6 + 3 = 9$ (c) $d = 7 − 2 = 5$

12.3   Find the lengths of the line segments connecting the following pairs of points:

(a) (6, 5) and (2, −3)   (b) (4, 5) and (−1, 1)   (c) (7, 3) and (−1, −2)

Solution

We substitute into the distance formula, Eq. 128, in each case taking the first given point as P_2 and the second point as P_1.

(a) $d = \sqrt{(6 − 2)^2 + [5 − (−3)]^2} = \sqrt{80} \cong 8.9$ (c) $d = \sqrt{[7 − (−1)]^2 + [3 − (−2)]^2} = \sqrt{89} \cong 9.4$

(b) $d = \sqrt{[4 − (−1)]^2 + (5 − 1)^2} = \sqrt{41} \cong 6.4$

12.4   Find the lengths of bridge girders A and B in Fig. 12-11.

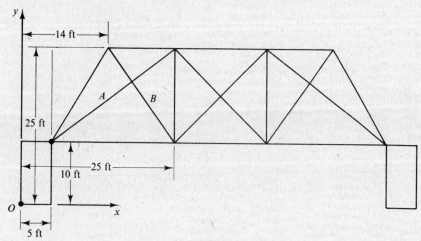

Fig. 12-11

Solution

Letting the origin be at O and taking the x axis in the horizontal direction, the ends of girder A have the coordinates $(5, 10)$ and $(25, 25)$. Taking the first of these points as P_1 and the second as P_2, we get, by Eq. 128,

$$\text{Length of } A = \sqrt{(25 - 5)^2 + (25 - 10)^2} = \sqrt{20^2 + 15^2} = \sqrt{625} = 25 \text{ ft}$$

for girder B, taking P_1 as $(14, 25)$ and P_2 as $(25, 10)$,

$$\text{Length of } B = \sqrt{(25 - 14)^2 + (10 - 25)^2} = \sqrt{11^2 + (-15)^2} = \sqrt{346} \cong 18.6 \text{ ft}$$

12.5   Find the distance between the holes in the template in Fig. 12-12.

Solution

By the distance formula, Eq. 128,

$$d^2 = (3.625 - 0.807)^2 + (1.575 - 0.552)^2$$

$$= 7.941 + 1.047$$

$$= 8.988$$

Thus,

$$d = 2.998 \text{ cm}$$

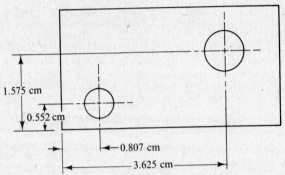

Fig. 12-12

DIRECTED DISTANCE

12.6   Find the directed distances PQ and QP between the points $P(0, 2)$ and $Q(0, -3)$.

Solution
$$PQ = -3 - 2 = -5$$
$$QP = 2 - (-3) = 5$$

SLOPE

12.7   Find the slope of the line passing through the points $(6, 5)$ and $(2, -3)$.

Solution

From Eq. 129, taking $(2, -3)$ as P_1 and $(6, 5)$ as P_2,

$$m = \frac{5 - (-3)}{6 - 2} = \frac{8}{4} = 2$$

12.8   Find the slope of the line passing through the points $(4, 5)$ and $(-1, 1)$.

Solution

From Eq. 129, taking $(-1, 1)$ as P_1 and $(4, 5)$ as P_2,

$$m = \frac{5 - 1}{4 - (-1)} = \frac{4}{5}$$

12.9   Find the slope of the line passing through the points $(7, 3)$ and $(-1, -2)$.

Solution

From Eq. 129, taking $(-1, -2)$ as P_1 and $(7, 3)$ as P_2,

$$m = \frac{3 - (-2)}{7 - (-1)} = \frac{5}{8}$$

12.10 Derive Eq. 136,

$$m_1 = -\frac{1}{m_2}$$

Solution

In Fig. 12-13, line L_1 has an angle of inclination of θ_1 and a slope

$$m_1 = \tan \theta_1 = \frac{a}{b}$$

The perpendicular L_2 to L_1 has an angle of inclination of θ_2 and a slope

$$m_2 = \tan \theta_2 = \frac{b}{-a} = -\frac{1}{m_1}$$

Therefore, $m_1 = -\dfrac{1}{m_2}$

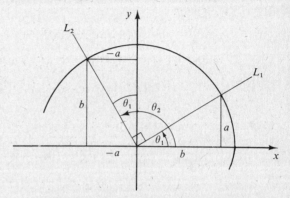

Fig. 12-13   Slopes of perpendicular lines.

12.11 A line has a slope of 3.58.   Find the slope of a line perpendicular to it.

Solution

By Eq. 136,

$$m = -\frac{1}{3.58} = -0.279$$

12.12 Find the slopes of girders A and B in Problem 12.4.

Solution

Using the definition of slope, Eq. 129,

$$\text{Slope of } A = \frac{15}{20} = \frac{3}{4}$$

and

$$\text{Slope of } B = -\frac{15}{11}$$

12.13 A vertical concentrated load is applied to a roof, Fig. 12-14, having a pitch of 3/4.   Find the slopes of the components of the load normal and tangential to the roof.

Solution

The tangential component is parallel to the roof and thus has the same slope of 3/4.   The normal component is perpendicular to the roof, so its slope will be the negative reciprocal of 3/4, or $-4/3$.

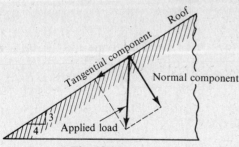

Fig. 12-14

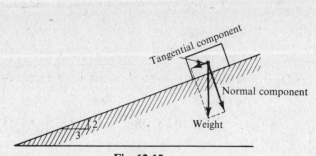

Fig. 12-15

12.14 A carton is placed on a material-handling chute, Fig. 12-15, which is inclined with a slope of 2/3. Find the slopes of the tangential and normal components of the weight of the carton.

Solution

The tangential component is parallel to the chute, and so it will have the same slope, 2/3. The normal component will have a slope equal to the negative reciprocal of 2/3, or $-3/2$.

ANGLE OF INCLINATION

12.15 The angle of inclination of a line is 32°. Find its slope.

Solution

From Eq. 130,

$$m = \tan 32° = 0.62$$

12.16 The slope of a line is 3. Find the angle of inclination.

Solution

From Eq. 130,

$$\tan \theta = 3$$

So,

$$\theta = \arctan 3 = 71.6°$$

12.17 The slope of a line is -2. Find the angle of inclination.

Solution

From Eq. 130,

$$\tan \theta = -2$$

So,

$$\theta = \arctan(-2) = 116.6°$$

GRADE AND GRADIENT

In surveying, the *grade* of a surface (also called *gradient* and *rate of grade*) is the vertical rise or fall of the surface per hundred units of horizontal run. Thus, a ten percent grade has a change in elevation of 10 feet in a horizontal distance of 100 feet. Ascending grades are positive and descending grades are negative.

The grade, since it is really a percentage, is therefore 100 times the rise divided by the run, or 100 times the slope.

$$\text{Grade} = \frac{\text{rise}}{\text{run}} \times 100 = \text{slope} \times 100 = 100m$$

12.18 A road surface rises 2.74 ft in a horizontal distance of 58.3 ft. Find the slope and the grade.

Solution

By Eq. 129,

$$\text{Slope} = m = \frac{2.74}{58.3} = 0.0470$$

So,

$$\text{Grade} = 100m = 4.70\%$$

12.19 A section of a road is to have a grade of $+3\%$. How much should it rise in a run of 258 m?

Solution

By Eq. 129,

$$\text{Grade} = 3 = \frac{\text{rise}}{258} \times 100$$

So,

$$\text{Rise} = \frac{3(258)}{100} = 7.74 \text{ m}$$

12.20 A section of roadway, Fig. 12-16, has a grade of $+4.50\%$. To what vertical angle must a transit be set so that its line of sight is parallel to the road surface?

Solution

By Eq. 129,

$$\text{Slope } m = \frac{4.50}{100} = 0.0450$$

By Eq. 130,

$$m = \tan \theta$$

$$= 0.0450$$

$$\theta = +2.58° = +2°35'$$

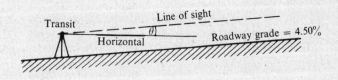

Fig. 12-16

12.21 A certain fork lift cannot climb a grade greater than 10%. How long a ramp is required to enable the lift to climb to a balcony 5 m above the floor?

Solution

Since

$$\text{Grade} = \text{slope} \times 100$$

$$\text{Slope} = \frac{\text{grade}}{100} = \frac{10}{100} = 0.1$$

By Eq. 129,

$$\text{Run} = \frac{\text{rise}}{\text{slope}} = \frac{5}{0.1} = 50 \text{ m}$$

12.22 What is the angle of inclination of the ramp of Problem 12.21?

Solution

By Eq. 130,

$$m = \tan \theta = 0.1$$

$$\theta = 5.7°$$

12.23 Find the angle between two lines having angles of inclination of $20°$ and $140°$.

Solution

From Eq. 137,

$$A = 140° - 20° = 120°$$

12.24 Find the angle between the two lines having slopes of -2 and $3/5$.

Solution

From Eq. 130,

$$\tan\theta_1 = \frac{3}{5}$$

$$\theta_1 = 31.0°$$

and

$$\tan\theta_2 = -2$$

$$\theta_2 = 117°$$

From Eq. 137,

$$A = 117° - 31° = 86°$$

EQUATION OF A LINE

12.25 A line has a slope of -2 and passes through the point $(1, -2)$. Write its equation in general form.

Solution

From Eq. 134,

$$-2 = \frac{y - (-2)}{x - 1}$$

Multiplying by $x - 1$,

$$-2x + 2 = y + 2$$

Transposing,

$$2x + y = 0$$

12.26 Write the equation in general form of a line having a slope of -1 and a y intercept of 3.

Solution

From Eq. 132, with $m = -1$ and $b = 3$,

$$y = (-1)x + 3$$

Transposing,

$$x + y - 3 = 0$$

12.27 Find the slope and y intercept of the line

$$6x + 4y - 7 = 0$$

Solution

Solving for y,

$$4y = -6x + 7$$

$$y = -\frac{3}{2}x + \frac{7}{4}$$

So,

$$\text{Slope} = -\frac{3}{2} \qquad y \text{ intercept} = \frac{7}{4}$$

12.28 Write the equation in general form of the line passing through the points $(3, 5)$ and $(-1, 2)$.

Solution

From Eq. 133, taking P_1 as $(3, 5)$,

$$\frac{y - 5}{x - 3} = \frac{2 - 5}{-1 - 3} = \frac{-3}{-4}$$

Cross-multiplying, $\qquad 4y - 20 = 3x - 9$

and transposing, $\qquad 3x - 4y + 11 = 0$

12.29 Write an equation, in general form, of the line passing through the points $(-2, 5)$ and $(3, -4)$.

Solution

From Eq. 133, taking P_1 as $(-2, 5)$ and P_2 as $(3, -4)$,

$$\frac{y - 5}{x - (-2)} = \frac{-4 - 5}{3 - (-2)}$$

$$\frac{y - 5}{x + 2} = \frac{-9}{5}$$

Cross-multiplying, $\qquad -9x - 18 = 5y - 25$

Transposing, $\qquad -9x - 5y + 7 = 0$

which may also be written,

$$9x + 5y - 7 = 0$$

12.30 Write the equation, in general form, of the line having an x intercept of 4 and a y intercept of -3.

Solution

Substituting in Eq. 135, with $a = 4$ and $b = -3$,

$$\frac{x}{4} + \frac{y}{-3} = 1$$

Multiplying by 12,

$$3x - 4y = 12$$

Transposing, $\qquad 3x - 4y - 12 = 0$

12.31 Write the equation, in general form, of the line which crosses the x axis at $x = -3$ and crosses the y axis at $y = 5$.

Solution

Substituting into Eq. 135, with $a = -3$ and $b = 5$,

$$\frac{x}{-3} + \frac{y}{5} = 1$$

Multiplying by -15,

$$5x - 3y = -15$$

Transposing, $\qquad 5x - 3y + 15 = 0$

12.32 A voltage increases linearly from an initial value of 12.4 V to a value of 36.8 V in an elapsed time of 10.6 s. (*a*) Find the slope of the voltage ramp, if *time* is taken as the abscissa and *voltage* as the ordinate. (*b*) Write the equation of the line in slope-intercept form.

Solution

(*a*) From Eq. 129,

$$\text{Slope} = \frac{V_2 - V_1}{10.6} = \frac{36.8 - 12.4}{10.6} = 2.30 \text{ volts per millisecond (V/ms)}$$

(*b*) Using the two-point form, Eq. 133, with the points (0, 12.4) and (10.6, 36.8),

$$\frac{V - 12.4}{t - 0} = \frac{36.8 - 12.4}{10.6 - 0} = 2.30$$

Solving for *V*,

$$V - 12.4 = 2.30t$$

$$V = 2.30t + 12.4$$

or, alternatively, substitute into Eq. 132, with $m = 2.30$ and $b = 12.4$ (the *y* intercept equals 12.4 because that is the value of the ordinate, voltage, when the abscissa, time, equals zero),

$$V = 2.30t + 12.4$$

12.33 The slope of the ramp portions of the waveform, Fig. 12-17, is 1/2 V/ms. (*a*) Find the maximum voltage reached. (*b*) Write the equation of the first ramp. (*c*) Find the voltage at 2.00 ms. (*d*) Find the time at which the voltage is 2.20 V.

Solution

(*a*) From Eq. 129,

$$\text{Rise} = \text{run} \times \text{slope}$$

$$V - 1.85 = 3.36\left(\frac{1}{2}\right) = 1.68$$

So, $V = 3.53 \text{ V}$

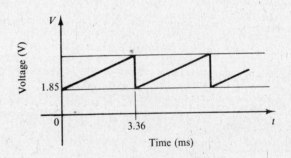

Fig. 12-17

(*b*) Using the slope-intercept form, Eq. 132, with $m = 1/2$ and $b = 1.85$,

$$V = \frac{1}{2}t + 1.85$$

(*c*) Substituting 2 ms into the equation of our line,

$$V = \frac{1}{2}(2) + 1.85 = 2.85 \text{ V}$$

(*d*) Substituting $V = 2.20$ into our equation and solving for *t*,

$$\frac{1}{2}t = 2.20 - 1.85 = 0.35$$

$$t = 0.70 \text{ ms}$$

TEMPERATURE COEFFICIENT OF RESISTANCE

12.34   The resistance R of a metal conductor at some temperature t is given by the formula

$$\boxed{R = R_1[1 + \alpha(t - t_1)]} \quad \textbf{208}$$

Where R_1 is the resistance at a temperature of t_1 degrees Celsius, and α is the temperature coefficient of resistance at temperature t_1. (a)   Write this linear equation in slope-intercept form. (b)   Find the slope. (c)   Find the y intercept.

Solution

(a)   Removing the brackets and parentheses,

$$R = R_1(1 + \alpha t - \alpha t_1) = R_1 + R_1\alpha t - R_1\alpha t_1$$

Regrouping,

$$R = (R_1\alpha)t + (R_1 - R_1\alpha t_1)$$

which is in slope-intercept form, with $m = R_1\alpha$ and $b = (R_1 - R_1\alpha t_1)$.

(b)   The slope is the coefficient of t,

$$\text{Slope} = R_1\alpha$$

(c)   The y intercept is the constant term

$$y \text{ intercept} = R_1 - R_1\alpha t_1$$

12.35   The temperature coefficient of resistance for copper is

$$\alpha = \frac{1}{234.5t_1}$$

(Note here that t is temperature.)
   If the resistance of a copper coil at 20.0°C is 46.8 Ω, find its resistance at 50.0°C.

Solution

The coefficient of resistance at 20.0°C is

$$\alpha = \frac{1}{234.5(20.0)} = 0.000\,213$$

From Eq. 208,

$$R = 46.8[1 + 0.000\,213(50.0 - 20.0)] = 47.1\ \Omega$$

12.36   How fast is the resistance of the coil of Problem 12.35 increasing with temperature?

Solution

The rate of increase is given by the slope of the line, which is, from Problem 12.34, $R_1\alpha$.

$$\text{Rate of increase} = R_1\alpha = 46.8(0.000\,213) = 0.009\,97\ \Omega/°C$$

TEMPERATURE GRADIENTS

12.37 If one side of a uniform wall, Fig. 12-18, is at a temperature t_1 and the other side at temperature t_2, the temperatures within the wall will plot as a straight line between t_1 and t_2. Write the equation of this straight line, taking the origin as shown in Fig. 12-18.

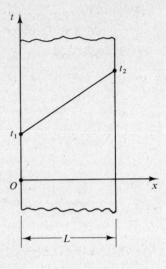

Fig. 12-18   Temperature gradient in a uniform wall.

Solution

The rise of the line is $t_2 - t_1$, in a run of L, so the slope is, by Eq. 129,

$$\text{Slope} = \frac{t_2 - t_1}{L}$$

and the y intercept is t_1. Using the slope-intercept form, Eq. 132,

$$t = \frac{t_2 - t_1}{L} x + t_1$$

where t is the temperature at a point located a distance x from the side having a temperature t_1.

12.38 The slope of the temperature line, Fig. 12-18, is called the *temperature gradient, G*.

$$\text{Temperature gradient } G = \frac{t_2 - t_1}{L}$$

Find the temperature gradient in a 1.50-ft-thick concrete wall that has one face at a temperature of 80°F and the other face at 10°F.

Solution

Using the formula for temperature gradient,

$$G = \frac{80 - 10}{1.50} = 46.7°F/\text{ft}$$

12.39 A 30.0-cm-thick chimney wall has its inside face at 365°C and its outside face at 75°C. Find the temperature gradient through the wall.

Solution

Using the formula for temperature gradient,

$$G = \frac{365 - 75}{30} = 9.67°C/\text{cm}$$

12.40 For the wall in Problem 12.38, find the temperature at a point 0.500 ft from the cooler side.

Solution

Using the equation of the temperature line, and substituting G for $(t_2 - t_1)L$, with $t_1 = 10°F$ and $G = 46.7°F/\text{ft}$,

$$t = Gx + t_1 = 46.7x + 10$$

When $x = 0.500$ ft,

$$t = 46.7(0.500) + 10 = 33.4°F$$

12.41 The inside face of a brick oven wall is at 350°C, and the temperature gradient is −250°C/m. At what distance from the inside face is the temperature 200°C?

Solution

Using the equation for the temperature line,

$$t = Gx + t_1$$

$$200 = -250x + 350$$

Solving for x,

$$-250x = 200 - 350 = -150$$

$$x = \frac{-150}{-250} = 0.60 \text{ m}$$

HYDRAULIC GRADE LINE

Figure 12-19 shows liquid flowing in a straight pipe of uniform cross section. If a hole were drilled in the pipe and a vertical tube inserted, liquid would rise in the pipe to a height depending on the pressure in the pipe. This height is called the *pressure head* and is given by

$$\text{Pressure head (ft)} = \frac{144p}{\gamma}$$

where p is the pressure in the pipe (lb/in²) and γ is the density of the liquid (lb/ft³).

A plot of the pressure head versus distance along the pipe is a straight line called the *hydraulic grade line*. The slope of the hydraulic grade line, measured from the horizontal, is called the *hydraulic gradient*.

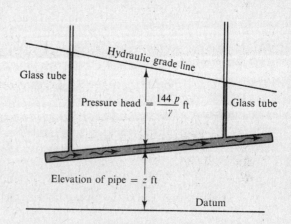

Fig. 12-19 Hydraulic grade line for a straight pipe.

12.42 A straight uniform pipe through which water is flowing has a pressure of 58 lb/in² at one end and 24 lb/in² at the other, as shown in Fig. 12-20. The low-pressure end is 100 ft from the other end, horizontally, and 8 ft higher. Find the hydraulic gradient.

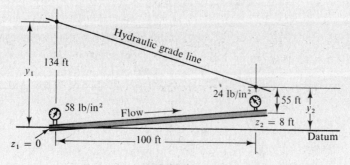

Fig. 12-20

Solution

Using a density $\gamma = 62.4$ lb/ft^3 for water, we convert the pressures to pressure head,

$$\frac{144(58)}{62.4} = 134 \text{ ft} \qquad \frac{144(24)}{62.4} = 55 \text{ ft}$$

To obtain the height of the grade line, we add to each pressure head the height z of the pipe above some datum. Choosing the datum at the elevation of the lower end of the pipe,

$$z_1 = 0 \qquad z_2 = 8$$

So the height of the grade line is

$$134 + 0 = 134 \text{ ft}$$

at the low end, and

$$55 + 8 = 63 \text{ ft}$$

at the high end. So,

$$\text{Hydraulic gradient} = \frac{\text{rise}}{\text{run}} = \frac{134 - 63}{100} = 0.71$$

12.43 A straight pipe is pitched downward with a slope of $-1/4$. Water ($\gamma = 62.4$ lb/ft^3) flows in the pipe. The pressure at the high end is 98 lb/in^2 and at the low end, 280 ft away, it is 73 lb/in^2. Find the hydraulic gradient.

Solution

The grade line will be at a height of

$$\frac{144(98)}{62.4} = 226 \text{ ft}$$

above the high end of the pipe, and at a height of

$$\frac{144(73)}{62.4} = 168 \text{ ft}$$

above the low end. Finding the difference in elevation of the two ends of the pipe,

$$\text{Rise} = \text{run(slope)} = 280\left(-\frac{1}{4}\right) = -70 \text{ ft}$$

Taking the low end of the pipe as the datum, the height of the grade line is 168 ft at the low end and $226 + 70 = 296$ ft at the high end. Then,

$$\text{Hydraulic gradient} = \text{slope} = \frac{296 - 168}{280} = 0.46$$

12.44 A certain horizontal water main has a hydraulic gradient of 0.02. Find the pressure required to deliver water at 30 lb/in^2 to a community 4.60 miles away.

Solution

$$\text{Run (in feet)} = 4.6(5280) = 24\,300 \text{ ft}$$

The height of the grade line at the community is

$$\frac{144(30)}{62.4} = 69.2 \text{ ft}$$

The amount by which it rises toward the pump end is

$$\text{Rise} = \text{run(slope)} = 24\,300(0.02) = 486 \text{ ft}$$

The height of the hydraulic grade line at the pump end is

$$69.2 + 486 = 555 \text{ ft}$$

Then, $$\text{Pressure head} = 555 = \frac{144p}{62.4}$$

Solving for p,

$$p = \frac{555(62.4)}{144} = 241 \text{ lb/in}^2$$

12.45 A straight water line slopes upward in the direction of flow with a slope of 0.4. The pressure at the low end is 65 lb/in² and at the high end, 143 ft away, it is 4.9 lb/in². Find the hydraulic gradient.

Solution

The grade line will be at a height of

$$\frac{144(65)}{62.4} = 150 \text{ ft}$$

above the low end, and at a height of

$$\frac{144(4.9)}{62.4} = 11.3 \text{ ft}$$

above the high end.   The rise in the pipe itself is, by Eq. 129,

$$\text{Rise} = \text{run(slope)} = 143(0.4) = 57.2 \text{ ft}$$

Taking the low end as the datum, the height of the grade line is 150 ft at the low end and $11.3 + 57.2 = 68.5$ ft above the datum at the high end.   So,

$$\text{Hydraulic gradient} = \text{slope} = \frac{150 - 68.5}{143} = 0.57$$

12.46 A horizontal pipe carrying oil ($\gamma = 60.2$ lb/ft³) has a hydraulic gradient of 0.150.   What pressure is required to deliver the oil at 55.0 lb/in² to a machine located 125 ft away?

Solution

The height of the grade line at the machine is

$$\frac{144(55.0)}{60.2} = 132 \text{ ft}$$

and it rises toward the pump end by an amount,

$$\text{Rise} = \text{run(slope)} = 125(0.150) = 18.8 \text{ ft}$$

So the height of the grade line at the pump end is

$$132 + 18.8 = 151 \text{ ft}$$

Then, $$\text{Pressure head} = \frac{144p}{60.2} = 151$$

Solving for P,

$$p = \frac{151(60.2)}{144} = 63.1 \text{ lb/in}^2$$

THE LOAD LINE

A basic diode circuit, consisting of a voltage source, a diode, and a resistor in series in a closed loop, is shown in Fig. 12-21.   We wish to find the current I that will flow for any given input voltage E.

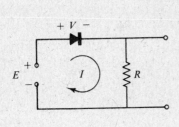

Fig. 12-21   Basic diode circuit.

Fig. 12-22   Load line.

Summing the voltages around the closed loop, by Kirchhoff's Law, Eq. 206, we know the sum must equal zero, so,

$$E - V - IR = 0$$

Transposing,

$$IR = -V + E$$

Dividing by R,

$$I = \left(-\frac{1}{R}\right)V + \frac{E}{R}$$

which is seen to be the equation of a straight line with a slope of $-1/R$ and a y intercept of E/R, Fig. 12-22.   This line is called the *load line*.

12.47   Graph three load lines for the following values of load resistance and input voltage.

	Load resistance (Ω)	Input voltage (V)
1	100	0.8
2	100	1.6
3	160	1.6

Solution

We compute the intercept E/R with the current axis for each load line,

$$\frac{E}{R} = \frac{0.8}{100} = 0.008 \text{ A} = 8 \text{ mA} \tag{1}$$

$$\frac{E}{R} = \frac{1.6}{100} = 0.016 \text{ A} = 16 \text{ mA} \tag{2}$$

$$\frac{E}{R} = \frac{1.6}{160} = 0.010 \text{ A} = 10 \text{ mA} \tag{3}$$

The intercepts with the voltage axis are just the given input voltages, 0.8, 1.6, and 1.6 V. These six points are plotted in Fig. 12-23 and the load lines drawn.

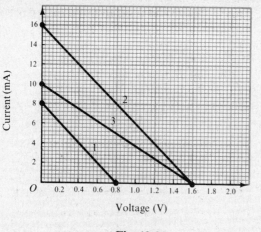

Fig. 12-23

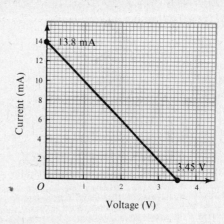

Fig. 12-24

12.48   Graph the load line for a device having a load resistance of 250 Ω and an input voltage of 3.45 V.

Solution

The equation of the load line is

$$I = -\frac{V}{R} + \frac{E}{R}$$

When V is 0,

$$I = \frac{E}{R} = \frac{3.45}{250} = 0.0138 \text{ A} = 13.8 \text{ mA}$$

which is the intercept on the "current" axis.   When I is zero,

$$V = E = 3.45 \text{ V}$$

which is the intercept on the voltage axis.   These two points are plotted in Fig. 12-24 and connected to obtain the load line.

12.49   Graph the load lines for a device having a load resistance of 250 Ω and input voltages of 1, 2, and 3 V.

Solution

Since the slope of the load line depends only upon the resistance, the slope of these lines will be equal and the same as the load line in Problem 12.48.   We thus draw lines parallel to the line of Problem 12.48, intersecting the voltage axis at 1, 2, and 3 V, as shown in Fig. 12-25.

For any given input voltage E and load resistance R, a certain current I will flow through the diode and a voltage V will exist across its terminals.   This pair of values (V, I) is called the *operating point*.   This point

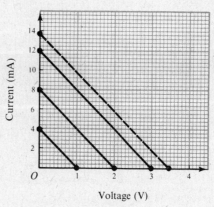

Fig. 12-25

must lie on the load line, and it must also lie on the *characteristic curve* of the diode (a typical characteristic curve for a silicon diode is shown in Fig. 12-26). The operating point is therefore at the intersection of the load line and the characteristic curve, Fig. 12-27.

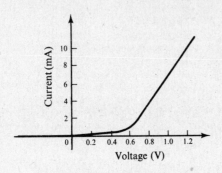

Silicon diode characteristic curve.
Fig. 12-26

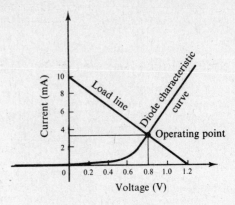

Locating the operating point.
Fig. 12-27

12.50 For the diode whose characteristic curve is given in Fig. 12-28, operating with a load of 100 Ω, find the operating points for input voltages of 0.8, 1.2, and 1.6 V.

Solution

Two of the load lines have already been graphed, Fig. 12-23, and the third is added in Fig. 12-28. The coordinates of the three points of intersection with the characteristic curve are read off the graph. They are

Input voltage (V)	Operating point	
	Current (mA)	Voltage (V)
0.8	1.5	0.66
1.2	3.6	0.83
1.6	6.0	1.0

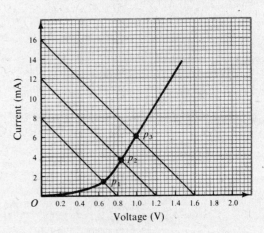

Fig. 12-28

STRAIGHT-LINE DEPRECIATION

Computations of depreciation for tax purposes may be done by several methods. The *straight-line method* assumes that the book value of a capital asset drops the same amount A every year, starting at the purchase price P, and ending at the salvage value S, after T years, as in Fig. 12-29. The total depreciation D is the total drop in book value.

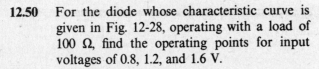

Total Depreciation	$D = P - S$	**169**

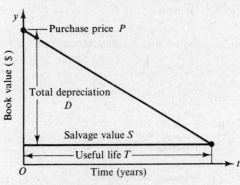

Fig. 12-29 Straight-line depreciation.

The depreciation A in any one year is equal to the slope of the line.

| Annual Depreciation | $A = \dfrac{D}{T} = \dfrac{P-S}{T}$ | **170** |

12.51 Write an equation for the straight line in Fig. 12-29.

Solution

The equation of the line in Fig. 12-29 may be obtained from Eq. 132, noting that the slope is $-A$ and the y intercept is P.

$$y = -At + P$$

or

| Book Value | $y = P - At$ | **171** |

This equation gives the *book value* y of the asset after a period of t years has elapsed.

12.52 A minicomputer is purchased for \$40,000. It is estimated to have a salvage value of \$15,000 after a useful life of 15 years. Find the annual depreciation, using the straight-line method.

Solution

From Eq. 169,

$$\text{Total depreciation} = \$40{,}000 - \$15{,}000 = \$25{,}000$$

and from Eq. 170,

$$\text{Annual depreciation} = \frac{\$25{,}000}{15} = \$1666.67$$

12.53 A punch press is purchased for \$18,000, and is depreciated at the rate of \$800 per year for 20 years. Find the salvage value.

Solution

From Eq. 169,

$$\text{Total depreciation} = 20(800) = \$16{,}000$$

So,

$$\text{Salvage value} = \$18{,}000 - \$16{,}000 = \$2000$$

12.54 A truck is purchased for \$5000. It is expected to have a salvage value of \$800 after a useful life of 10 years. Find the annual straight-line depreciation.

Solution

From Eq. 169,

$$\text{Total depreciation} = \$5000 - \$800 = \$4200$$

From Eq. 170,

$$\text{Annual depreciation} = \frac{\$4200}{10} = \$420 \text{ per year}$$

12.55 A machine is purchased for $7667 and is expected to have a salvage value of $1000 after 10 years. A second machine costs $6091 and will be scrapped for $2000 after 15 years. After how many years will the book values of the two machines be equal?

Solution

From Eq. 170, the annual depreciation is

$$A_1 = \frac{\$7667 - \$1000}{10} = \$666.70 \text{ for machine 1}$$

and

$$A_2 = \frac{\$6091 - \$2000}{15} = \$272.73 \text{ for machine 2}$$

From Eq. 171, the book values are

$$y_1 = 7667 - 666.70t \text{ for machine 1}$$

and

$$y_2 = 6091 - 272.73t \text{ for machine 2}$$

Since $y_1 = y_2$,

$$7667 - 666.70t = 6091 - 272.73t$$

Solving for t,

$$393.97t = 1576$$

$$t = 4 \text{ years}$$

Supplementary Problems

12.56 Find the lengths of the line segments connecting the following pairs of points:

(a) $(0, 3)$ and $(3, 0)$ (c) $(1, 9)$ and $(0, 2)$ (e) $(-3, -1)$ and $(4, 0)$

(b) $(2, 5)$ and $(6, 1)$ (d) $(-1, 2)$ and $(3, -4)$

12.57 Find the directed distances AB and BA between the points $A(1, 4)$ and $B(1, -6)$.

12.58 Find the slope of the lines passing through the following pairs of points:

(a) $(2, 1)$ and $(1, 2)$ (b) $(0, 3)$ and $(2, -4)$ (c) $(-1, -1)$ and $(1, 1)$

12.59 Find the slope of a line perpendicular to a line having a slope of

(a) 2.76   (b) 1.39   (c) 4.37

12.60 Find the angle of inclination of lines with the following slopes:

(a) 2.74   (b) 1.25   (c) −3.14

12.61 Find the slope of lines with the following angles of inclination:

(a) 27°   (b) 78°   (c) 111°

12.62 Write the equation of the following lines in general form:

(a) slope = 2, y intercept = 4   (c) slope = −1, passes through $(2, -1)$

(b) passes through $(2, 1)$ and $(-3, 4)$

12.63   Find the equation of the line having an x intercept of -1 and a y intercept of 4.

12.64   Find the slopes and y intercepts of the following lines:

(a) $3x - 2y + 3 = 0$ (b) $x + 3y - 2 = 0$ (c) $-7x - y - 5 = 0$

12.65   Find the angle between the lines having the following angles of inclination:

(a) $13°$ and $90°$ (b) $-60°$ and $30°$ (c) $240°$ and $190°$

12.66   Find the angle between the two lines having the following slopes:

(a)   1 and $\dfrac{2}{3}$ (b) -3 and 2   (c)   0 and $\dfrac{4}{7}$

Answers to Supplementary Problems

12.56   (a)   4.24   (b)   5.66   (c)   7.07   (d)   7.21   (e)   7.07

12.57 $AB = -10,\ BA = 10$

12.58   (a) -1 (b) $-\dfrac{7}{2}$ (c)   1

12.59   (a) $m = -0.362$ (b) $m = -0.719$ (c) $m = -0.229$

12.60   (a) $69.9°$ (b) $51.3°$ (c) $-72.3°$

12.61   (a) $m = 0.510$ (b) $m = 4.70$ (c) $m = -2.61$

12.62   (a) $2x - y + 4 = 0$ (b) $3x + 5y - 11 = 0$ (c) $x + y - 1 = 0$

12.63 $4x - y + 4 = 0$

12.64   (a) $m = \dfrac{3}{2},\ y$ intercept $= \dfrac{3}{2}$ (b) $m = -\dfrac{1}{3},\ y$ intercept $= \dfrac{2}{3}$ (c) $m = -7,\ y$ intercept $= -5$

12.65   (a) $77°$ (b) $90°$ (c) $50°$

12.66   (a) $11.3°$ (b) $135°$ (c) $29.7°$

Chapter 13

Quadratic Equations

13.1 TERMINOLOGY

A *quadratic function* is a polynomial whose highest degree term is of second degree.

Example 13.1:   The functions

$$f(x) = 3x^2 - 2x + 5 \qquad f(x) = 5 - 3x^2 \qquad f(x) = x(x - 4) \qquad f(x) = x - 6 + 8x^2$$

are quadratic functions.

A *quadratic equation* is a polynomial equation whose highest degree term is of second degree.

Example 13.2

$$2x^2 + 3x - 5 = 0 \qquad x^2 = 37 \qquad 5x^2 - 6x = 0 \qquad 9x^2 - 5 = 0$$

are all quadratic equations.

A quadratic equation is said to be in *general form* when it is written in the form,

Quadratic in General Form	$ax^2 + bx + c = 0$

138

where a, b, and c are constants.

Example 13.3:   Write the quadratic equation

$$4 - \frac{3x^2}{2} = 5x$$

in general form.

Solution: Transposing the $5x$, and arranging the terms in descending order of the exponents,

$$-\frac{3}{2}x^2 - 5x + 4 = 0$$

The equation is now in general form. However, quadratics in general form are usually written without fractional coefficients, and with the coefficient of the x^2 term positive. So multiplying both sides by -2,

$$3x^2 + 10x - 8 = 0$$

A quadratic equation always has *two solutions* or *roots*, although the two roots are sometimes equal. The roots are not always real numbers, but are sometimes imaginary or complex numbers.

13.2 SOLUTION OF PURE QUADRATICS

A quadratic is called *pure* if the x term is missing. Thus, $x^2 - 4 = 0$ is a pure quadratic.

Example 13.4:   Solve the equation $2x^2 - 50 = 0$.

Solution: Transposing,

$$2x^2 = 50$$

Dividing by 2,

$$x^2 = 25$$

Taking the square root,

$$x = \pm\sqrt{25} = \pm 5$$

Example 13.5:   Solve the equation $3x^2 - 11 = 0$.

Solution

$$3x^2 = 11$$
$$x^2 = \frac{11}{3}$$

$$x = \pm\sqrt{\frac{11}{3}} = \pm\frac{\sqrt{33}}{3} \cong \pm 1.91$$

Example 13.6:   Solve the equation $x^2 + 16 = 0$.

Solution

$$x^2 = -16$$

$$x = \pm\sqrt{-16} = \pm 4i$$

where $i = \sqrt{-1}$ (see Section 10.7).

13.3 SOLUTION OF INCOMPLETE QUADRATICS

A quadratic equation is called *incomplete* when the constant term is zero.

Example 13.7:   The equations

$$x^2 + 3x = 0 \qquad 2x^2 = 7x \qquad 3x - 5x^2 = 0$$

are incomplete quadratics.

To solve an incomplete quadratic, factor by removing the common factor x from the two terms containing it, and then equate each factor to zero.

Example 13.8:   Solve the equation $x^2 + 3x = 0$.

Solution: Factoring,

$$x(x + 3) = 0$$

This equation is satisfied if either or both of the factors is equal to zero.   Equating each factor to zero,

$$x = 0$$
and
$$x + 3 = 0$$
Transposing,
$$x = -3$$

So the two roots are $x = 0$ and $x = -3$.

Common Error	Do not cancel an x from the terms of an incomplete quadratic, as this results in the loss of a root.

Example 13.9: The incomplete quadratic of the previous example should *not* be solved in the following way:

Transpose the $3x$,

$$x^2 = -3x$$

Divide by x,

$$x = -3$$

which is correct, but the other root, $x = 0$, has been lost.

13.4 SOLVING QUADRATICS BY FACTORING

To solve a quadratic by factoring, write it in general form, factor it by Eqs. 35 to 38, and set each factor equal to zero.

Example 13.10: Solve by factoring.

$$x^2 + 5x - 6 = 0$$

Solution: Factoring the left-hand side by Eq. 35,

$$(x + 6)(x - 1) = 0$$

Setting each factor equal to zero,

$$x + 6 = 0 \qquad \text{and} \qquad x - 1 = 0$$
$$x = -6 \qquad\qquad\qquad x = 1$$

Thus, $x = -6$ and $x = 1$ are the two roots of the given quadratic.

Check: Substituting each root back into the original equation, we get

$$(-6)^2 + 5(-6) - 6 = 36 - 30 - 6 = 0 \qquad \text{checks}$$
$$(1)^2 + 5(1) - 6 = 1 + 5 - 6 = 0 \qquad \text{checks}$$

Example 13.11

$$x^2 - 4x + 4 = 0$$

Solution:

$$(x - 2)(x - 2) = 0$$
$$x - 2 = 0 \qquad x - 2 = 0$$
$$x = 2 \qquad\qquad x = 2$$

Thus, our equation has the double root, $x = 2$.

Check: $$(2)^2 - 4(2) + 4 = 4 - 8 + 4 = 0 \qquad \text{checks}$$

Example 13.12

$$3x^2 - 16x - 12 = 0$$

Solution:

$$(3x + 2)(x - 6) = 0$$
$$3x + 2 = 0 \qquad\qquad x - 6 = 0$$
$$x = -\frac{2}{3} \qquad\qquad x = 6$$

Check

$$3\left(-\frac{2}{3}\right)^2 - 16\left(-\frac{2}{3}\right) - 12 = \frac{12}{9} + \frac{32}{3} - 12 = \frac{12 + 96 - 108}{9} = 0 \quad \text{checks}$$

and

$$3(6)^2 - 16(6) - 12 = 108 - 96 - 12 = 0 \quad \text{checks}$$

13.5 SOLVING QUADRATICS BY COMPLETING THE SQUARE

To *complete the square* means to manipulate an expression into the form of a perfect square trinomial such as in Eqs. 37 and 38, which is then factored as in the following examples.

Example 13.13:   Solve $x^2 - 4x - 5 = 0$ by completing the square.

Solution: Transpose the constant term.

$$x^2 - 4x = 5$$

Complete the square by adding the square of half the coefficient of the x term to both sides. The coefficient of the x term is -4. We take half of -4 and square it, getting $(-2)^2 = 4$. Then add this 4 to *both sides* of the equation.

$$x^2 - 4x + 4 = 5 + 4$$

Factoring by Eq. 38,

$$(x - 2)^2 = 9$$

Take the square root of both sides

$$x - 2 = \pm 3$$

Transposing,

$$x = 2 \pm 3$$

So

$$x = 2 + 3 = 5$$

and

$$x = 2 - 3 = -1$$

Check

$$(5)^2 - 4(5) - 5 = 25 - 20 - 5 = 0 \quad \text{checks}$$
$$(-1)^2 - 4(-1) - 5 = 1 + 4 - 5 = 0 \quad \text{checks}$$

Common Error	When adding the quantity needed to complete the square to the left-hand side, it is easy to forget to add the same quantity to the right-hand side. $$x^2 - 4x \boxed{+4} = 5 \boxed{+4}$$ ⬆ ──── don't forget

Common Error	After completing the square and factoring the perfect square trinomial, it is tempting to set each factor equal to the RHS in an *incorrect* imitation of the method used in Section 13.4. The reasoning used there, that if either or both factors were equal to zero the equation would be satisfied, is obviously valid only if the RHS is zero.

Example 13.14:   Solve $3x^2 + 2x - 8 = 0$.

Solution: To complete the square, the coefficient of x^2 must be 1; so we start by dividing through by 3, the coefficient of the x^2 term.

$$x^2 + \frac{2}{3}x - \frac{8}{3} = 0$$

Transposing,

$$x^2 + \frac{2}{3}x = \frac{8}{3}$$

We complete the square by adding to each side the square of half the middle coefficient, or

$$\left[\frac{1}{2}\left(\frac{2}{3}\right)\right]^2 = \left[\frac{1}{3}\right]^2 = \frac{1}{9}$$

$$x^2 + \frac{2}{3}x + \frac{1}{9} = \frac{8}{3} + \frac{1}{9}$$

Factoring by Eq. 37,

$$\left(x + \frac{1}{3}\right)^2 = \frac{25}{9}$$

Taking the square root,

$$x + \frac{1}{3} = \pm\frac{5}{3}$$

Transposing,

$$x = -\frac{1}{3} \pm \frac{5}{3} = -2 \text{ and } \frac{4}{3}$$

Example 13.15:   Solve $x^2 - 4x + 9 = 0$.

Solution: Transposing,

$$x^2 - 4x = -9$$

Completing the square,

$$x^2 - 4x + 4 = -9 + 4$$

Factoring, by Eq. 38,

$$(x - 2)^2 = -5$$

Taking the square root,

$$x - 2 = \pm\sqrt{-5}$$

Transposing,

$$x = 2 \pm \sqrt{5}i$$

We thus get two complex roots,

$$x = 2 + \sqrt{5}i$$

and

$$x = 2 - \sqrt{5}i$$

13.6 SOLVING QUADRATICS BY FORMULA

The *quadratic formula* is obtained by solving the general quadratic

$$ax^2 + bx + c = 0$$

by completing the square (see Problem 13.5e).

| Quadratic Formula | $x = \dfrac{-b \pm \sqrt{b^2 - 4ac}}{2a}$ | **139** |

Its use in solving quadratic equations is shown in the following examples.

Example 13.16:   Solve $3x^2 - 4x - 2 = 0$ by formula.

Solution: We substitute $a = 3$, $b = -4$, and $c = -2$ into the quadratic formula,

$$x = \frac{-(-4) \pm \sqrt{(-4)^2 - 4(3)(-2)}}{2(3)} = \frac{4 \pm \sqrt{16 + 24}}{6} = \frac{4 \pm \sqrt{40}}{6} = \frac{4 \pm 2\sqrt{10}}{6} = \frac{2 \pm \sqrt{10}}{3}$$

$$\cong 1.72 \quad \text{and} \quad -0.387$$

Example 13.17:   Solve $2x^2 - 4x + 3 = 0$.

Solution: From Eq. 139, with $a = 2$, $b = -4$, $c = 3$,

$$x = \frac{-(-4) \pm \sqrt{(-4)^2 - 4(2)(3)}}{2(2)} = \frac{4 \pm \sqrt{-8}}{4} = \frac{4}{4} \pm \frac{2\sqrt{-2}}{4} = 1 \pm \frac{\sqrt{2}}{2}i$$

13.7   PREDICTING THE NATURE OF THE ROOTS

The part of the quadratic formula under the radical sign, $b^2 - 4ac$, is called the *discriminant*. Calculation of the discriminant tells us whether the roots are real or complex, and if they are equal.

| The Discriminant | if $b^2 - 4ac > 0$,   the roots are real and unequal.
if $b^2 - 4ac = 0$,   the roots are real and equal.
if $b^2 - 4ac < 0$,   the roots are complex. | **140** |

Example 13.18:   Determine the nature of the roots in the equation $2x^2 - 3x + 5 = 0$.

Solution: Computing the discriminant,

$$b^2 - 4ac = 9 - 4(2)(5) = 9 - 40 = -31$$

Therefore since $-31 < 0$, the roots are complex.

13.8   GRAPHICAL SOLUTION OF QUADRATICS

To graphically solve the quadratic equation

$$ax^2 + bx + c = 0$$

make a graph of the quadratic *function*

$$y = f(x) = ax^2 + bx + c$$

and note where the curve crosses the x axis.   These x intercepts are the roots of the quadratic equation because they are the values of x at which y is equal to zero.   If the curve does not touch the x axis, there are no real roots.   If it touches the x axis at just one point, the two roots are equal.

Example 13.19: Graphically determine the approximate value of the roots of the quadratic equation,

$$x^2 - 6x + 7 = 0$$

Solution: We plot the quadratic function,

$$y = x^2 - 6x + 7$$

by choosing values of x, substituting them one by one into the function, and solving for y.

When $x = 0$,

$$y = 0 - 0 + 7 = 7$$

When $x = 1$,

$$y = 1^2 - 6(1) + 7 = 2$$

and so on, until we have enough point pairs to make a graph. Our table of point pairs is

x	0	1	2	3	4	5	6
y	7	2	-1	-2	-1	2	7

Plotting these points, we obtain the curve in Fig. 13-1. It crosses the x axis at $x \cong 1.6$ and $x \cong 4.4$.

Check: Substituting these values into the given equation,

$$f(1.6) = (1.6)^2 - 6(1.6) + 7 = -0.04$$
$$f(4.4) = (4.4)^2 - 6(4.4) + 7 = -0.04$$

Since substitution of these values gives answers which differ slightly from zero, we see that the roots obtained do not check exactly, emphasizing the approximate nature of the graphical method.

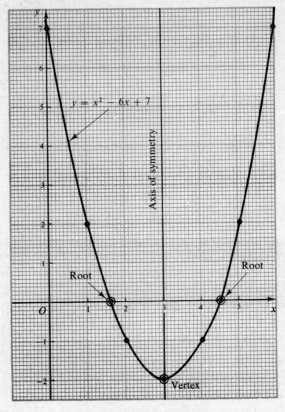

Fig. 13-1 The parabola.

The graph of a quadratic function is called a *parabola*, a curve which has many interesting properties and applications in technology.

Note that the parabola in Fig. 13-1 is symmetrical about the line $x = 3$. This line is called the *axis of symmetry*. The point where the parabola crosses the axis of symmetry is called the *vertex*.

13.9 EQUATIONS OF QUADRATIC TYPE

Certain equations, with degrees other than second, can be reduced to quadratics and solved by the methods of this chapter.

Example 13.20: Solve the equation $x^4 - 5x^2 + 4 = 0$.

Solution: Make the substitution; let $w = x^2$. The original equation then becomes the quadratic

$$w^2 - 5w + 4 = 0$$

Factoring, $\qquad\qquad (w - 1)(w - 4) = 0$

So, $\qquad\qquad w = 1 \quad$ and $\quad w = 4$

But, $\qquad\qquad w = x^2 = 1 \quad$ and $\quad w = x^2 = 4$

So, $\qquad\qquad x = \pm 1 \quad$ and $\quad x = \pm 2$

Check: When $x = 1$,

$$1^4 - 5(1)^2 + 4 \overset{?}{=} 0$$
$$1 - 5 + 4 = 0 \qquad \text{checks}$$

When $x = -1$,

$$(-1)^4 - 5(-1)^2 + 4 \overset{?}{=} 0$$
$$1 - 5 + 4 = 0 \qquad \text{checks}$$

When $x = 2$,

$$2^4 - 5(2)^2 + 4 \overset{?}{=} 0$$
$$16 - 20 + 4 = 0 \qquad \text{checks}$$

When $x = -2$,

$$(-2)^4 - 5(-2)^2 + 4 \overset{?}{=} 0$$
$$16 - 20 + 4 = 0 \qquad \text{checks}$$

An equation may be solved by this method if the power of the unknown in one term is twice the power of the unknown in another term.   For example, x^2 and x^4, x^3 and x^6, $x^{1/2}$ and x, and so on.

Example 13.21:   Solve the equation $2x^{-1/3} + x^{-2/3} + 1 = 0$.

Solution: Inspecting the powers of x, we see that one of them, $-2/3$, is twice the other, $-1/3$. We make the substitution

$$w = x^{-1/3}$$

and our equation becomes

$$2w + w^2 + 1 = 0$$

or

$$w^2 + 2w + 1 = 0$$

which factors into

$$(w + 1)(w + 1) = 0$$

Setting each factor equal to zero,

$$w + 1 = 0 \qquad \text{and} \qquad w + 1 = 0$$

So,

$$w = -1 = x^{-1/3}$$

Cubing both sides,

$$x^{-1} = (-1)^3 = -1$$

By Eq. 14,

$$x^{-1} = \frac{1}{x} = -1$$

So,

$$x = \frac{1}{-1} = -1$$

Check: Substituting $x = -1$ into the original equation,

$$2(-1)^{-1/3} + (-1)^{-2/3} + 1 \overset{?}{=} 0$$

$$\frac{2}{\sqrt[3]{-1}} + \frac{1}{\sqrt[3]{(-1)^2}} + 1 \overset{?}{=} 0$$

$$\frac{2}{-1} + \frac{1}{1} + 1 \overset{?}{=} 0$$

$$-2 + 1 + 1 = 0 \qquad \text{checks}$$

13.10 WORD PROBLEMS INVOLVING QUADRATICS

The procedure to be followed when setting up a quadratic word problem is no different than that used in Chapter 5. Find the roots of the resulting quadratic equation by one of the methods in this chapter. Often one of the roots will not make sense in the physical problem and it should be discarded. Other times, the second root may give another unexpected, but equally valid, solution.

Example 13.22:   A driver travels 80 miles from city A to city B, then continues on to city C, 110 miles from B, at an average rate 12 mi/h greater than on the first part of the trip. Find his average speed on the first part of the trip if the total traveling time is 5.5 h.

Solution: Let $x =$ speed from A to B, mi/h. Then by Eq. 177,

$$\frac{80}{x} = \text{time from A to B, h}$$

Similarly,

$$\frac{110}{x + 12} = \text{time from B to C, h}$$

The total time is the sum of the two individual times; so,

$$\frac{80}{x} + \frac{110}{x + 12} = 5.5$$

Multiplying by $x(x + 12)$,

$$80(x + 12) + 110x = 5.5x(x + 12)$$

Simplifying, $80x + 960 + 110x = 5.5x^2 + 66x$

or $5.5x^2 - 124x - 960 = 0$

By the quadratic formula,

$$x = \frac{124 \pm \sqrt{(-124)^2 - 4(5.5)(-960)}}{2(5.5)} = \frac{124 \pm 191}{11}$$

So, $x = 29$ mi/h

and $x = -6.1$ mi/h   (which we discard)

Solved Problems

GENERAL FORM

13.1   Write the following quadratics in general form and identify the constants a, b, and c.

(a) $5x = 3x^2 - 2$ (c) $x(x - 5) = 3$

(b) $\frac{3x}{2} = 5 - x^2$ (d) $2x(x - 3) - (x - 1)(x + 2) = 0$

Solution

(a) $5x = 3x^2 - 2$

Transposing the $5x$,

$$3x^2 - 5x - 2 = 0$$

So, $a = 3$ $b = -5$ and $c = -2$

(b) $\dfrac{3x}{2} = 5 - x^2$

Transposing,
$$x^2 + \dfrac{3x}{2} - 5 = 0$$

Multiplying by 2,
$$2x^2 + 3x - 10 = 0$$

So, $a = 2 \qquad b = 3 \qquad$ and $\qquad c = -10$

(c) $x(x - 5) = 3$

Clearing parentheses,
$$x^2 - 5x = 3$$

Transposing, $x^2 - 5x - 3 = 0$

So, $a = 1 \qquad b = -5 \qquad$ and $\qquad c = -3$

(d) $2x(x - 3) - (x - 1)(x + 2) = 0$

Clearing parentheses,
$$2x^2 - 6x - (x^2 + x - 2) = 0$$

Collecting terms,
$$x^2 - 7x + 2 = 0$$

So, $a = 1 \qquad b = -7 \qquad$ and $\qquad c = 2$

PURE QUADRATICS

13.2   Solve the following pure quadratics:

(a) $3x^2 + 6 = 18$ (c) $x^2 + 49 = 0$ (e) $2x^2 - 3 = 0$

(b) $5x^2 - 20 = 160$ (d) $2x^2 + 18 = 0$ (f) $4x^2 + 506 = 0$

Solution

(a)
$$3x^2 + 6 = 18$$
$$3x^2 = 18 - 6 = 12$$
$$x^2 = 4$$
$$x = \pm 2$$

Check: Substituting $x = 2$ into the original equation,
$$3(2)^2 + 6 \overset{?}{=} 18$$
$$12 + 6 = 18 \qquad \text{checks}$$

For $x = -2$,
$$3(-2)^2 + 6 \overset{?}{=} 18$$
$$12 + 6 = 18 \qquad \text{checks}$$

(b)
$$5x^2 - 20 = 160$$
$$5x^2 = 160 + 20 = 180$$
$$x^2 = 36$$
$$x = \pm 6$$

(c)
$$x^2 + 49 = 0$$
$$x^2 = -49$$
$$x = \pm\sqrt{-49} = \pm 7i$$

Check: When $x = 7i$,

$$(7i)^2 + 49 \overset{?}{=} 0$$
$$7^2 i^2 + 49 \overset{?}{=} 0$$

Since $i^2 = -1$,

$$-49 + 49 = 0 \qquad \text{checks}$$

When $x = -7i$,

$$(-7i)^2 + 49 \overset{?}{=} 0$$
$$49i^2 + 49 \overset{?}{=} 0$$
$$-49 + 49 = 0 \qquad \text{checks}$$

(d)
$$2x^2 + 18 = 0$$
$$2x^2 = -18$$
$$x^2 = -9$$
$$x = \pm\sqrt{-9} = \pm 3i$$

(e)
$$2x^2 - 3 = 0$$
$$2x^2 = 3$$
$$x^2 = \frac{3}{2}$$
$$x = \pm\sqrt{\frac{3}{2}} \cong \pm 1.22$$

(f)
$$4x^2 + 506 = 0$$
$$4x^2 = -506$$
$$x^2 = -\frac{506}{4}$$
$$x = \pm\sqrt{-\frac{506}{4}} \cong \pm 11.25i$$

INCOMPLETE QUADRATICS

13.3   Solve the following incomplete quadratics:

(a) $3x = 4x^2$ (c) $4x(x - 1) = 2x(x + 2)$ (e) $2x(x - 4) = x(x + 5)$

(b) $3x - 30x^2 = 0$ ·   (d) $x(x + 3) = 3x(x - 2)$

Solution

(a) $3x = 4x^2$

Transposing, $4x^2 - 3x = 0$

Factoring, $x(4x - 3) = 0$

So, $x = 0$ and $4x - 3 = 0$

$$x = \frac{3}{4}$$

Check: For $x = 0$,

$$3(0) \overset{?}{=} 4(0)^2$$

$$0 = 0 \quad \text{checks}$$

For $x = 3/4$,

$$3\left(\frac{3}{4}\right) \overset{?}{=} 4\left(\frac{3}{4}\right)^2$$

$$\frac{9}{4} \overset{?}{=} 4\left(\frac{9}{16}\right)$$

$$\frac{9}{4} = \frac{9}{4} \quad \text{checks}$$

(b) $3x - 30x^2 = 0$

Factoring, $x(3 - 30x) = 0$

So, $x = 0$ and $3 - 30x = 0$

$$x = \frac{1}{10}$$

Check: For $x = 0$,

$$3(0) - 30(0)^2 \overset{?}{=} 0$$

$$0 = 0 \quad \text{checks}$$

For $x = 1/10$,

$$3\left(\frac{1}{10}\right) - 30\left(\frac{1}{10}\right)^2 \overset{?}{=} 0$$

$$\frac{3}{10} - \frac{30}{100} \overset{?}{=} 0$$

$$\frac{3}{10} - \frac{3}{10} = 0 \quad \text{checks}$$

(c) $4x(x - 1) = 2x(x + 2)$

Clearing parentheses,

$$4x^2 - 4x = 2x^2 + 4x$$

Transposing and collecting terms,

$$2x^2 - 8x = 0$$

or $x^2 - 4x = 0$

Factoring, $x(x - 4) = 0$

So, $x = 0$ and $x - 4 = 0$

$$x = 4$$

(d) $x(x + 3) = 3x(x - 2)$

Clearing parentheses,

$$x^2 + 3x = 3x^2 - 6x$$

Transposing and collecting terms,

$$2x^2 - 9x = 0$$

Factoring, $$x(2x - 9) = 0$$

So, $x = 0$ and $2x - 9 = 0$

$$x = \frac{9}{2}$$

(e) $2x(x - 4) = x(x + 5)$

Clearing parentheses,

$$2x^2 - 8x = x^2 + 5x$$

Transposing and collecting terms,

$$x^2 - 13x = 0$$

Factoring, $$x(x - 13) = 0$$

So, $x = 0$ and $x = 13$

SOLUTION BY FACTORING

13.4   Solve the following quadratics by factoring:

(a) $x^2 + 9x + 18 = 0$ (c) $x^2 + 6x + 9 = 0$ (e) $2x^2 + 8x + 6 = 0$

(b) $x^2 + 6x - 16 = 0$ (d) $x^2 - 8x + 16 = 0$ (f) $4x^2 + 5x + 1 = 0$

Solution

(a) $x^2 + 9x + 18 = 0$ (d) $x^2 - 8x + 16 = 0$

           $(x + 3)(x + 6) = 0$ $(x - 4)(x - 4) = 0$

    $x + 3 = 0$ $x + 6 = 0$ $x - 4 = 0$ $x - 4 = 0$

       $x = -3$ $x = -6$ $x = 4$ $x = 4$

(b) $x^2 + 6x - 16 = 0$ (e) $2x^2 + 8x + 6 = 0$

           $(x + 8)(x - 2) = 0$ $x^2 + 4x + 3 = 0$

    $x + 8 = 0$ $x - 2 = 0$ $(x + 3)(x + 1) = 0$

       $x = -8$ $x = 2$ $x = -3$ $x = -1$

(c) $x^2 + 6x + 9 = 0$ (f) $4x^2 + 5x + 1 = 0$

           $(x + 3)(x + 3) = 0$ $(4x + 1)(x + 1) = 0$

    $x + 3 = 0$ $x + 3 = 0$ $4x + 1 = 0$ $x + 1 = 0$

       $x = -3$ $x = -3$ $x = -\dfrac{1}{4}$ $x = -1$

COMPLETING THE SQUARE

13.5   Solve the following quadratics by completing the square:

(a) $x^2 - 10x - 2 = 0$ (c) $x^2 - 4x - 1 = 0$ (e) $ax^2 + bx + c = 0$

(b) $2x^2 - 4x - 9 = 0$ (d) $3x^2 - 4x - 1 = 0$

Solution

(a) $x^2 - 10x - 2 = 0$

Transposing,
$$x^2 - 10x = 2$$

Completing the square,
$$x^2 - 10x + 25 = 2 + 25$$

Factoring, by Eq. 38,
$$(x - 5)^2 = 27$$

Taking the square root,
$$x - 5 = \pm\sqrt{27}$$

Transposing,
$$x = 5 \pm \sqrt{27} = 5 \pm 3\sqrt{3}$$

(b) $2x^2 - 4x - 9 = 0$

Transposing,
$$2x^2 - 4x = 9$$

Dividing by 2,
$$x^2 - 2x = \frac{9}{2}$$

Completing the square,
$$x^2 - 2x + 1 = \frac{9}{2} + 1$$

Factoring, by Eq. 38,
$$(x - 1)^2 = \frac{11}{2}$$

Taking the square root,
$$x - 1 = \pm\sqrt{\frac{11}{2}} = \pm\frac{\sqrt{22}}{2}$$

Transposing,
$$x = 1 \pm \frac{\sqrt{22}}{2}$$

(c) $x^2 - 4x - 1 = 0$

Transposing,
$$x^2 - 4x = 1$$

Completing the square,
$$x^2 - 4x + 4 = 1 + 4$$

Factoring, by Eq. 38,
$$(x - 2)^2 = 5$$

Taking the square root,
$$x - 2 = \pm\sqrt{5}$$

Transposing,
$$x = 2 \pm \sqrt{5}$$

(d) $3x^2 - 4x - 1 = 0$

Transposing and dividing by 3,
$$x^2 - \frac{4x}{3} = \frac{1}{3}$$

Completing the square,

$$x^2 - \frac{4x}{3} + \left(\frac{2}{3}\right)^2 = \frac{1}{3} + \frac{4}{9}$$

Factoring, by Eq. 38,

$$\left(x - \frac{2}{3}\right)^2 = \frac{7}{9}$$

Taking the square root,

$$x - \frac{2}{3} = \pm \frac{\sqrt{7}}{3}$$

Transposing,

$$x = \frac{2}{3} \pm \frac{\sqrt{7}}{3}$$

(e) $ax^2 + bx + c = 0$

Transposing the c and dividing by a,

$$x^2 + \frac{b}{a}x = -\frac{c}{a}$$

Completing the square,

$$x^2 + \frac{b}{a}x + \left(\frac{b}{2a}\right)^2 = \frac{b^2}{4a^2} - \frac{c}{a}$$

Factoring, by Eq. 37,

$$\left(x + \frac{b}{2a}\right)^2 = \frac{b^2 - 4ac}{4a^2}$$

Taking the square root,

$$x + \frac{b}{2a} = \pm \frac{\sqrt{b^2 - 4ac}}{2a}$$

Transposing,

$$\boxed{\text{Quadratic Formula} \quad x = \frac{-b \pm \sqrt{b^2 - 4ac}}{2a}} \qquad \mathbf{139}$$

PREDICTING THE NATURE OF THE ROOTS

13.6   Determine if the roots of the following quadratics are real or complex, and whether they are equal or unequal:

(a) $x^2 - 6x - 11 = 0$ (c) $x^2 - 8x + 16 = 0$

(b) $2x^2 + 3x + 5 = 0$ (d) $3x^2 + 4x - 1 = 0$

Solution

(a)   From the discriminant, Eq. 140, with $a = 1$, $b = -6$,   and $c = -11$,

$$b^2 - 4ac = (-6)^2 - 4(1)(-11) = 36 + 44 = 80$$

Since the discriminant is positive, the roots are real and unequal.

(b) From Eq. 140, with $a = 2$, $b = 3$,   and $c = 5$,

$$b^2 - 4ac = 3^2 - 4(2)(5) = 9 - 40 = -31$$

A negative discriminant means the roots are complex.

(c) From Eq. 140, with $a = 1$, $b = -8$,   and $c = 16$,

$$b^2 - 4ac = (-8)^2 - 4(1)(16) = 64 - 64 = 0$$

Therefore, the roots are real and equal.

(d) From Eq. 140, with $a = 3$, $b = 4$,   and $c = -1$,

$$b^2 - 4ac = 4^2 - 4(3)(-1) = 16 + 12 = 28$$

The roots are real and unequal.

SOLUTION BY QUADRATIC FORMULA

13.7   Solve by quadratic formula:

(a) $x^2 - 6x - 11 = 0$ (d) $2x^2 - 3x + 5 = 0$

(b) $x^2 = 4x + 13$ (e) $px^2 - sx - r = 0$ (p, s, and r are constants)

(c) $5x^2 - 12x - 12 = 0$

Solution

(a) $x^2 - 6x - 11 = 0$

By Eq. 139,

$$x = \frac{-(-6) \pm \sqrt{(-6)^2 - 4(1)(-11)}}{2(1)} = \frac{6 \pm \sqrt{80}}{2} = 3 \pm 2\sqrt{5} \cong 7.47 \quad \text{and} \quad -1.47$$

(b) Writing the equation in general form, Eq. 138,

$$x^2 - 4x - 13 = 0$$

By Eq. 139,

$$x = \frac{4 \pm \sqrt{16 - 4(-13)}}{2} = \frac{4 \pm \sqrt{68}}{2} = 2 \pm \sqrt{17} \cong 6.12 \quad \text{and} \quad -2.12$$

(c) By Eq. 139,

$$x = \frac{12 \pm \sqrt{144 - 4(5)(-12)}}{2(5)} = \frac{12 \pm \sqrt{384}}{10} = \frac{6 \pm 4\sqrt{6}}{5} \cong 3.16 \quad \text{and} \quad -0.760$$

(d) By Eq. 139,

$$x = \frac{3 \pm \sqrt{9 - 4(2)(5)}}{2(2)} = \frac{3 \pm \sqrt{-31}}{4} \cong 0.75 \pm 1.39i$$

(e) By Eq. 139,

$$x = \frac{s \pm \sqrt{s^2 - 4p(-r)}}{2p} = \frac{s \pm \sqrt{s^2 + 4pr}}{2p}$$

GRAPHICAL SOLUTION OF QUADRATICS

13.8 Plot the following quadratics and graphically find the approximate values of the roots:

(a) $y = x^2 - 6x + 8$ (c) $y = 5x^2 - 6x + 1$

(b) $y = 9x^2 - 6x + 1$

Solution

(a) Substituting values of x into the quadratic, we obtain the following table of point pairs:

x	0	1	2	3	4	5	6
y	8	3	0	-1	0	3	8

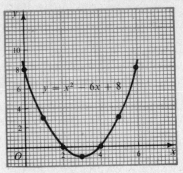

We see from the table that the roots are $x = 2$ and $x = 4$. The curve is plotted in Fig. 13-2 and is seen to cross the x axis at those points.

Fig. 13-2

(b) The following point pairs are obtained by substitution into the equation:

x	-1	0	0.25	0.5	1	2
y	16	1	0.06	0.25	4	25

The resulting parabola has a single root somewhere in the neighborhood of $x \cong 0.3$, Fig. 13-3.

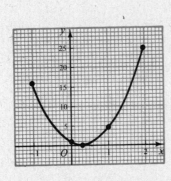

Fig. 13-3

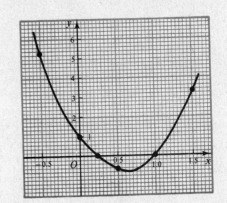

Fig. 13-4

(c) We obtain the following point pairs:

x	-0.5	0	0.5	1.0	1.5
y	5.25	1	-0.75	0	3.25

The plot, Fig. 13-4, shows roots at $x \cong 0.2$ and $x \cong 1.0$.

EQUATIONS OF QUADRATIC TYPE

13.9 Solve the following equations for all values of x:

(a) $x^4 - 10x^2 + 9 = 0$ (c) $9x^4 - 6x^2 + 1 = 0$ (e) $x^{-2/3} - x^{-1/3} = 0$

(b) $x^6 - 6x^3 + 8 = 0$ (d) $x^{-1} - 2x^{-1/2} + 1 = 0$

Solution

(a) Make the substitution, $w = x^2$. So,

$$w^2 - 10w + 9 = 0$$

Factoring,
$$(w - 1)(w - 9) = 0$$
$$w = x^2 = 1 \qquad \text{and} \qquad w = x^2 = 9$$

Taking the square root,
$$x = \pm 1 \qquad \text{and} \qquad x = \pm 3$$

(b) Letting $w = x^3$,

$$w^2 - 6w + 8 = 0$$

Factoring,
$$(w - 2)(w - 4) = 0$$

So,
$$w = x^3 = 2 \qquad \text{and} \qquad w = x^3 = 4$$

Taking the cube root,
$$x = \sqrt[3]{2} \cong 1.26 \qquad \text{and} \qquad x = \sqrt[3]{4} \cong 1.59$$

(c) Letting $w = x^2$,

$$9w^2 - 6w + 1 = 0$$

Factoring,
$$(3w - 1)^2 = 0$$

Solving for w,

$$3w - 1 = 0$$

$$w = x^2 = \frac{1}{3}$$

Taking the square root,

$$x = \pm \sqrt{\frac{1}{3}} \cong \pm 0.577$$

(d) Letting $w = x^{-1/2}$, then $x^{-1} = w^2$. So,

$$w^2 - 2w + 1 = 0$$

Factoring,
$$(w - 1)^2 = 0$$
$$w = x^{-1/2} = 1$$

Raising both sides to the -2 power,

$$x = 1^{-2} = 1$$

(e) Letting $w = x^{-1/3}$, then $x^{-2/3} = w^2$. So,

$$w^2 - w = 0$$

Factoring,
$$w(w - 1) = 0$$
$$w = x^{-1/3} = 0 \qquad \text{and} \qquad w = x^{-1/3} = 1$$

Cubing both sides,

$$x^{-1} = 0^3 = 0 \qquad \text{and} \qquad x^{-1} = 1^3 = 1$$

$$\frac{1}{x} = 0 \qquad\qquad\qquad \frac{1}{x} = 1$$

$$x = \frac{1}{0} \qquad\qquad\qquad x = 1$$

$$\text{(undefined)}$$

WORD PROBLEMS INVOLVING QUADRATICS

13.10   The angle iron in Fig. 13-5 has a cross-sectional area of 4.45 cm². Find the thickness t.

Solution

Dividing the given area into two rectangles as shown, one of which has an area of $9t$ and the other an area of $t(5 - t)$, we get

$$9t + t(5 - t) = 4.45$$
$$9t + 5t - t^2 - 4.45 = 0$$
$$t^2 - 14t + 4.45 = 0$$

By the quadratic formula,

$$t = \frac{14 \pm \sqrt{(14)^2 - 4(4.45)}}{2} = \frac{14 \pm 13.3}{2}$$

$$= 0.35 \text{ cm} \qquad \text{(and 13.7 cm, which we discard because it is obviously too large)}$$

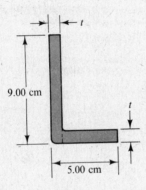

9.00 cm

5.00 cm

Fig. 13-5

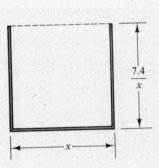

7.4 / x

x

Fig. 13-6

13.11   A long strip of copper, 12 in. in width, is to have its edges bent up at right angles to form an open trough, Fig. 13-6, having a cross-sectional area of 7.4 in². Find the width and depth of the trough. Disregard the thickness of the copper.

Solution

Let

$$x = \text{width of trough, in.}$$

Then,

$$\frac{7.4}{x} = \text{depth of trough, in.}$$

The width of the copper strip equals the width of the trough plus twice the depth,

$$12 = x + 2\left(\frac{7.4}{x}\right)$$

Multiplying by x,

$$12x = x^2 + 14.8$$

or

$$x^2 - 12x + 14.8 = 0$$

By the quadratic formula,

$$x = \frac{12 \pm \sqrt{144 - 4(14.8)}}{2} = \frac{12 \pm 9.21}{2} = 11 \text{ in.} \quad \text{and} \quad 1.4 \text{ in.} = \text{width}$$

So, $\text{Depth} = \frac{7.4}{11} = 0.67 \text{ in.}$ and $\frac{7.4}{1.4} = 5.3 \text{ in.}$

Thus, we have two possible solutions.

13.12 A metal frame of uniform thickness has outside dimensions of 10.3 cm and 13.5 cm as in Fig. 13-7. Find the thickness x of the frame, if it encloses an area of 90.0 cm².

Solution

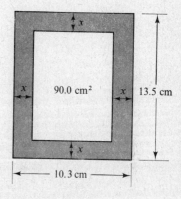

The inside width $= 10.3 - 2x$, and the inside height $= 13.5 - 2x$. By Eq. 90,

$$(10.3 - 2x)(13.5 - 2x) = 90.0$$

Simplifying, $(10.3)(13.5) - 20.6x - 27x + 4x^2 = 90.0$

or $4x^2 - 47.6x + 49.1 = 0$

By the quadratic formula,

$$x = \frac{47.6 \pm \sqrt{(47.6)^2 - 4(4)(49.1)}}{2(4)} = \frac{47.6 \pm 38.5}{8} = 1.14 \text{ cm}$$

Fig. 13-7

(discarding the larger root)

13.13 Write an equation for the wall thickness t of a cylindrical tube of volume V, inside radius r, and length L.

Solution

By Eq. 109, the volume of a hollow cylinder is

$$V = [\pi(r + t)^2 - \pi r^2]L$$

since $\pi(r + t)^2 - \pi r^2$ is the surface area of the cylinder base.
Solving for t, we divide by πL,

$$(r + t)^2 - r^2 = \frac{V}{\pi L}$$

Removing parentheses,

$$r^2 + 2rt + t^2 - r^2 = \frac{V}{\pi L}$$

Transposing, $t^2 + 2rt - \frac{V}{\pi L} = 0$

By the quadratic formula, with $a = 1$, $b = 2r$,   and $c = -\frac{V}{\pi L}$,

$$t = \frac{-2r \pm \sqrt{4r^2 - 4\left(-\frac{V}{\pi L}\right)}}{2} = -r \pm \sqrt{r^2 + \frac{V}{\pi L}}$$

Since t must be positive, we may discard one root. So

$$t = -r + \sqrt{r^2 + \frac{V}{\pi L}}$$

13.14   Find the depth of cut h needed to produce the flat of Fig. 13-8.

Solution

Draw line OP, the perpendicular bisector of the flat, and radius OQ. By the Pythagorean Theorem, Eq. 72, in triangle OPQ,

$$(r - h)^2 + \left(\frac{W}{2}\right)^2 = r^2$$

Removing parentheses,

$$r^2 - 2rh + h^2 + \frac{W^2}{4} = r^2$$

Transposing, $$h^2 - 2rh + \frac{W^2}{4} = 0$$

or $$4h^2 - 8rh + W^2 = 0$$

By the quadratic formula, with $a = 4$, $b = -8r$,   and $c = W^2$,

$$h = \frac{8r \pm \sqrt{64r^2 - 4(4)W^2}}{8} = r \pm \frac{1}{2}\sqrt{4r^2 - W^2}$$

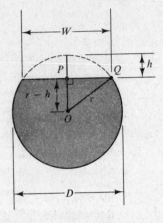

Fig. 13-8

13.15   What two resistances will give a total resistance of 525 Ω when wired in series and 124 Ω when wired in parallel?

Solution

Let R = one resistance, Ω.   Then, by Eq. 201, $525 - R$ = other resistance, Ω.   By Eq. 202,

$$\frac{1}{124} = \frac{1}{R} + \frac{1}{525 - R}$$

Multiplying by $124R(525 - R)$,

$$R(525 - R) = 124(525 - R) + 124R$$

$$525R - R^2 = 65\,100 - 124R + 124R$$

$$R^2 - 525R + 65\,100 = 0$$

By the quadratic formula,

$$R = \frac{525 \pm \sqrt{(525)^2 - 4(1)(65\,100)}}{2} = \frac{525 \pm 123}{2} = 324\ \Omega \quad \text{and} \quad 201\ \Omega$$

13.16   A 4.50-m-diameter cylindrical tank fits snugly into the right angle formed where a wall meets a floor. Find the maximum radius r that another cylindrical tank may have that will fit between the corner and the 4.50-m tank. See Fig. 13-9.

Solution

In right triangle AOB,

$$(OB)^2 = (OA)^2 + (AB)^2 = 2(2.25)^2 = 10.1$$

So, $OB = \sqrt{10.1} = 3.18$ m

Then, $BC = 3.18 - 2.25 - r = 0.93 - r$

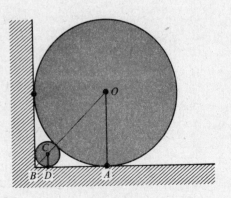

Fig. 13-9

In right triangle BCD,

$$(BC)^2 = (CD)^2 + (BD)^2$$

But $CD = BD = r$ and $BC = 0.93 - r$,   so,

$$(0.93 - r)^2 = 2r^2$$

$$0.87 - 1.86r + r^2 = 2r^2$$

$$r^2 + 1.86r - 0.87 = 0$$

By the quadratic formula,

$$r = \frac{-1.86 \pm \sqrt{(1.86)^2 - 4(1)(-0.87)}}{2} = \frac{-1.86 \pm 2.63}{2} = 0.385 \text{ m} \quad \text{and} \quad -2.25 \text{ m} \quad \text{(discard)}$$

13.17   *Sight Distance:*   The sight distance C on a circular highway curve of radius R and roadway width of $2m$, Fig. 7-75, is given by,

$$C^2 = 8Rm - 4m^2$$

How wide would a roadway have to be to have a sight distance of 600 ft for a curve with a 2250-ft radius?

Solution

Substituting $C = 600$ ft   and $R = 2250$ ft   into the given equation,

$$(600)^2 = 8(2250)m - 4m^2$$

or

$$m^2 - 4500m + 90\,000 = 0$$

By the quadratic formula,

$$m = \frac{4500 \pm \sqrt{(4500)^2 - 4(1)(90\,000)}}{2} = \frac{4500 \pm 4460}{2} = 4480 \text{ ft} \quad \text{and} \quad 20 \text{ ft}$$

so, discarding the larger root, the road width = 40 ft.

13.18   Crew A takes 3 days longer to erect a certain type of house than does crew B.   After crew A has been working on such a house for 5 days, they are joined by crew B and together finish it in 3 additional days.   How long would it take each crew to finish a house, working alone?

Solution

Let x = time for crew B, days.   Then, $x + 3$ = time for crew A, days.   By Eq. 161,

$$\frac{1}{x} = \text{crew B's rate, houses/day}$$

and

$$\frac{1}{x + 3} = \text{crew A's rate, houses/day}$$

By Eq. 163,

$$\frac{1}{x}(3) + \frac{1}{x + 3}(8) = 1$$

since crew B works for only 3 days but crew A works for a total of 8 days.

Multiplying by $x(x + 3)$,

$$3(x + 3) + 8x = x(x + 3)$$

$$3x + 9 + 8x = x^2 + 3x$$

or

$$x^2 - 8x - 9 = 0$$

Factoring, $(x - 9)(x + 1) = 0$

So, $x = 9$ and $x = -1$ (which we discard)

Time for crew B $= x = 9$ days

Time for crew A $= x + 3 = 12$ days

13.19   *The Golden Ratio:* When the ratio of a smaller number x to a larger number y equals the ratio of the larger number to their sum,

$$\frac{x}{y} = \frac{y}{x + y}$$

we say the two numbers are in the golden ratio (see Problem 4.16).   Solve the above equation for x.

Solution

Cross-multiplying,

$$x^2 + yx = y^2$$

Transposing, $x^2 + yx - y^2 = 0$

which is now a quadratic in x, with $a = 1$, $b = y$,   and $c = -y^2$.

Substituting into the quadratic formula,

$$x = \frac{-y \pm \sqrt{y^2 - 4(1)(-y^2)}}{2(1)} = \frac{-y \pm y\sqrt{1 + 4}}{2}$$

Factoring out y,

$$x = \left(\frac{-1 \pm \sqrt{5}}{2}\right)y$$

Evaluating the quantity in parentheses to get a decimal value for the golden ratio,

$$x \cong 0.618y$$

and $x \cong -1.618y$

Dropping the negative value, the golden ratio is then

$$\frac{x}{y} \cong 0.618$$

13.20   *Freely Falling Body:* The vertical distance s traveled by a body given an initial velocity V_0 and falling for a time t is given by the quadratic equation

$$\boxed{s = V_0 t + \frac{at^2}{2}} \quad \textbf{178}$$

where a is the acceleration due to gravity (approximately 32 ft/s^2 or 980 cm/s^2).   Find the distance traveled by an object in 2.0 s if it is thrown downward with a velocity of 25 cm/s.

Solution

Choosing the downward direction as positive,

$V_0 = 25$ cm/s   and $a = 980$ cm/s^2

Substituting into Eq. 178,

$$s = 25(2.0) + \frac{980(2.0)^2}{2} = 2\overline{0}00 \text{ cm}$$

13.21 An object is projected upward with a velocity of $2\overline{0}0$ ft/s. When will it be 75 ft above the ground?

Solution

Taking the upward direction as positive, we substitute in Eq. 178, with $V_0 = 200$, $s = 75$, and $g = -32$,

$$75 = 200t - 16t^2$$

or

$$16t^2 - 200t + 75 = 0$$

By the quadratic formula,

$$t = \frac{200 \pm \sqrt{(200)^2 - 4(16)(75)}}{2(16)} = \frac{200 \pm 188}{32}$$

$$= 0.38 \text{ s} \quad \text{and} \quad 12 \text{ s}$$

representing the times the object is 75 ft above the ground on its upward flight and then coming down.

13.22 *The Parabolic Arch:* The equation of the parabolic bridge arch in Fig. 13-10 is $y = 0.0625x^2 - 5.00x + 100$. Find the distances a, b, c, and d.

Solution

Be sure to refer to point O on Fig. 13-10 as the origin.

$y = a$ when $x = 0$. Substituting $x = 0$ in the given equation,

$$a = 0 - 0 + 100 = 100 \text{ ft}$$

$x = c$ when $y = 50$. Substituting,

$$y = 50 = 0.0625c^2 - 5.00c + 100$$

$$0.0625c^2 - 5.00c + 50 = 0$$

$$c^2 - 80c + 800 = 0$$

By the quadratic formula,

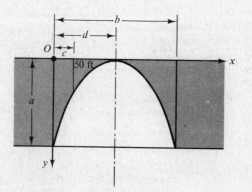

Fig. 13-10

$$c = \frac{80 \pm \sqrt{(80)^2 - 4(800)}}{2} = \frac{80 \pm 56.6}{2} = 68.3 \text{ ft} \quad \text{and} \quad 11.7 \text{ ft}$$

the distances to either side of the centerline.

$x = d$ when $y = 0$. Substituting,

$$y = 0 = 0.0625d^2 - 5.00d + 100$$

$$d^2 - 80d + 1600 = 0$$

Factoring,

$$(d - 40)^2 = 0$$

$$d = 40 \text{ ft}$$

$$b = 2d = 2(40) = 80 \text{ ft}$$

13.23 *Parabolic Reflector:* Spotlight and searchlight reflectors, microwave and radar antennas, solar collectors, and radio and optical telescopes all make use of the focusing property of the parabola; rays running parallel to the axis of symmetry will all pass through a single point, the *focus*, after being reflected from the parabola. Conversely, rays originating at the focus will be reflected from the parabola in such a way as to emerge parallel to the axis of symmetry, Fig. 13-11.

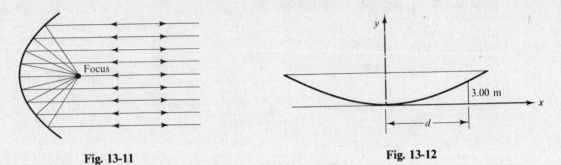

Fig. 13-11   Fig. 13-12

The equation of the parabolic surface of the 20.0-m-diameter antenna dish in Fig. 13-12 is $25y = x^2$. Find (a) the depth of the dish and (b) the distance d at which the support struts are 3.00 m long.

Solution

(a) When $x = 10.0$ m,

$$y = \frac{x^2}{25} = \frac{100}{25} = 4.00 \text{ m} = \text{depth of dish}$$

(b) When $y = 3.00$ m,

$$x^2 = 25y = 25(3.00) = 75.0$$

$$x = 8.66 \text{ m}$$

13.24 A parabolic antenna dish is 20 m in diameter. If the vertex is at the origin and the y axis is taken along the axis of symmetry, the equation of the cross section is $x^2 = 50y$. Make a sketch of the cross section, giving coordinates on the curve for every meter of radius, to two decimal places.

Solution

Substituting values of the radius x from 0 to 10 into the equation, we get the following table of point pairs.

x	0	1	2	3	4	5	6	7	8	9	10
y	0	0.02	0.08	0.18	0.32	0.50	0.72	0.98	1.28	1.62	2.00

These points are plotted in Fig. 13-13.

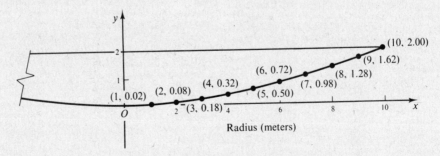

Radius (meters)

Fig. 13-13

13.25 *Square-Law Devices:* An electrical device whose output is proportional to the square of the input is often called a *square-law device*. Included in this category are bolometers and field effect transistors.   For each of these, the curve of output vs. input is a *parabola*.

The output current I_D resulting when an input voltage V_{GS} is applied to a junction field effect transistor (JFET) is given by

$$I_D = I_{DSS}\left[1 - \frac{V_{GS}}{V_{GS(off)}}\right]^2$$

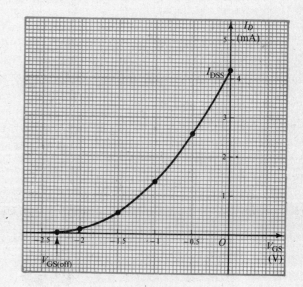

where I_{DSS} = the drain saturation current
$V_{GS(off)}$ = the gate source pinchoff voltage

(a) Plot I_D vs. V_{GS} (the *transconductance curve*) for a range of V_{GS} from -2.3 to 0 V, if $I_{DSS} = 4.2$ mA $V_{GS(off)} = -2.3$ V.

(b) Express V_{GS} as a function of I_D.

(c) Compute V_{GS} when $I_D = -2$ mA.

Fig. 13-14

Solution

(a) Substituting the given values into the equation for I_D, and converting 4.2 mA to 0.0042 A,

$$I_D = 0.0042\left(1 + \frac{V_{GS}}{2.3}\right)^2$$

Substituting values of V_{GS} into this equation gives the following table of point pairs.

V_{GS} (V)	-2.3	-2	-1.5	-1	-0.5	0
I_D (mA)	0	0.07	0.51	1.34	2.57	4.20

Plotting these points, we get a transconductance curve, Fig. 13-14, which is parabolic in shape.

(b) Starting with the expression,

$$I_D = 0.0042(1 + 0.43V_{GS})^2$$

Squaring, $$I_D = 0.0042(1 + 0.86V_{GS} + 0.18V_{GS}^2)$$

Removing parentheses,

$$I_D = 0.0042 + 0.0036V_{GS} + 0.00076V_{GS}^2$$

Multiplying by 10^4 and transposing,

$$7.6V_{GS}^2 + 36V_{GS} + 42 - 10^4 I_D = 0$$

By the quadratic formula,

$$V_{GS} = \frac{-36 \pm \sqrt{(36)^2 - 4(7.6)(42 - 10^4 I_D)}}{2(7.6)} = \frac{-36 \pm 551\sqrt{I_D}}{15.2} = -2.37 \pm 36.3\sqrt{I_D}$$

(c)   When $I_D = 2$ mA $= 0.002$ A,

$$V_{GS} = -2.37 \pm 36.3\sqrt{0.002} = -0.75 \text{ V}$$

which agrees with the graphical value fairly well, and

$$V_{GS} = -3.99 \text{ V}$$

which we discard because it is outside the operating range of the device.

13.26   If we let $x = V_{GS}/V_{GS(off)}$ and $y = I_D/I_{DSS}$,
we may rewrite the equation for a JFET as

$$y = (1 - x)^2$$

(a)   Make a graph of x vs. y.

(b)   Find x when $y = \dfrac{1}{2}$.

Solution

(a)   Making a table of point pairs,

x	0	$\frac{1}{4}$	$\frac{1}{2}$	$\frac{3}{4}$	1
y	1	$\frac{9}{16}$	$\frac{1}{4}$	$\frac{1}{16}$	0

These points are plotted in Fig. 13-15.

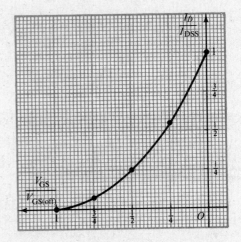

Fig. 13-15

(b)   When $y = \dfrac{1}{2}$,

$$\frac{1}{2} = (1 - x)^2$$

Taking the square root,

$$1 - x = \sqrt{\frac{1}{2}} = 0.707$$

Transposing, $x = 1 - 0.707 = 0.293 = \dfrac{V_{GS}}{V_{GS(off)}}$

13.27   *Vertical Curves:*   Where there is a change
in the slope of a highway surface, such as
at the top of a hill or the bottom of a dip,
the roadway is often made parabolic in
shape to provide a smooth transition
between the two different grades. To construct the required curve, we make use of
the following property of the parabola:

> *The offset C from a tangent to a parabola
> is proportional to the squares of the
> distance x from the point of tangency.*

or $C = kx^2$, where k is a constant of
proportionality, found from a known offset.

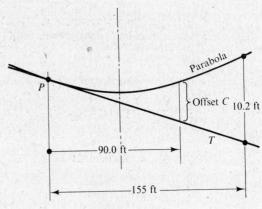

Fig. 13-16

In Fig. 13-16, the offset from the tangent T is 10.2 ft at a distance of 155 ft from the point of tangency P. Find the offset C at 90.0 ft.

Solution

Since $C = 10.2$ when $x = 155$,

So,
$$k = \frac{C}{x^2} = \frac{10.2}{(155)^2} = 4.25 \times 10^{-4}$$

When $x = 90.0$,
$$C = kx^2 = (4.25 \times 10^{-4})(90.0)^2 = 3.44 \text{ ft}$$

13.28 Two grade lines having grades of -2 and $+5$ percent meet at an elevation of 1000 ft. The grade lines are to be joined by a parabolic curve, with the horizontal distance between the two points of tangency being 800 ft. Grade stakes are to be placed every 50 ft along the roadway. Find the required elevation at each grade stake.

Solution

In Fig. 13-17, we locate the two tangent points PVC (point of vertical curve) and PVT (point of vertical tangency) 400 ft to either side of the point of intersection PI of the two grade lines. Since the -2 percent grade line drops 2 ft per hundred, or 8 ft in 400 ft, the elevation of the PVC is 1008.00 ft. Similarly, the PVT is at 1020.00 ft. Now extend the back tangent to Q.

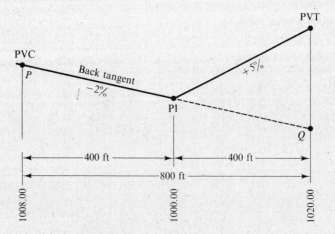

Fig. 13-17 Finding the elevations of the PVC and PVT (vertical scale exaggerated).

Along the back tangent and its extension, lay off stations, or stages, horizontally spaced 50 ft (see Fig. 13.18). The offset C, measured vertically from PQ to the parabola, will be
$$C = kx^2$$

where x is the horizontal distance of the station from the PVC. We evaluate k by noting that the offset at station 16 must be equal to the elevation of the PVT minus the elevation of Q. Since the -2 percent grade line drops 8 ft in 400 ft, the elevation of Q is 992 ft. Therefore,
$$C_{16} = 1020 \text{ ft} - 992 \text{ ft} = 28 \text{ ft}$$

Since x at this station is 800 ft,
$$28 = k(800)^2$$

$$k = \frac{28}{(800)^2} = 4.375 \times 10^{-5}$$

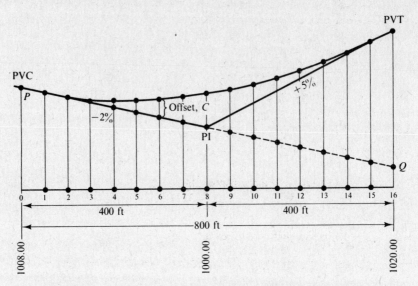

Fig. 13-18   Drawing the parabolic sag curve (vertical scale exaggerated).

The remainder of the offsets can now be computed by the equation

$$C = 4.375 \times 10^{-5} x^2$$

This is done in tabular form.

Station	x	Tangent elevation	Offset C	Curve elevation
PVC	0	1008.00	0	1008.00
1	50	1007.00	0.11	1007.11
2	100	1006.00	0.44	1006.44
3	150	1005.00	0.98	1005.98
4	200	1004.00	1.75	1005.75
5	250	1003.00	2.73	1005.73
6	300	1002.00	3.94	1005.94
7	350	1001.00	5.36	1006.36
PI	400	1000.00	7.00	1007.00
9	450	999.00	8.86	1007.86
10	500	998.00	10.94	1008.94
11	550	997.00	13.23	1010.23
12	600	996.00	15.75	1011.75
13	650	995.00	18.48	1013.48
14	700	994.00	21.44	1015.44
15	750	993.00	24.61	1017.61
PVT	800	992.00	28.00	1020.00

The curve is plotted in Fig. 13-18, with an exaggerated vertical scale.

Supplementary Problems

13.29   Write the following quadratics in general form.

   (a) $3 = 2x - 5x^2$ (b) $3x^2 - 9 = 6x$ (c) $0 = 2 - 4x + 7x^2$ (d) $x - 3 = x^2$

13.30   Solve the following pure quadratics.

(a) $2x^2 + 6 = 24$ (b) $3x^2 - 147 = 0$ (c) $x^2 - 8 = 17$ (d) $4x^2 - 31 = -6$

13.31   Solve the following incomplete quadratics.

(a) $2x^2 = x$ (b) $x(x - 2) = 2x(x + 4)$ (c) $6x + 17x^2 = 0$

13.32   Solve the following quadratics by factoring.

(a) $x^2 + 7x + 12 = 0$ (b) $x^2 - 2x - 15 = 0$ (c) $x^2 + 5x + 4 = 0$ (d) $9x^2 + 12x - 5 = 0$

13.33   Solve the following quadratics by completing the square.

(a) $x^2 - 3x - 9 = 0$ (b) $2x^2 - 4x - 18 = 0$ (c) $x^2 - x - 1 = 0$ (d) $4x^2 - 7x - 6 = 0$

13.34   Solve the following by the quadratic formula.

(a) $x^2 + 4x - 11 = 0$ (b) $5x^2 - 3x - 15 = 0$ (c) $7x^2 + 5x - 1 = 0$ (d) $x^2 - 2x - 7 = 0$

13.35   Plot the following quadratics.   Graphically determine the roots.

(a) $y = x^2 - 2x + 4$ (b) $y = 4x^2 + x - 3$ (c) $y = x^2 + 3x - 6$ (d) $y = 2x^2 - x - 7$

13.36   Solve for x.

(a) $x^4 - 11x^2 + 28 = 0$ (b) $x^{-4} - 9x^{-2} = 0$ (c) $x^{-1} - 2x^{-1/2} + 1 = 0$

(d) $x^{-2/3} + 2x^{-1/3} + 1 = 0$

13.37   Find the dimensions of a field that has a perimeter of 858 m and an area of 45 200 m^2.

13.38   Two technicians working together can wire a receiver in 6.2 h.   Working alone, the slower technician takes 3 h longer to wire a receiver than the faster technician.   How long would it take the faster technician, working alone, to wire one receiver?

13.39   A motorist drives 270 mi to another town.   She then makes the return trip in 1 h less time by increasing her average speed by 9.0 mi/h.   Find her original speed.

Answers to Supplementary Problems

13.29   (a) $5x^2 - 2x + 3 = 0$ (b) $3x^2 - 6x - 9 = 0$ (c) $7x^2 - 4x + 2 = 0$ (d) $x^2 - x + 3 = 0$

13.30   (a) $x = \pm 3$ (b) $x = \pm 7$ (c) $x = \pm 5$ (d) $x = \pm \dfrac{5}{2}$

13.31   (a) $x = 0, x = \dfrac{1}{2}$ (b) $x = 0, x = -10$ (c) $x = 0, x = -\dfrac{6}{17}$

13.32   (a) $x = -3, x = -4$ (b) $x = -3, x = 5$ (c) $x = -1, x = -4$ (d) $x = \dfrac{1}{3}, x = -\dfrac{5}{3}$

13.33   (a) $x \cong 4.854, x \cong -1.854$ (b) $x \cong 4.162, x \cong -2.162$ (c) $x \cong 1.618, x \cong -0.6180$

(d) $x \cong 2.380, x \cong 0.6302$

13.34   (a) $x \cong 1.87, x \cong -5.87$ (b) $x \cong 2.058, x \cong -1.458$ (c) $x \cong 0.1629, x \cong -0.8772$

(d) $x \cong 3.828, x \cong -1.828$

13.35   The equations are plotted in Fig. 13-19.

(a) no real roots   (b) $x \cong -1, x \cong 0.75$ (c) $x \cong 1.4, x \cong -4.4$ (d) $x \cong 2.15, x \cong -1.65$

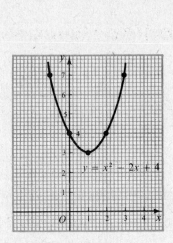

$y = x^2 - 2x + 4$

(a)

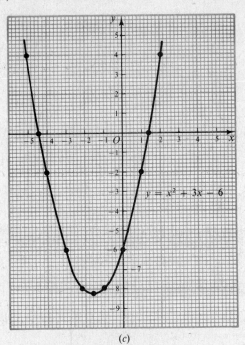

$y = x^2 + 3x - 6$

(c)

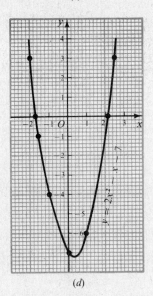

$y = 4x^2 + x - 3$

(b)

$y = 2x^2 - x - 7$

(d)

Fig. 13-19

13.36   (a) $x = \pm 2, x = \pm \sqrt{7}$ (b) $x = \pm \dfrac{1}{3}$ (c) $x = 1$ (d) $x = -1$

13.37   length = 243 m, width = 186 m

13.38   11 h

13.39   45 mi/h

Chapter 14

Logarithms and Exponential Functions

14.1 DEFINITIONS

The *logarithm* of a number N is the exponent to which a base b must be raised to obtain N.

Example 14.1: To what power must the base 10 be raised to obtain 100?

Solution: Since 10 must be raised to the power 2 to obtain 100, we say the logarithm of 100 to the base 10 is 2 and it is written

$$\log_{10} 100 = 2$$

In general, if

then,

$$\boxed{\begin{array}{c} b^x = N \\ \log_b N = x \end{array}} \quad \textbf{142}$$

14.2 CONVERTING BETWEEN LOGARITHMIC AND EXPONENTIAL FORM

It is often necessary to convert an expression from logarithmic to exponential form, or vice versa. This is done by means of Eqs. 142. It is helpful to keep in mind the key phrase, "The logarithm ... is the exponent ..."

Example 14.2: Change $2^3 = 8$ to logarithmic form.

Solution

$$\log_2 8 = 3$$

Example 14.3: Change $\log_{10} x = 4$ to exponential form.

Solution

$$10^4 = x$$

Example 14.4: Change $e^x = 26$ to logarithmic form.

Solution

$$\log_e 26 = x$$

14.3 SOLVING SIMPLE LOGARITHMIC EQUATIONS

Example 14.5: Find x if

$$x = \log_3 9$$

Solution: Changing to exponential form,

$$3^x = 9 = 3^2$$

Since the bases are equal, we may equate the exponents, so,

$$x = 2$$

Example 14.6:   Solve for x, where

$$3 = \log_x 8$$

Solution: Changing to exponential form,

$$x^3 = 8 = 2^3$$

Since the exponents are equal, we may equate the bases, so,

$$x = 2$$

Example 14.7:   Solve for x,

$$\log_2 x = 7$$

Solution: Changing to exponential form,

$$2^7 = x$$

So, $$x = 128$$

14.4 LAWS OF LOGARITHMS

(a) Products. The log of a product is equal to the sum of the logs of the factors.

$$\boxed{\log_b MN = \log_b M + \log_b N} \quad \mathbf{143}$$

Example 14.8

$$\log_{10}(12.5)(57.3) = \log_{10} 12.5 + \log_{10} 57.3$$

Common Error	The log of a sum is *not* equal to the sum of the logs. $$\log(M + N) \neq \log M + \log N$$

Common Error	The product of two logs in *not* equal to the sum of the logs. $$(\log M)(\log N) \neq \log M + \log N$$

(b) Quotients. The log of a quotient is equal to the log of the dividend minus the log of the divisor.

$$\boxed{\log_b \frac{M}{N} = \log_b M - \log_b N} \quad \mathbf{144}$$

Example 14.9

$$\log_{10} \frac{18.5}{11.3} = \log_{10} 18.5 - \log_{10} 11.3$$

Common Error	Similar errors are made with quotients as with products. $$\log (M - N) \neq \log M - \log N$$ $$\frac{\log M}{\log N} \neq \log M - \log N$$

Example 14.10:   Express as a single logarithm

$$x = \log_{10} 3 + \log_{10} 8 - \log_{10} 4$$

Solution: By Eqs. 143 and 144,

$$x = \log_{10} \frac{3(8)}{4} = \log_{10} 6$$

(c)   Powers.   The log of a number raised to a power is equal to the power times the log of the number.

$$\boxed{\log_b M^p = p \log_b M} \quad \mathbf{145}$$

Example 14.11

$$\log_{10} 4.55^3 = 3 \log_{10} 4.55$$

Example 14.12:   Express as a single logarithm,

$$x = 3 \log_{10} 4 - \log_{10} 8$$

Solution: By Eq. 145,

$$x = \log_{10} 4^3 - \log_{10} 8 = \log_{10} 64 - \log_{10} 8$$

and by Eq. 144,

$$x = \log_{10} \frac{64}{8} = \log_{10} 8$$

(d)   Roots

$$\boxed{\log_b \sqrt[q]{M} = \frac{1}{q} \log_b M} \quad \mathbf{146}$$

Example 14.13

$$\log_{10} \sqrt{57} = \frac{1}{2} \log_{10} 57$$

Example 14.14:   Express as a single logarithm

$$\frac{\log x}{3} + \log 4$$

Solution: By Eq. 146,

$$\frac{\log x}{3} + \log 4 = \frac{1}{3}\log x + \log 4 = \log \sqrt[3]{x} + \log 4$$

and by Eq. 143,

$$\frac{\log x}{3} + \log 4 = \log 4 \sqrt[3]{x}$$

(e) Log of the Base

$$\boxed{\log_b b = 1} \quad \textbf{148}$$

Example 14.15

$$\log_{10} 10 = 1$$

Example 14.16

$$\log_5 5 = 1$$

Example 14.17

$$\log_e e = 1$$

(f) Log of the Base Raised to a Power

$$\boxed{\log_b b^n = n} \quad \textbf{151}$$

Example 14.18

$$\log_{10} 10^2 = 2$$

Example 14.19

$$\log_{10} 10\,000 = \log_{10} 10^4 = 4$$

Example 14.20

$$\log_{10} 0.001 = \log_{10} 10^{-3} = -3$$

14.5 SYSTEMS OF LOGARITHMS

Two systems are in common use.

(a) *Common* **Logarithms, or the** *Briggsian* **System.**   The base used is the number 10.   They are usually written with the base omitted.   Thus, log 4.5 is taken to mean $\log_{10} 4.5$.

(b) *Natural* **Logarithms, or the** *Naperian* **System.**   The base used is the number $e \cong 2.718 \ldots$ (see Section 14.9).   Natural logs are usually written ln x, which is understood to mean $\log_e x$.

14.6 COMMON LOGARITHMS

(a) Common Logarithms by Calculator. Simply enter the number and depress the *log x* key, or the *log* key.

Example 14.21:   Find log 35.74 by calculator.

Solution: We enter the number 35.74, and depress the *log x* key, obtaining

$$\log 35.74 = 1.553$$

(b) Common Logarithms from the Tables. Tables are seldom used now for obtaining logarithms, but should it become necessary, the following procedure can be used.

Write the number in scientific notation, in the form

$$N \times 10^p$$

Then by Eq. 143,

$$\log (N \times 10^p) = \log N + \log 10^p$$

The logarithm is thus composed of two parts.

1. The *mantissa*, log N, found in the log table, Appendix D.
2. The *characteristic*, log 10^p, which, by Eq. 151, is equal to p.

Example 14.22:   Find log 185.

Solution

$$\log 185 = \log (1.85 \times 10^2) = \log 1.85 + \log 10^2$$

The characteristic = log 10^2 = 2.

The mantissa is read from the log table. Locate the first two digits (18) at the left of the table and the third digit (5) along the top. Where this row and column intersect, read the mantissa 2672. So, the mantissa = 0.2672. (Note that decimal points are omitted from the table, and should be placed before the first digit of the mantissa.) Adding the characteristic and mantissa,

$$\log 185 = 0.2672 + 2 = 2.2672$$

Third digit

Common Logarithms of Numbers

N	0	1	2	3	4	5	6	7	8	9
10	0000	0043	0086	0128	0170	0212	0253	0294	0334	0374
11	0414	0453	0492	0531	0569	0607	0645	0682	0719	0755
12	0792	0828	0864	0899	0934	0969	1004	1038	1072	1106
13	1139	1173	1206	1239	1271	1303	1335	1367	1399	1430
14	1461	1492	1523	1553	1584	1614	1644	1673	1703	1732
15	1761	1790	1818	1847	1875	1903	1931	1959	1987	2014
16	2041	2068	2095	2122	2148	2175	2201	2227	2253	2279
17	2304	2330	2355	2380	2405	2430	2455	2480	2504	2529
18	2553	2577	2601	2625	2648	2672	2695	2718	2742	2765
19	2788	2810	2833	2856	2878	2900	2923	2945	2967	2989
20	3010	3032	3054	3075	3096	3118	3139	3160	3181	3201
21	3222	3243	3263	3284	3304	3324	3345	3365	3385	3404
22	3424	3444	3464	3483	3502	3522	3541	3560	3579	3598

First two digits (18)

Example 14.23:   Find log 57 700.

Solution

$$\log 57\,700 = \log\,(5.77 \times 10^4) = \log 5.77 + \log 10^4$$

From the table, log 5.77 = 0.7612; so

$$\log 57\,700 = 0.7612 + 4 = 4.7612$$

Example 14.24:   Find log 0.006 38.

Solution

$$\log 0.006\,38 = \log\,(6.38 \times 10^{-3}) = \log 6.38 + \log 10^{-3} = 0.8048 - 3$$

This answer is also commonly written as

$$\log 0.006\,38 = 7.8048 - 10$$

a form convenient for logarithmic computation. It was obtained by adding 7 to the mantissa and subtracting 7 from the characteristic. Thus

$$0.8048 - 3 = 0.8048 \boxed{+7} - 3 \boxed{-7} = 7.8048 - 10$$

A third way of writing the same answer is obtained by subtracting the characteristic from the mantissa,

$$\log 0.006\,38 = 0.8048 - 3 = -2.1952$$

A calculator will give this form of the log of a number less than one.

(c) Interpolating. When finding the mantissa of a four-digit number, use linear interpolation as described in Section 8.11.

Example 14.25:   Find log 1548.

Solution

$$\log 1548 = \log\,(1.548 \times 10^3) = \log 1.548 + \log 10^3$$

From the table, we find that

$$\log 1.540 = 0.1875$$

and

$$\log 1.550 = 0.1903$$

N	0	1	2	3	④	⑤	6	7	8	9
10	0000	0043	0086	0128	0170	0212	0253	0294	0334	0374
11	0414	0453	0492	0531	0569	0607	0645	0682	0719	0755
12	0792	0828	0864	0899	0934	0969	1004	1038	1072	1106
13	1139	1173	1206	1239	1271	1303	1335	1367	1399	1430
14	1461	1492	1523	1553	1584	1614	1644	1673	1703	1732
⑮	1761	1790	1818	1847	1875	1903	1931	1959	1987	2014
16	2041	2068	2095	2122	2148	2175	2201	2227	2253	2279
17	2304	2330	2355	2380	2405	2430	2455	2480	2504	2529
18	2553	2577	2601	2625	2648	2672	2695	2718	2742	2765
19	2788	2810	2833	2856	2878	2900	2923	2945	2967	2989

Finding the tabular differences,

$$
\begin{array}{c|c}
\text{Number} & \text{Log} \\
\hline
1.540 & 0.1875 \\
0.010\ \ 0.008\ \ 1.548 & M \ \ x \ \ 0.0028 \\
1.550 & 0.1903 \\
\end{array}
$$

Forming a proportion and solving for x,

$$\frac{0.008}{0.010} = \frac{x}{0.0028}$$

$$x = \frac{8(0.0028)}{10} = 0.0022$$

So, the mantissa M is

$$0.1875 + 0.0022 = 0.1897$$

Adding the characteristic, 3,

$$\log 1548 = 0.1897 + 3 = 3.1897$$

14.7 ANTILOGARITHMS

If the logarithm of some number N is L, then N is called the *antilogarithm* of L. If

$$L = \log N$$

then,

$$N = \text{antilog } L = 10^L$$

(a) Antilogs by Calculator. If your calculator has a 10^x key, enter the number of which you want the antilog, and depress this key.

Example 14.26:   Find the antilog of 2.47 by calculator, to three significant figures.

Solution

>   Enter 2.47
>   Depress the 10^x key
>   Read 295

In other words,

$$\text{Antilog } 2.47 = 10^{2.47} = 295$$

If your calculator has no 10^x key but has a y^x key, you must first enter the number 10.

Example 14.27:   Find the antilog of 2.82 by calculator, to four significant figures.

Solution

>   Enter 10
>   Depress the y^x key
>   Enter 2.82
>   Depress [=] key
>   Read 660.7

(b) Antilogs from the Tables. To find antilogs from the tables, we reverse the procedure used for finding logarithms, as shown in the following example.

Example 14.28:   If $\log N = 2.7723$,   find N.

Solution

$$\log N = 2.7723 = 2 + 0.7723 = \text{characteristic} + \text{mantissa}$$

We find the antilogs of the characteristic and mantissa separately.   From Eq. 151,

$$\text{Characteristic} = 2 = \log 10^2$$

and $$\text{Mantissa} = 0.7723 = \log 5.92 \text{ (from table)}$$

N	0	1	②	3	4	5	6	7	8	9
55	7404	7412	7419	7427	7435	7443	7451	7459	7466	7474
56	7482	7490	7497	7505	7513	7520	7528	7536	7543	7551
57	7559	7566	7574	7582	7589	7597	7604	7612	7619	7627
58	7634	7642	7649	7657	7664	7672	7679	7686	7694	7701
59	7709	7716	7723	7731	7738	7745	7752	7760	7767	7774
60	7782	7789	7796	7803	7810	7818	7825	7832	7839	7846
61	7853	7860	7868	7875	7882	7889	7896	7903	7910	7917

So, $$\log N = \log 10^2 + \log 5.92 = \log (5.92 \times 10^2)$$

$$N = 592$$

Example 14.29:   If $\log N = -3.6840$,   find N.

Solution: We must have a *positive* mantissa in order to use the tables.   To obtain a positive mantissa, we add 4 to our number, and write -4 after the number to preserve its value.

$$-3.6840 + 4 - 4 = 0.3160 - 4 = \text{mantissa} + \text{characteristic}$$

So, $$\log N = 0.3160 - 4.$$

Then find the antilogs of the mantissa and characteristic separately.   From the table,

$$0.3160 = \log 2.07$$

and by Eq. 151,

$$-4 = \log 10^{-4}$$

So, $$\log N = \log 2.07 + \log 10^{-4}$$

and by Eq. 143,

$$\log N = \log (2.07 \times 10^{-4})$$

So, $$N = 2.07 \times 10^{-4} = 0.000\,207$$

(c) Interpolation. Interpolate if the given logarithm is not an exact table value.

Example 14.30:   If $\log N = 2.1635$,   find N.

Solution

$$\log N = 2.1635 = 2 + 0.1635$$

The antilog B of the mantissa 0.1635 lies between 1.45 and 1.46. Taking tabular differences,

$$\begin{array}{ll} \log 1.45 = 0.1614 \\ x \quad \log B = 0.1635 \quad .0021 \\ 0.01 \quad \log 1.46 = 0.1644 \quad .0030 \end{array}$$

Forming a proportion,

$$\frac{x}{0.01} = \frac{21}{30}$$

Solving for x,

$$x = 0.007$$

So,

$$\log B = 1.45 + 0.007 = 1.457$$

which is the antilog of the mantissa.

N	0	1	2	3	4	5	6	7	8	9
10	0000	0043	0086	0128	0170	0212	0253	0294	0334	0374
11	0414	0453	0492	0531	0569	0607	0645	0682	0719	0755
12	0792	0828	0864	0899	0934	0969	1004	1038	1072	1106
13	1139	1173	1206	1239	1271	1303	1335	1367	1399	1430
14	1461	1492	1523	1553	1584	1614	1644	1673	1703	1732
15	1761	1790	1818	1847	1875	1903	1931	1959	1987	2014
16	2041	2068	2095	2122	2148	2175	2201	2227	2253	2279

1.457

0.1635

The antilog of the characteristic 2 is 10^2. Thus,

$$N = \text{antilog } 2.1635 = 1.457 \times 10^2 = 145.7$$

(d) Negative Numbers. The *logarithm of a negative number* is not defined. If we were to try to find, for example,

$$x = \log(-5)$$

we would have to find some number x such that

$$10^x = -5$$

and it is evident that there is no such number.

14.8 LOGARITHMIC COMPUTATION

We can make use of the laws of logarithms to perform arithmetic computations.

(a) Multiplication. Use Eq. 143,

$$\log MN = \log M + \log N$$

to reduce a multiplication problem to one of addition.

Example 14.31:   Multiply $(3750)(4.45)$.

Solution

$$\log\left[(3750)(4.45)\right] = \log 3750 + \log 4.45 = \log (3.750 \times 10^3) + \log (4.45 \times 10^0)$$

$$= 3.5740 + 0.6484 = 4.2224$$

Taking the antilog,

$$(3750)(4.45) = \text{antilog } 4.2224$$

$$= 16\,700 \qquad \text{rounded to three significant figures}$$

(b) Division. Use Eq. 144,

$$\log \frac{M}{N} = \log M - \log N$$

to change a division computation to one of subtraction.

Example 14.32:   Divide $842/233$.

Solution

$$\log \frac{842}{233} = \log 842 - \log 233 = 2.9253 - 2.3674 = 0.5579$$

$$\frac{842}{233} = \text{antilog } 0.5579$$

$$= 3.61 \qquad \text{rounded to three significant figures}$$

(c) Powers. Use Eq. 145,

$$\log M^p = p \log M$$

to reduce the operation of raising to a power to one of multiplication.

Example 14.33:   Find $(7.38)^5$.

Solution

$$\log\left[(7.38)^5\right] = 5 \log 7.38 = 5(0.8681) = 4.3405$$

$$(7.38)^5 = \text{antilog } 4.3405$$

$$= 2.19 \times 10^4$$

(d) Roots. Use Eq. 146,

$$\log \sqrt[q]{M} = \frac{1}{q} \log M$$

to reduce the operation of taking a root to one of division.

Example 14.34:   Find $\sqrt[3]{285}$.

Solution

$$\log \sqrt[3]{285} = \frac{1}{3} \log 285 = \frac{2.4548}{3} = 0.8183$$

$$\sqrt[3]{285} = \text{antilog } 0.8183$$

$$= 6.58$$

14.9   NATURAL LOGARITHMS

(a)   Natural Logarithms on the Calculator.   Simply enter the number and depress the *ln* key, or *ln x* key. Remember that natural logs are usually written ln *x*, which is understood to mean $\log_e x$.

Example 14.35:   Find ln 48.3 by calculator, to four significant figures.

Solution

> Enter 48.3
> Depress the *ln* key
> Read 3.877

So,
$$\ln 48.3 = 3.877$$

(b)   Natural Logarithms from Tables.   Write the number in scientific notation.   Find its logarithm in the table in Appendix E.   Then find the logarithm of the power of ten in *Multiples of $\log_e$ 10* given below the table.

Example 14.36:   Find ln 144.

Solution

$$\ln 144 = \ln (1.44 \times 10^2) = \ln 1.44 + \ln 10^2 = \ln 1.44 + 2 \ln 10$$

From the table,

Natural Logarithms

N	0	1	2	3	4	5	6	7	8	9
1.0	0.0 000	100	198	296	392	488	583	677	770	862
1.1	953	*044	*133	*222	*310	*398	*484	*570	*655	*740
1.2	0.1 823	906	989	*070	*156	*231	*311	*390	*469	*546
1.3	0.2 624	700	776	852	937	*001	*075	*148	*221	*293
1.4	0.3 365	436	507	577	646	716	784	853	920	988
1.5	0.4 055	121	187	253	348	383	447	511	574	637
1.6	700	762	824	886	947	*008	*068	*128	*188	*247
1.7	0.5 306	365	423	481	539	596	653	710	766	822
1.8	878	933	988	*043	*098	*152	*206	*259	*313	*366
4.7	476	497	518	539	560	581	602	623	644	665
4.8	686	707	728	748	769	790	810	831	851	872
4.9	892	913	933	953	974	994	*014	*034	*054	*074
5.0	1.6 094	114	134	154	174	194	214	233	253	273

Multiples of $\log_e$ 10

$\log_e 10 = 2.3026$	$-\log_e 10 = 7.6974 - 10$
$2 \log_e 10 = 4.6052$	$-2 \log_e 10 = 5.3948 - 10$
$3 \log_e 10 = 6.9078$	$-3 \log_e 10 = 3.0922 - 10$
$4 \log_e 10 = 9.2103$	$-4 \log_e 10 = 0.7897 - 10$
$5 \log_e 10 = 11.5129$	$-5 \log_e 10 = 9.4871 - 20$

So, $\ln 144 = 0.3646 + 4.6052 = 4.9698$

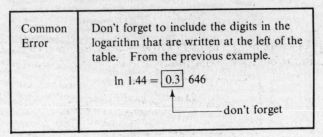

| Common Error | Don't forget to include the digits in the logarithm that are written at the left of the table.   From the previous example. |

$$\ln 1.44 = \boxed{0.3}\ 646$$

don't forget

Example 14.37:   Find $\ln 0.00607$.

Solution

$$\ln 0.00607 = \ln (6.07 \times 10^{-3}) = \ln 6.07 + (-3 \ln 10)$$
$$= 1.8034 + 3.0922 - 10 = 4.8956 - 10 = -5.1044$$

(c) Antilogs from the Log Table. Reverse the procedure for finding logarithms. If the number for which you want the antilog is greater than 2.3115 (which is the last number in the table, Appendix E), subtract multiples of $\ln 10$ to break the large number down into two numbers, one of which is a multiple of $\ln 10$ and the other is less than or equal to 2.3115.

Example 14.38:   If $\ln N = 11.1169$,   find N.

Solution

$$\ln N = 11.1169$$

Subtracting 4 $\ln 10$, (9.2103),

$$11.1169 - 9.2103 = 1.9066$$

Therefore, $$\ln N = 9.2103 + 1.9066 = \ln 10^4 + \ln 6.73 = \ln (6.73 \times 10^4)$$

$$N = 6.73 \times 10^4$$

(d) Antilogs by Calculator. If $\ln N = x$,   then

$$N = e^x$$

which can be evaluated by using the e^x key on your calculator.

Example 14.39:   If $\ln N = 5.86$,   find N to three significant figures, by calculator.

Solution: Changing to exponential form,

$$N = e^{5.86}$$

      Enter 5.86

      Depress the e^x key

      Read 351

(e) Antilogs from the Table of Powers of e. Values of e^n and e^{-n} are given in Appendix F. They may be used to find natural antilogs.

Example 14.40:   Find N if $\ln N = 0.65$.

Solution: Converting to exponential form,

$$N = e^{0.65}$$

From the table,

$$N = e^{0.65} = 1.916$$

(f) Converting Natural Logarithms to Common Logarithms. Simply divide the natural logarithm by ln 10 ($= 2.3026$) to obtain the common logarithm.   This is referred to as *change of base*.

$$\log N = \frac{\ln N}{\ln 10} = \frac{\ln N}{2.3026} \qquad \boxed{\textbf{153}}$$

(See Problem 14.12)

Example 14.41:   Find $\log N$ if $\ln N = 5.3874$.

Solution

$$\log N = \frac{\ln N}{2.3026} = \frac{5.3874}{2.3026} = 2.3397$$

Example 14.42:   Find $\ln N$ if $\log N = 3.9474$.

Solution

$$\ln N = \log N \,(2.3026) = 3.9474(2.3026) = 9.0893$$

14.10   SOLVING EXPONENTIAL EQUATIONS WITH LOGS

Exponential equations may be solved by *taking the logarithms of both sides.*

Example 14.43:   Solve for x: $5^x = 23$.

Solution: Taking the common log (natural logs would work as well) of both sides

$$\log 5^x = \log 23$$

Using Eq. 145 to remove x from the exponent,

$$\log 5^x = x \log 5 = \log 23$$

Dividing,

$$x = \frac{\log 23}{\log 5} \cong \frac{1.36}{0.699} = 1.95$$

Example 14.44:   Solve for x: $18.5 = 3.21e^{2x}$.

Solution: Dividing by 3.21,

$$5.76 = e^{2x}$$

Taking natural logs of both sides,

$$\ln 5.76 = \ln e^{2x}$$

By Eq. 145,

$$\ln 5.76 = 2x \ln e = 2x$$

since by Eq. 148, $\ln e = 1$.   Solving for x,

$$x = \frac{\ln 5.76}{2} \cong \frac{1.75}{2} = 0.875$$

Example 14.45:   Solve for x: $5^{2x-3} = 117^x$.

Solution: Taking common logs of both sides,

$$\log 5^{2x-3} = \log 117^x$$

By Eq. 145,

$$(2x - 3) \log 5 = x \log 117$$

Removing parentheses,

$$2x \log 5 - 3 \log 5 = x \log 117$$

Transposing, $2x \log 5 - x \log 117 = 3 \log 5$

Factoring, $x(2 \log 5 - \log 117) = 3 \log 5$

Dividing, $x = \dfrac{3 \log 5}{2 \log 5 - \log 117} \cong \dfrac{3(0.6990)}{2(0.6990) - 2.068} = -3.129$

14.11 LOGARITHMIC EQUATIONS

Use the laws of logarithms, Eqs. 143 to 148, to simplify and solve logarithmic equations.

Example 14.46:   Solve $\log (3x - 1) + \log 8 = \log (x + 15)$.

Solution: Use Eq. 143 to obtain a single logarithm on the left-hand side,

$$\log [(3x - 1)8] = \log (x + 15)$$

Taking the antilog of both sides,

$$(3x - 1)8 = x + 15$$

Solving for x,

$$24x - 8 = x + 15$$

$$23x = 23$$

$$x = 1$$

Check: $\log [3(1) - 1] + \log 8 \overset{?}{=} \log (1 + 15)$

$$\log 2 + \log 8 \overset{?}{=} \log 16$$

$$\log 16 = \log 16 \qquad \text{checks}$$

Check all solutions to logarithmic equations, as extraneous roots are sometimes obtained.

14.12 THE POWER FUNCTION

1. A power function is a relationship of the form,

Power Function	$y = ax^n$	**156**

where the constants a and n can be any positive or negative number.

Example 14.47

$$y = 3x^2 \qquad y = 20x^{8.7} \qquad y = -8x^{-6} \qquad y = -2x^{-\sqrt{5}}$$

are examples of power functions.

2. When n is positive, the graph of the power function is called *parabolic*.

Example 14.48:   Plot $y = 3x^3$ for values of x from 0 to 5.

Solution: Making a table of point pairs,

x	0	1	2	3	4	5
y	0	3	24	81	192	375

These points are plotted in Fig. 14-1.

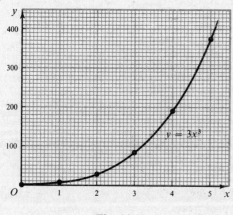

Fig. 14-1

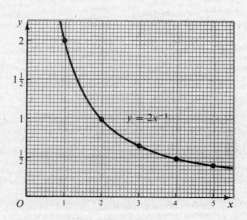

Fig. 14-2

3. When n is negative, the graph of the power function is called *hyperbolic*.

Example 14.49:   Plot $y = 2x^{-1}$ for values of x from 1 to 5.

Solution: Making a table of point pairs,

x	1	2	3	4	5
y	2	1	$\frac{2}{3}$	$\frac{1}{2}$	$\frac{2}{5}$

These points are plotted in Fig. 14-2.

4. Logarithms may be used to evaluate or solve power functions.

Example 14.50:   For the power function

$$y = 2.41x^{1.57}$$

find x when $y = 15.4$.

Solution: Dividing by 2.41,

$$x^{1.57} = \frac{15.4}{2.41} = 6.39$$

Taking logs of both sides,

$$\log x^{1.57} = \log 6.39$$

By Eq. 145,

$$1.57 \log x = \log 6.39$$

So,
$$\log x = \frac{\log 6.39}{1.57} = 0.513$$

Taking the antilog,
$$x = 3.26$$

14.13 EXPONENTIAL GROWTH AND DECAY

(a) Exponential Growth. When some quantity grows exponentially from an initial amount a, the final amount y obtained after a period of time x has passed is

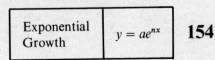

where n is the growth rate per unit time and e is the base of the natural logarithms.

Example 14.51: The population of a certain city is 100 000. If it grows exponentially at the rate of 3 % per year, find the population after 10 years.

Solution: Substituting into Eq. 154, with $a = 100\,000$, $x = 10$ years, and $n = 0.03/\text{year}$,
$$y = 100\,000e^{10(0.03)} = 100\,000e^{0.3} = 135\,000$$

(b) Exponential Decay. The formula for *exponential decay* is the same as that for exponential growth, except that the exponent is *negative*.

| Exponential Decay | $y = ae^{-nx}$ | **155** |

Example 14.52: The temperature of an object, initially at $20\overline{0}0°C$, is dropping exponentially at the rate of 25 % per min. Find the temperature after 45 s.

Solution: The units of n and x must agree; so we convert
$$x = 45 \text{ s} = 0.75 \text{ min}$$

Substituting into Eq. 155, with $a = 2000$, $n = 0.25$, and $x = 0.75$,
$$y = 2000e^{-0.25(0.75)} = 2000e^{-0.188} = 2000(0.829) = 1660°C$$

(c) Half-Life. The *half-life* of a material is the time it takes for it to decay exponentially to half its original amount.

Example 14.53: Find the half-life of a material that decays exponentially at the rate of 0.11 %/year.

Solution: From Eq. 155, with $y = a/2$ and $n = 0.0011$,
$$\frac{a}{2} = ae^{-0.0011x}$$

$$e^{-0.0011x} = \frac{1}{2}$$

By Eq. 142, $$-0.0011x = \ln\frac{1}{2} = -0.693$$

$$x = \frac{0.693}{0.0011} = 630 \text{ years}$$

(d) Time Constant. If we let $T = 1/n$,   Eq. 155 becomes

$$y = ae^{-x/T}$$

The quantity T is called the *time constant*, and it has the units of time.

Example 14.54:   A quantity decreases exponentially from an initial amount of 255 units to an amount of 184 units in 1.44 s.   Find the time constant T.

Solution: Substituting,

$$184 = 255e^{-1.44/T}$$

$$e^{-1.44/T} = \frac{184}{255} = 0.722$$

Taking natural logs,

$$\ln e^{-1.44/T} = \ln 0.722 = -0.326$$

By Eq. 145,

$$\frac{-1.44}{T} \ln e = -0.326$$

Since $\ln e = 1$,

$$T = \frac{-1.44}{-0.326} = 4.41 \text{ s}$$

Solved Problems

CONVERTING BETWEEN LOGARITHMIC AND EXPONENTIAL FORM

14.1   Convert to logarithmic form.

(a) $3^5 = 243$ (b) $10^5 = 100\,000$ (c) $4^6 = 4096$ (d) $e^3 = 20.09$

Solution

By Eq. 142,

(a) $\log_3 243 = 5$ (b) $\log_{10} 100\,000 = 5$ (c) $\log_4 4096 = 6$ (d) $\log_e 20.09 = 3$

14.2   Convert to exponential form.

(a) $\log_{10} 10\,000 = 4$ (b) $\log_5 25 = 2$ (c) $\log_2 256 = 8$ (d) $\log_8 512 = 3$

Solution

By Eq. 142,

(a) $10^4 = 10\,000$ (b) $5^2 = 25$ (c) $2^8 = 256$ (d) $8^3 = 512$

SOLVING SIMPLE LOGARITHMIC EQUATIONS

14.3   Solve for x.

(a) $x = \log_3 81$ (c) $3 = \log_x 64$ (e) $\log_3 x = 6$

(b) $x = \log_5 3125$ (d) $3 = \log_x 216$ (f) $\log_4 x = 5$

Solution

(a) By Eq. 142,

$$3^x = 81$$

but $81 = 3^4$, so,

$$3^x = 3^4$$

Equating the exponents,

$$x = 4$$

(b) By Eq. 142,

$$5^x = 3125 = 5^5$$

Equating the exponents,

$$x = 5$$

(c) By Eq. 142,

$$x^3 = 64$$

Taking the cube root,

$$x = 4$$

(d) By Eq. 142,

$$x^3 = 216 = 6^3$$

So,

$$x = 6$$

(e) By Eq. 142,

$$x = 3^6 = 729$$

(f) By Eq. 142,

$$x = 4^5 = 1024$$

LAWS OF LOGARITHMS

14.4   Express the following as single logarithms:

(a) $x = \log 4 - \log 6 + \log 3$ (c) $x = 2\log 2 + 2\log 4 - 4\log 2$

(b) $x = 3\log 2 + \log 8 - \log 4$

Solution

(a) $x = \log\left[\dfrac{4(3)}{6}\right] = \log 2$ (b) $x = \log 2^3 + \log 8 - \log 4$ (c) $x = \log 2^2 + \log 4^2 - \log 2^4$

$$= \log\left[\frac{8(8)}{4}\right] = \log 16 \qquad = \log\left[\frac{4(16)}{16}\right] = \log 4$$

COMMON LOGARITHMS

14.5   Find the common logarithms of the following numbers to four decimal places.   Use a calculator or the tables.

(a) 91   (b) 12.5   (c) 1090   (d) 55   (e) 907   (f) 0.012

Solution

(*a*) log 91 = log (9.1×10^1)

$\qquad$ = log 9.1 + log 10^1

$\qquad$ = 0.9590 + 1 = 1.9590

(*b*) log 12.5 = log (1.25×10^1)

$\qquad$ = log 1.25 + log 10^1

$\qquad$ = 0.0969 + 1 = 1.0969

(*c*) log 1090 = log (1.090×10^3)

$\qquad$ = log 1.090 + log 10^3

$\qquad$ = 0.0374 + 3 = 3.0374

(*d*) log 55 = log (5.5×10^1)

$\qquad$ = log 5.5 + log 10^1

$\qquad$ = 0.7404 + 1 = 1.7404

(*e*) log 907 = log (9.07×10^2)

$\qquad$ = log 9.07 + log 10^2

$\qquad$ = 0.9576 + 2 = 2.9576

(*f*) log 0.012 = log (1.2×10^{-2})

$\qquad$ = log 1.2 + log 10^{-2}

$\qquad$ = 0.0792 − 2

$\qquad$ = 8.0792 − 10   or   −1.9208

14.6 Find the common logarithms of the following numbers to four decimal places by calculator or by interpolation in the tables:

(*a*) 1.235   (*b*) 1009   (*c*) 2755   (*d*) 0.1096

Solution

(*a*) log 1.235 = log (1.235×10^0) = log 1.235 + log 10^0

From the table,

$$
0.01 \left[\; 0.005 \left[\begin{array}{l} \text{log } 1.23 \; = 0.0899 \\ \text{log } 1.235 = M \quad\rule{1cm}{0.4pt}\; x \\ \text{log } 1.24 \; = 0.0934 \end{array} \right] 0.0035 \right.
$$

Writing a proportion and solving for *x*,

$$\frac{5}{10} = \frac{x}{0.0035}$$

$$x = 0.001\,75$$

So the mantissa is

$$M = 0.0899 + 0.001\,75 = 0.0917$$

So,$\qquad$ log 1.235 = 0.0917

(*b*) log 1009 = log (1.009×10^3) = log 1.009 + log 10^3

$$
0.010 \left[\; 0.009 \left[\begin{array}{l} \text{log } 1.000 = 0 \\ \text{log } 1.009 = M \quad\rule{1cm}{0.4pt}\; x \\ \text{log } 1.010 = 0.0043 \end{array} \right] 0.0043 \right.
$$

$$\frac{9}{10} = \frac{x}{0.0043}$$

$$x = 0.0039$$

$$M = 0 + 0.0039 = 0.0039$$

So,$\qquad$ log 1009 = 0.0039 + 3 = 3.0039

(c) $\log 2755 = \log (2.755 \times 10^3) = \log 2.755 + \log 10^3$

$$0.010 \; \begin{array}{|c} 0.005 \left[\begin{array}{l} \log 2.75 \; = 0.4393 \\ \log 2.755 = M \\ \log 2.76 \; = 0.4409 \end{array} \right] x \end{array} \; 0.0016$$

$$\frac{5}{10} = \frac{x}{0.0016}$$

$$x = 0.0008$$

$$M = 0.4393 + 0.0008 = 0.4401$$

So, $\log 2755 = 0.4401 + 3 = 3.4401$

(d) $\log 0.1096 = \log (1.096 \times 10^{-1}) = \log 1.096 + \log 10^{-1}$

$$0.010 \; \begin{array}{|c} 0.006 \left[\begin{array}{l} \log 1.09 \; = 0.0374 \\ \log 1.096 = M \\ \log 1.10 \; = 0.0414 \end{array} \right] x \end{array} \; 0.0040$$

$$\frac{6}{10} = \frac{x}{0.0040}$$

$$x = 0.0024$$

$$M = 0.0374 + 0.0024 = 0.0398$$

So, $\log 0.1096 = 0.0398 - 1 = 9.0398 - 10 = -0.9602$

ANTILOGARITHMS

14.7   Find the antilogarithms of the following to three significant figures by calculator or tables:

(a) $\log N = 2.5119$ (c) $\log N = 3.7825$ (e) $\log N = 2.6053$

(b) $\log N = -1.9101$ (d) $\log N = 8.3139 - 10$

Solution

(a) $\log N = 2.5119 = 2 + 0.5119$ (d) $\log N = 8.3139 - 10 = 0.3139 - 2$

$\qquad\qquad = \log 10^2 + \log 3.25$ $= \log 2.06 + \log 10^{-2}$

$\qquad\qquad = \log (3.25 \times 10^2)$ $= \log (2.06 \times 10^{-2})$

$\qquad N = 325$ $N = 0.0206$

(b) $\log N = -1.9101 = 0.0899 - 2$ (e) $\log N = 2.6053 = 2 + 0.6053$

$\qquad\qquad = \log 1.23 + \log 10^{-2}$ $= \log 10^2 + \log 4.03$

$\qquad\qquad = \log (1.23 \times 10^{-2})$ $= \log (4.03 \times 10^2)$

$\qquad N = 0.0123$ $N = 403$

(c) $\log N = 3.7825 = 3 + 0.7825$

$\qquad\qquad = \log 10^3 + \log 6.06$

$\qquad\qquad = \log (6.06 \times 10^3)$

$\qquad N = 6060$

14.8 Find N to four significant figures by calculator or by interpolating in the tables.

(a) $\log N = 1.6296$ (b) $\log N = 3.8776$ (c) $\log N = -1.9488$ (d) $\log N = 0.5506$

Solution

(a) Finding the mantissa 0.6296 in the table, we obtain,

$$0.01 \left[\; x \left[\begin{array}{l} \log 4.26 = 0.6294 \\ \log B \;\; = 0.6296 \\ \log 4.27 = 0.6304 \end{array} \right. 2 \right] 10$$

Forming a proportion,

$$\frac{x}{0.01} = \frac{2}{10}$$

$$x = 0.002$$

$$B = 4.26 + 0.002 = 4.262$$

Since the characteristic is 1,

$$N = 4.262 \times 10^1 = 42.62$$

(b) From the table,

$$\log 7.54 = 0.8774$$

$$\log B = 0.8776$$

$$\log 7.55 = 0.8779$$

So,

$$\frac{x}{0.01} = \frac{2}{5}$$

$$x = 0.004$$

$$B = 7.54 + 0.004 = 7.544$$

and

$$N = 7.544 \times 10^3 = 7544$$

(c) $\log N = -1.9488 = 0.0512 - 2$

From the table,

$$\log 1.12 = 0.0492$$

$$\log B = 0.0512$$

$$\log 1.13 = 0.0531$$

So,

$$\frac{x}{0.01} = \frac{20}{39}$$

$$x = 0.005$$

and

$$B = 1.12 + 0.005 = 1.125$$

So,

$$N = 1.125 \times 10^{-2} = 0.01125$$

(d) From the table,

$$\log 3.55 = 0.5502$$

$$\log B = 0.5506$$

$$\log 3.56 = 0.5514$$

So,

$$\frac{x}{0.01} = \frac{4}{12}$$

$$x = 0.003$$

$$B = 3.55 + 0.003 = 3.553$$

$$N = 3.553 \times 10^0 = 3.553$$

LOGARITHMIC COMPUTATION

14.9   Perform the following computations using logarithms:

(a) $x = (475)(3.25)$ (c) $x = (6.95)^3$ (e) $x = \dfrac{(7.16)^4(34.1)}{2.18}$

(b) $x = \dfrac{239}{16.0}$ (d) $x = \sqrt[4]{113}$ (f) $x = \dfrac{6.78}{(3.33)^2}(2.34)$

Solution

(a) $x = (475)(3.25)$

$\log x = \log 475 + \log 3.25$

$\quad = \log (4.75 \times 10^2) + \log (3.25 \times 10^0)$

$\quad = 2.6767 + 0.5119$

$\quad = 3.1886$

$x = \text{antilog } 3.1886 = 1540$

(b) $x = \dfrac{239}{16.0}$

$\log x = \log 239 - \log 16.0$

$\quad = \log (2.39 \times 10^2) - \log (1.60 \times 10^1)$

$\quad = 2.3784 - 1.2041$

$\quad = 1.1743$

$x = \text{antilog } 1.1743 = 14.9$

(c) $x = (6.95)^3$

$\log x = 3 \log 6.95$

$\quad = 3(0.8420)$

$\quad = 2.526$

$x = \text{antilog } 2.526 = 336$

(d) $x = \sqrt[4]{113}$

$\log x = \log (113)^{1/4}$

$\quad = \dfrac{1}{4} \log (1.13 \times 10^2)$

$\quad = \dfrac{1}{4} (2.0531)$

$\quad = 0.5133$

$x = \text{antilog } 0.5133 = 3.26$

(e) $x = \dfrac{(7.16)^4 \times 34.1}{2.18}$

$\log x = 4 \log 7.16 + \log 34.1 - \log 2.18$

$\quad = 4(0.8549) + 1.5328 - 0.3385$

$\quad = 4.6139$

$x = \text{antilog } 4.6139 = 41\,100$

(f) $x = \dfrac{6.78}{(3.33)^2}(2.34)$

$\log x = \log 6.78 - 2 \log 3.33 + \log 2.34$

$\quad = 0.8312 - 2(0.5224) + 0.3692$

$\quad = 0.1556$

$x = \text{antilog } 0.1556 = 1.43$

NATURAL LOGARITHMS

14.10 Find the natural logarithms of the following numbers to four decimal places, by calculator or tables:

(a) 3.2   (b) 294   (c) 0.036   (d) 4090   (s) 61.5   (f) 316.0

Solution

(a) $\ln 3.2 = \ln (3.2 \times 10^0)$
$= \ln 3.2 + \ln 10^0$
$= \ln 3.2 + 0 = 1.1632$

(b) $\ln 294 = \ln (2.94 \times 10^2)$
$= \ln 2.94 + 2 \ln 10$
$= 1.0784 + 4.6052 = 5.6836$

(c) $\ln 0.036 = \ln (3.6 \times 10^{-2})$
$= \ln 3.6 + (-2) \ln 10$
$= 1.2809 + 5.3948 - 10 = -3.3243$

(d) $\ln 4090 = \ln (4.090 \times 10^3)$
$= \ln 4.090 + 3 \ln 10$
$= 1.4085 + 6.9078 = 8.3163$

(e) $\ln 61.5 = \ln (6.15 \times 10^1)$
$= \ln 6.15 + \ln 10$
$= 1.8165 + 2.3026 = 4.1191$

(f) $\ln 316.0 = \ln (3.16 \times 10^2)$
$= \ln 3.16 + 2 \ln 10$
$= 1.1506 + 4.6052 = 5.7558$

14.11 Find the antilogarithms of each of the following natural logarithms by calculator or from the tables:

(a) 1.9359   (b) 9.0734   (c) 3.0634   (d) 5.5646   (e) −1.7451

Solution

(a) $\ln N = 1.9359$

From the table,

$$N = 6.93$$

(b) $\ln N = 9.0734$

Removing multiples of $\ln 10$,

$$\ln N = 6.9078 + 2.1656$$
$$= \ln 10^3 + \ln 8.72$$
$$= \ln (8.72 \times 10^3)$$

So, $N = 8720$

(c) $\ln N = 3.0634$
$= 2.3026 + 0.7608$
$= \ln 10^1 + \ln 2.14$
$= \ln (2.14 \times 10^1)$

$N = 21.4$

(d) $\ln N = 5.5646$
$= 4.6052 + 0.9594$
$= \ln 10^2 + \ln 2.61$
$= \ln (2.61 \times 10^2)$

$N = 261$

(e) $\ln N = -1.7451$
$= 0.5575 - 2.3026$
$= \ln 1.7463 - \ln 10$
$= \ln \left(\dfrac{1.7463}{10} \right)$

$N = 0.174\,63$

CHANGE OF BASE

14.12 Show that $\log N = \ln N / \ln 10$.

Solution

Let $x = \log N$

Changing to exponential form,

$$10^x = N$$

Taking natural logs of both sides,

$$\ln 10^x = x \ln 10 = \ln N$$

Dividing,

$$x = \frac{\ln N}{\ln 10}$$

but $x = \log N$, so

$$\boxed{\log N = \frac{\ln N}{\ln 10}} \quad \mathbf{153}$$

where $\ln 10 \cong 3.2026$.

14.13   Find $\log N$, if

(a) $\ln N = 1.8165$ (b) $\ln N = 2.1793$

Solution

By Eq. 153,

(a) $\log N = \dfrac{\ln N}{2.3026} = \dfrac{1.8165}{2.3026} = 0.7889$ (b) $\log N = \dfrac{\ln N}{2.3026} = \dfrac{2.1793}{2.3026} = 0.9465$

14.14   Find $\ln N$, if

(a) $\log N = 2.1461$ (b) $\log N = 1.8756$

Solution

By Eq. 153,

(a) $\ln N = 2.3026 \log N = 2.1461(2.3026) = 4.942$ (b) $\ln N = 2.3026 \log N = 1.8756(2.3026) = 4.319$

EXPONENTIAL EQUATIONS

14.15   Solve for x: $3^x = 5$.

Solution

Taking the logs of both sides,

$$\log 3^x = \log 5$$

By Eq. 145,

$$x \log 3 = \log 5$$

$$x = \frac{\log 5}{\log 3} \cong \frac{0.699}{0.477} = 1.47$$

14.16   Solve for x: $(8.95)^x = 27.3$.

Solution

Taking logs,

$$\log (8.95)^x = \log 27.3$$

By Eq. 145,

$$x \log 8.95 = \log 27.3$$

$$x = \frac{\log 27.3}{\log 8.95} \cong \frac{1.44}{0.952} = 1.51$$

14.17 Solve for x: $(9.92)^{x+1} = 18.3$.

Solution

Taking logs,

$$\log (9.92)^{x+1} = \log 18.3$$

$$(x+1) \log 9.92 = \log 18.3$$

$$x + 1 = \frac{\log 18.3}{\log 9.92} \cong 1.267$$

$$x = 1.267 - 1 = 0.267$$

14.18 Solve for x: $(0.552)^{x^2} = (1.55)^x$.

Solution

Taking logs,

$$\log (0.552)^{x^2} = \log (1.55)^x$$

By Eq. 145,

$$x^2 \log 0.552 = x \log 1.55$$

$$-0.258x^2 - 0.190x = 0$$

Factoring,

$$x(0.258x + 0.190) = 0$$

$$x = 0 \qquad \text{and} \qquad x = -\frac{0.190}{0.258} = -0.736$$

14.19 Solve for x: $(142)^{\sqrt{x}} = 528$.

Solution

Taking logs,

$$\log (142)^{\sqrt{x}} = \log 528$$

By Eq. 145,

$$\sqrt{x} \log 142 = \log 528$$

$$\sqrt{x} = \frac{\log 528}{\log 142} \cong 1.26$$

Squaring both sides,

$$x = 1.60$$

14.20 Solve for x: $e^{3x} = 18.9$.

Solution

Taking natural logs,

$$\ln e^{3x} = \ln 18.9 \cong 2.94$$

By Eq. 145,

$$3x \ln e = 2.94$$

By Eq. 148,

$$3x = 2.94$$
$$x = 0.980$$

14.21   Solve for x: $11.3e^{5x} = 27.5$.

Solution

Dividing by 11.3,

$$e^{5x} = \frac{27.5}{11.3} = 2.43$$

Taking natural logs,

$$\ln e^{5x} = \ln 2.43$$
$$5x \ln e = \ln 2.43$$
$$5x = \ln 2.43$$
$$x = \frac{\ln 2.43}{5} \cong \frac{0.888}{5} = 0.178$$

14.22   Solve for x: $115e^{2x+1} = 577$.

Solution

Dividing,

$$e^{2x+1} = \frac{577}{115} = 5.02$$

Taking natural logs,

$$\ln e^{2x+1} = (2x+1)\ln e = \ln 5.02$$
$$2x + 1 = \ln 5.02$$
$$x = \frac{\ln 5.02 - 1}{2} \cong 0.307$$

14.23   Solve for x: $e^{3x+5} = 2e^{x-1}$.

Solution

$$\frac{e^{3x+5}}{e^{x+1}} = 2$$

By Eq. 9,

$$e^{(3x+5)-(x+1)} = e^{2x+4} = 2$$

Taking natural logs,

$$\ln e^{2x+4} = \ln 2$$
$$(2x+4)\ln e = \ln 2 \cong 0.693$$
$$2x = 0.693 - 4 = -3.31$$
$$x = -1.66$$

14.24   Solve for x: $5.52e^{2x^2} = 9.73$.

Solution

Dividing,
$$e^{2x^2} = \frac{9.73}{5.52} = 1.76$$

Taking natural logs,

$$2x^2 \ln e = \ln 1.76 \cong 0.565$$
$$x^2 = 0.283$$
$$x = \pm 0.532$$

14.25   *Flow over Weirs:*   The flow rate Q (ft^3/s) over a rectangular weir L feet wide is given by

$$Q = 3.33LH^{3/2}$$

where H is the water level above the weir, in feet, as in Fig. 14-3.   Find the flow rate when the water level is 1.25 ft above a 4.00-ft-wide weir.

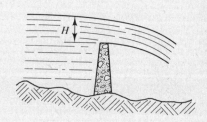

Fig. 14-3

Solution

Substituting, with $H = 1.25$ and $L = 4.00$,

$$Q = 3.33(4.00)(1.25)^{3/2}$$
$$= 3.33(4.00)(1.40) = 18.6 \text{ ft}^3/\text{s}$$

14.26   The approximate density d (lb/ft^3) of sea water at a depth of h miles is given by the exponential equation,

$$d = 64e^{0.00676h}$$

At what depth will the density be 65 lb/ft^3?

Solution

Substituting $d = 65$,

$$65 = 64e^{0.00676h}$$

$$e^{0.00676h} = \frac{65}{64} = 1.02$$

Taking natural logs of both sides,

$$\ln e^{0.00676h} = 0.00676h \ln e = \ln 1.02$$

Since $\ln e = 1$,

$$0.00676h = \ln 1.02 = 0.0198$$

$$h = 2.9 \text{ mi}$$

14.27   A weight W hangs by a rope which passes over a rough cylindrical beam, Fig. 14-4.   The force F needed to hold the weight in equilibrium is given by the exponential equation

$$F = We^{\mu\theta}$$

where θ is the angle of contact in radians and μ is the coefficient of friction between the rope and the beam. If a force of 150 lb will support a weight of 210 lb with $1\frac{1}{4}$ turns of wrap, how many turns will be needed for the same force to hold 500 lb?

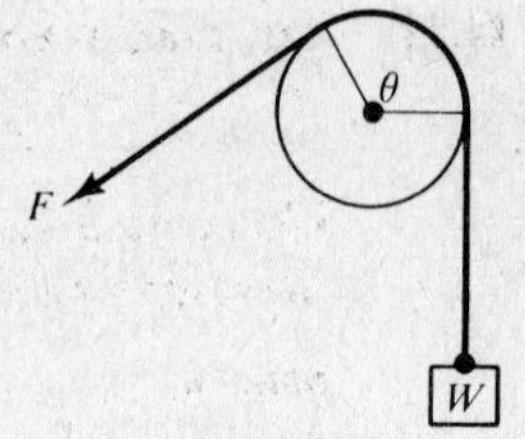

Fig. 14-4

Solution

First evaluate μ by substituting into the equation, $F = 150$ lb, $W = 210$ lb, and $\theta = 1.25$ turns $= 7.85$ rad. Then,

$$150 = 210e^{7.85\mu}$$

So,

$$e^{7.85\mu} = \frac{150}{210} = 0.714$$

Taking natural logs of both sides,

$$7.85\mu = \ln 0.714 = -0.336$$

$$\mu = -0.0429$$

So,

$$F = We^{-0.0429\theta}$$

Substituting the *new* values, $W = 500$ and $F = 150$,

$$150 = 500e^{-0.0429\theta}$$

$$e^{-0.0429\theta} = \frac{150}{500} = 0.300$$

Taking natural logs,

$$-0.0429\theta = \ln 0.300 = -1.20$$

$$\theta = 28.1 \text{ rad} = 4.47 \text{ turns}$$

14.28 The efficiency E of an Otto cycle automobile engine having a compression ratio r is

$$E = \left(1 - \frac{1}{r^{0.4}}\right) \times 100$$

Find the efficiency for engines having compression ratios of 7, 8, and 9.

Solution

Letting $r = 7, 8$, and 9,

$$7^{0.4} = 2.18 \qquad 8^{0.4} = 2.30 \qquad \text{and} \qquad 9^{0.4} = 2.41$$

For $r = 7$,

$$E = \left(1 - \frac{1}{2.18}\right)100 = 54\%$$

For $r = 8$,

$$E = \left(1 - \frac{1}{2.30}\right)100 = 57\%$$

For $r = 9$,

$$E = \left(1 - \frac{1}{2.41}\right)100 = 59\%$$

14.29   The maximum pressure p in a cylinder having a compression ratio r is approximately

$$p = p_m r^{1.33}$$

where p_m is the intake manifold pressure. What compression ratio is needed to produce a maximum pressure 16 times as great as the manifold pressure?

Solution

Letting $p = 16p_m$,

$$16p_m = p_m r^{1.33}$$

$$r^{1.33} = 16$$

Taking natural logs of both sides,

$$\ln r^{1.33} = \ln 16$$

By Eq. 145,

$$1.33 \ln r = \ln 16 = 2.77$$

$$\ln r = 2.08$$

Taking the antilog,

$$r = e^{2.08} = 8.04$$

LOGARITHMIC EQUATIONS

14.30   Solve for x: $\log 5 + \log (x - 3) = \log (2x - 3)$.

Solution

By Eq. 143,

$$\log [5(x - 3)] = \log (2x - 3)$$

Taking antilogs,

$$5x - 15 = 2x - 3$$

$$3x = 12$$

$$x = 4$$

Check: $\log 5 + \log (4 - 3) \overset{?}{=} \log (8 - 3)$

$$0.6990 + 0 = 0.6990 \quad \text{checks}$$

14.31   Solve for x: $\log (x - 2) + \log 3 = \log (x + 4)$.

Solution

By Eq. 143,

$$\log [3(x - 2)] = \log (x + 4)$$

So, $3x - 6 = x + 4$

$$2x = 10$$

$$x = 5$$

Check: $\log (5 - 2) + \log 3 \overset{?}{=} \log (5 + 4)$

$$0.4771 + 0.4771 = 0.9542 \quad \text{checks}$$

14.32   Solve for x: $\ln x + \ln (x + 2) = 1$.

Solution

By Eq. 143,

$$\ln [x(x + 2)] = 1$$
$$\ln (x^2 + 2x) = 1$$

By Eq. 142,

$$x^2 + 2x = e^1 = e$$
$$x^2 + 2x - e = 0$$

By Eq. 139,

$$x = \frac{-2 \pm \sqrt{4 - 4(-e)}}{2} = \frac{-2 \pm 2\sqrt{1 + e}}{2} = -1 \pm \sqrt{1 + e} \cong -1 \pm 1.9283 = 0.9283 \quad \text{and} \quad -2.9283$$

Check: When $x = 0.9283$,

$$\ln (0.9283) + \ln (2.9283) \overset{?}{=} 1$$
$$-0.0744 + 1.0744 = 1 \quad \text{checks}$$

When $x = -2.9283$, we get the log of a negative number, which is not permitted.

14.33   Solve for x: $2 \log x - \log (30 - 2x) = 1$.

Solution

By Eqs. 144 and 145,

$$\log x^2 - \log (30 - 2x) = \log \frac{x^2}{30 - 2x} = 1$$

By Eq. 142,

$$\frac{x^2}{30 - 2x} = 10^1 = 10$$

Solving for x,

$$x^2 = 300 - 20x$$
$$x^2 + 20x - 300 = 0$$

Factoring,

$$(x + 30)(x - 10) = 0$$
$$x = -30 \quad \text{and} \quad x = 10$$

Check: When $x = -30$,

$$2 \log (-30) - \log (90) \overset{?}{=} 1$$

Does not check, as $\log (-30)$ is not permissible.

When $x = 10$,

$$2 \log 10 - \log 10 \overset{?}{=} 1$$
$$\log 10 = 1 \quad \text{checks}$$

14.34   Solve for x: $\ln 2x + \ln (3x - e) - 2 = \ln 8$.

Solution

Transposing,

$$\ln 2x + \ln (3x - e) - \ln 8 = 2$$

By Eqs. 143 and 144,

$$\ln \frac{2x(3x - e)}{8} = 2$$

By Eq. 142,

$$\frac{x(3x - e)}{4} = e^2$$

Solving for x,

$$3x^2 - ex = 4e^2$$

$$3x^2 - ex - 4e^2 = 0$$

By Eq. 139,

$$x = \frac{e \pm \sqrt{e^2 - 4(3)(-4e^2)}}{6} = \frac{e \pm \sqrt{e^2 + 48e^2}}{6} = \frac{e \pm \sqrt{49e^2}}{6} = \frac{e \pm 7e}{6} = \frac{4e}{3} \quad \text{and} \quad -e$$

Check: When $x = 4e/3$,

$$\ln \frac{8e}{3} + \ln (4e - e) - 2 \overset{?}{=} \ln 8$$

$$\ln 8 - \ln 3 + \ln e + \ln 3 + \ln e - 2 \overset{?}{=} \ln 8$$

$$\ln 8 + 2 \ln e - 2 \overset{?}{=} \ln 8$$

$$\ln 8 = \ln 8 \quad \text{checks}$$

When $x = -e$,   we get the illegal operation of taking the log of a negative number.

14.35   The difference in elevation h (ft) between two locations having barometer readings of B_1 and B_2 inches of mercury (in. Hg) is given by the logarithmic equation,

$$h = 60\,470 \log \left(\frac{B_2}{B_1} \right)$$

where B_1 is the reading at the higher location (and is therefore the *lower* reading).

Find the difference in elevation between two stations having barometer readings of 29.25 in. and 25.45 in.

Solution

Substituting, $h = 60\,470 \log \dfrac{29.25}{25.45} = 60\,470(0.060\,44) = 3655 \text{ ft}$

14.36   What will be the barometer reading 925.0 ft above a station having a reading of 28.74 in.?

Solution

Letting $h = 925.0$ and $B_2 = 28.74$,

$$925.0 = 60\,470 \log \left(\frac{28.74}{B_1} \right)$$

$$\log \left(\frac{28.74}{B_1} \right) = \frac{925.0}{60\,470} = 0.015\,30$$

Switching to exponential form,

$$\frac{28.74}{B_1} = 10^{0.01530} = 1.036$$

$$B_1 = \frac{28.74}{1.036} = 27.74 \text{ in.}$$

DECIBELS

If the power input to a network or device is P_1 and the power output is P_2, the amount of decibels gained or lost in the device is given by the logarithmic equation,

Decibels Gained or Lost	$dB = 10 \log_{10} \dfrac{P_2}{P_1}$	**221**

14.37 A certain amplifier gives a power output of 30 W for an input of 100 mW.   Find the dB gain.

Solution

From Eq. 221,

$$dB \text{ gain} = 10 \log \frac{30}{0.1} = 10 \log 300 = 25 \text{ dB}$$

14.38 A transmission line has a loss of 2.15 dB.   Find the power transmitted for an input of 1580 kW.

Solution

From Eq. 221,

$$dB \text{ loss} = -2.15 = 10 \log \frac{P_2}{1580}$$

$$\log \frac{P_2}{1580} = -0.215$$

By Eq. 142,

$$\frac{P_2}{1580} = 10^{-0.215} = 0.610$$

$$P_2 = 1580(0.610) = 964 \text{ kW}$$

14.39 What power input is needed to produce a 40-W output with an amplifier having a 30-dB gain?

Solution

From Eq. 221,

$$dB \text{ gain} = 30 = 10 \log \frac{40}{P_1}$$

$$\log \frac{40}{P_1} = 3$$

By Eq. 142,

$$\frac{40}{P_1} = 10^3 = 1000$$

$$P_1 = \frac{40}{1000} = 0.040 \text{ W} = 40 \text{ mW}$$

14.40 The heat loss q per foot of cylindrical pipe insulation (Fig. 14-5) having an inside radius r_1 and outside radius r_2 is given by the logarithmic equation,

$$q = \frac{2\pi k(t_1 - t_2)}{\ln\left(\dfrac{r_2}{r_1}\right)} \text{ British thermal units/hour (Btu/h)}$$

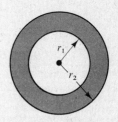

where t_1 and t_2 are the inside and outside temperatures (°F) and k is the conductivity of the insulation. Find q for a 2-in.-thick insulation having a conductivity of 0.025 wrapped around a 6-in.-diameter pipe at 400°F, if the surroundings are at 80°F.

Fig. 14-5

Solution

Substituting, with $k = 0.025$, $r_1 = 3$ in., $r_2 = 5$ in., $t_1 = 400°F$, and $t_2 = 80°F$,

$$q = \frac{2\pi(0.025)(400 - 80)}{\ln\left(\dfrac{5}{3}\right)} = \frac{50.3}{0.51} = 98 \text{ Btu/h per foot } [\text{Btu/(h)/(ft)}]$$

14.41 Find the heat loss if the insulation thickness in the previous problem were (a) halved, (b) doubled.

Solution

(a) Letting $r_2 = 4$ in.,

$$q = \frac{50.3}{\ln\left(\dfrac{4}{3}\right)} = \frac{50.3}{0.29} = 173 \text{ Btu/(h)/(ft)}$$

(b) Letting $r_2 = 7$ in.,

$$q = \frac{50.3}{\ln\left(\dfrac{7}{3}\right)} = \frac{50.3}{0.85} = 59 \text{ Btu/(h)/(ft)}$$

POWER FUNCTION

14.42 For the power function $y = 8.75x^{5.33}$, find x when $y = 573$.

Solution

Dividing,

$$x^{5.33} = \frac{573}{8.75} = 65.5$$

Taking logs of both sides,

$$\log x^{5.33} = \log 65.5$$

By Eq. 145,

$$5.33 \log x = \log 65.5 = 1.82$$

$$\log x = \frac{1.82}{5.33} = 0.341$$

Taking the antilog,

$$x = 10^{0.341} = 2.19$$

14.43   For the power function $y = 0.595x^{-1.54}$, find x when $y = 0.0113$.

Solution

Dividing,

$$x^{-1.54} = \frac{0.0113}{0.595} = 0.0190$$

Taking logs,

$$\log x^{-1.54} = -1.54 \log x = \log 0.0190$$

$$\log x = \frac{\log 0.0190}{-1.54} = \frac{-1.72}{-1.54} = 1.12$$

$$x = 13.2$$

EXPONENTIAL GROWTH AND DECAY

14.44   A quantity grows exponentially at the rate of 3% per year for 5 years, starting at an amount of 1000 units. Find the final amount.

Solution

From Eq. 154, with $a = 1000$, $n = 0.03$, and $x = 5$,

$$y = 1000e^{(0.03)(5)} = 1000(1.16) = 1160 \text{ units}$$

14.45   How long will it take for a quantity to triple when growing exponentially at an annual rate of 15%?

Solution

From Eq. 154, with $y = 3a$ and $n = 0.15$,

$$3a = ae^{0.15x}$$

$$e^{0.15x} = 3$$

By Eq. 142,

$$0.15x = \ln 3 = 1.10$$

$$x = 7.33 \text{ years}$$

14.46   Find the rate of growth to enable a quantity to increase exponentially from 10 000 units to 18 000 units in 90 days.

Solution

From Eq. 154, with $y = 18\,000$, $a = 10\,000$, and $x = 90$,

$$18\,000 = 10\,000e^{90n}$$

$$e^{90n} = 1.8$$

By Eq. 142,

$$90n = \ln 1.8 = 0.59$$

$$n = 0.0065 = 0.65\% \text{ per day}$$

14.47 How many units of some material must there be initially to have $50\overline{0}$ units left after the amount has decreased exponentially at the rate of 2.3% per h for 18 h?

Solution

From Eq. 155, with $y = 500$, $n = 0.023$, and $x = 18$,

$$500 = ae^{-0.023(18)} = ae^{-0.41} = 0.664a$$

$$a = \frac{500}{0.664} = 753 \text{ units}$$

14.48 Find the half-life of a material that decays exponentially at the rate of 1.15% per year.

Solution

From Eq. 155, with $y = a/2$ and $n = 0.0115$,

$$\frac{a}{2} = ae^{-0.0115x}$$

$$e^{-0.0115x} = \frac{1}{2}$$

From Eq. 142,

$$-0.0115x = \ln \frac{1}{2} = -0.693$$

$$x = 60.3 \text{ years}$$

TRANSIENT CURRENTS IN A REACTIVE CIRCUIT

14.49 When the switch in Fig. 14-6 is closed, the current i will grow exponentially according to the equation,

$$i = \frac{E(1 - e^{-Rt/L})}{R} \text{ A}$$

where L is the inductance in henries (H) and R is the resistance in ohms. Find the current at 0.050 s if $R = 125 \ \Omega$, $L = 2.48$ H, and $E = 6.15$ V.

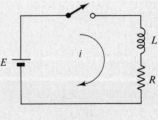

Fig. 14-6

Solution

Substituting into the given equation,

$$e^{-Rt/L} = e^{-125(0.050)/2.48} = e^{-2.52} = 0.0804$$

So,

$$i = \frac{6.15(1 - 0.0804)}{125} = 0.0452 \text{ A} = 45.2 \text{ mA}$$

14.50 When a fully discharged capacitor C, Fig. 14-7, is connected across a battery, the current i flowing into the capacitor will decay exponentially according to the equation,

$$i = \frac{E}{R} e^{-t/RC}$$

If $E = 55$ V, $R = 120\ \Omega$, and $C = 0.000\,40$ farad (F), find the time for the current to decrease to 0.20 A.

Solution

Substituting,
$$0.20 = \frac{55}{120} e^{-t/120(0.000\,40)}$$

$$e^{-t/0.048} = \frac{120(0.20)}{55} = 0.436$$

Taking natural logs of both sides,

$$-\frac{t}{0.048} = \ln 0.436 = -0.829$$

$$t = 0.048(0.829) = 0.040 \text{ s} = 40 \text{ ms}$$

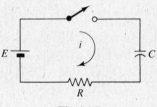

Fig. 14-7

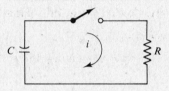

Fig. 14-8

14.51 When a capacitor C, charged to a voltage v_0, is discharged through a resistor R, Fig. 14-8, the current i will decay exponentially according to the equation,

$$i = \frac{v_0}{R} e^{-t/RC}$$

Find the current after 50 ms in a circuit where $v_0 = 115$ V, $C = 255$ microfarads (μF), and $R = 1500\ \Omega$.

Solution

Substituting,

$$i = \frac{115}{1500} e^{-0.050/1500(255 \times 10^{-6})} = 0.0767 e^{-0.131} = 0.0767(0.878) = 0.0673 \text{ A} = 67.3 \text{ mA}$$

14.52 The voltage in a certain circuit is seen to decay exponentially from an initial voltage of 45.8 V. After 25 ms, the voltage is 18.3 V.

(a) Find the time constant T.   (b) Write the equation $V = f(t)$.

Solution

(a) The formula for exponential decay, Eq. 155, written in terms of the variables in this problem, is

$$V = V_0 e^{-t/T}$$

Solving for the time constant T,

$$e^{-t/T} = \frac{V}{V_0}$$

$$-\frac{t}{T} = \ln\left(\frac{V}{V_0}\right)$$

$$T = \frac{-t}{\ln\left(\dfrac{V}{V_0}\right)}$$

Substituting our known values,

$$T = \frac{-0.025}{\ln\left(\dfrac{18.3}{45.8}\right)} = 0.0273 \text{ s}$$

(b) Substituting this value for T into our equation,

$$V = 45.8e^{-t/0.0273}$$

14.53 The speed of a motor armature is observed to decrease exponentially at the rate of 5 % per s after the power is shut off.   How long will it take to reach half its original speed?

Solution

Substituting in Eq. 155, with $y = a/2$ and $n = 0.05/s$,

$$\frac{a}{2} = ae^{-0.05x}$$

$$e^{-0.05x} = 0.5$$

Taking natural logs of both sides,

$$-0.05x = \ln 0.5 = -0.693$$

$$x = 14 \text{ s}$$

14.54 *Newton's Law of Cooling:* When a body, initially at a temperature t_0, is placed in cooler surroundings, its temperature will drop exponentially according to the formula,

$$t = t_0 e^{-kT}$$

where k is a constant and t is the temperature at any time T. If the temperature of a casting drops from 1500°C to 1000°C in 30 min, how long will it take to cool to 100°C?

Solution

First finding k, we substitute, with $T = 0.5$ h, $t_0 = 1500$°C,   and $t = 1000$°C,

$$1000 = 1500e^{-0.5k}$$

$$e^{-0.5k} = \frac{1000}{1500} = 0.667$$

Taking natural logs,

$$-0.5k = \ln 0.667 = -0.405$$

$$k = 0.81$$

Letting $t = 100°C$ and solving for T,

$$100 = 1500e^{-0.81T}$$

$$e^{-0.81T} = \frac{100}{1500} = 0.0667$$

$$-0.81T = \ln 0.0667 = -2.71$$

$$T = 3.3 \text{ h}$$

14.55   A piece of tool steel, initially at 1000°C, is quenched in oil and is seen to cool exponentially at the rate of 10% per s. Find its temperature after 10 s.

Solution

Substituting in Eq. 155, with $a = 1000$, $n = 0.10/s$, and $x = 10$ s,

$$\text{Temperature after 10 s} = y = 1000e^{-0.10(10)} = 1000e^{-1} = \frac{1000}{e} = 368°C$$

Supplementary Problems

14.56   Convert to logarithmic form.

     (a) $4^4 = 256$ (b) $9^5 = 59\,049$ (c) $e^5 = 148.4$ (d) $2^9 = 512$

14.57   Convert to exponential form.

     (a) $\log_{10} 1000 = 3$ (b) $\log_2 512 = 9$ (c) $\log_5 125 = 3$ (k) $\log_8 4096 = 4$

14.58   Solve for x.

     (a) $x = \log_4 64$ (b) $x = \log_6 7776$ (c) $\log_5 x = 3$ (d) $6 = \log_x 4096$

14.59   Express the following as single logarithms:

     (a) $x = \log 2 - \log 3 + \log 4$ (c) $x = 3 \log 7 - \log 9 - 3 \log 3$

     (b) $x = 6 \log 7 + \log 5 - 2 \log 2$

14.60   Find the common logarithms of the following numbers to four decimal places. Use a calculator or the tables.

     (a)   52   (b)   21.1   (c)   0.06   (d)   1020   (e)   1.23

14.61   Interpolate to find the common logarithms of the following numbers to four decimal places, or use a calculator.

     (a)   1013   (b)   2.317   (c)   0.1733

14.62   Find the antilogarithms of the following to three significant figures, by calculator or tables:

     (a) $\log N = 2.4200$ (b) $\log N = 1.798$ (c) $\log N = 2.2014$ (d) $\log N = 2.9782$

14.63   Interpolate to find the antilogarithms of the following, or use a calculator:

(a) $\log N = 1.6792$ (b) $\log N = 2.0988$ (c) $\log N = 0.4074$

14.64   Perform the following computations using logarithms:

(a) $x = (231)(7.0)$ (b) $x = \dfrac{9.2}{8.3}$ (c) $x = \sqrt{77}$ (d) $x = (3.25)^4$

14.65   Find the natural logarithms of the following numbers to five significant figures, by calculator or tables:

(a) 2.2   (b) 5   (c) 27.01   (d) 211.0

14.66   Find the antilogarithms of each of the following natural logarithms to two significant figures, by calculator or tables:

(a) 3.7377   (b) 1.7047   (c) 2.5657   (d) −3.1011

14.67   Find $\log N$.

(a) $\ln N = 2.9957$ (b) $\ln N = 1.4586$

14.68   Find $\ln N$.

(a) $\log N = 1.2553$ (b) $\log N = -1.0044$

14.69   Solve for x, to four significant figures.

(a) $2^x = 7$ (b) $(6.37)^x = 33.6$ (c) $(0.22)^{x^2} = (2.03)^x$

14.70   Solve for x, to four significant figures.

(a) $e^x = 7.11$ (b) $27e^x = 3e^{2x}$ (c) $2.2e^{3x^2} = 17$

14.71   Solve for x.

(a) $\log 3 - \log (x + 2) = \log x$ (c) $2 \log 2 + \log 7 - \log (x + 3) = \log (2x + 8)$

(b) $\ln x + \ln (3x + 2) = 4$

14.72   For the power function $y = 6.35x^{1.75}$, find x when $y = 3.63$.

14.73   For the power function $y = 0.019x^{6.90}$, find x when $y = 0.0035$.

14.74   How long will it take for a quantity to increase fourfold when growing exponentially at an annual rate of 13%?

14.75   Find the rate of growth, if 10 000 units increase exponentially to 15 000 units in 60 days.

14.76   What is the half-life of a material that decays exponentially at a rate of 3% per year?

Answers to Supplementary Problems

14.56   (a) $\log_4 256 = 4$ (b) $\log_9 59\,049 = 5$ (c) $\log_e 148.4 = 5$ (d) $\log_2 512 = 9$

14.57   (a) $10^3 = 1000$ (b) $2^9 = 512$ (c) $5^3 = 125$ (d) $8^4 = 4096$

14.58   (a) $x = 3$ (b) $x = 5$ (c) $x = 125$ (d) $x = 4$

14.59   (a) $\log \dfrac{8}{3}$ (b) $\log \left(\dfrac{5 \times 7^6}{4} \right)$ (c) $\log \left(\dfrac{7^3}{243} \right)$

14.60   (a) 1.7160   (b) 1.3243   (c) -1.2218 (d) 3.0086   (e) 0.0899

14.61   (a) 3.0056   (b) 0.3649   (c) -0.7612

14.62   (a) 263   (b) 62.8   (c) 159   (d) 951

14.63   (a) 47.77   (b) 125.5   (c) 2.555

14.64   (a) 1617   (b) 0.1108   (c) 8.7750   (d) 111.57

14.65   (a) 0.788 46   (b) 1.6094   (c) 3.2962   (d) 5.3519

14.66   (a) 42   (b) 5.5   (c) 13   (d) 0.045

14.67   (a) $\log N = 1.3010$ (b) $\log N = 0.6335$

14.68   (a) $\ln N = 2.8905$ (b) $\ln N = -2.3127$

14.69   (a) 2.807   (b) 1.898   (c) 0, -0.4676

14.70   (a) 1.962   (b) 2.197   (c) ± 0.8256

14.71   (a) $x = 1$ (b) $x = 3.95$ (c) 0.275

14.72 $x = 0.726\,47$

14.73 $x = 0.782\,57$

14.74 $x = 10.7$ years

14.75 $r = 0.68\,\%$ per day

14.76 $x = 23.1$ years

Chapter 15

Oblique Triangles

15.1 OBLIQUE TRIANGLES

Oblique triangles do not contain a right angle. The six trigonometric functions, Eqs. 73 to 78, cannot be used for their solution, nor can the Pythagorean Theorem, Eq. 72. We still use the fact that the sum of the angles is 180° (Eq. 60), as well as two new relationships, the *law of sines* and the *law of cosines*.

15.2 THE LAW OF SINES

In any triangle, the lengths of the sides are proportional to the sines of the opposite angles. In triangle ABC, Fig. 15-1,

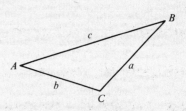

Law of Sines	$\dfrac{a}{\sin A} = \dfrac{b}{\sin B} = \dfrac{c}{\sin C}$	**61**

(See Problem 15.1 for the derivation.)

Fig. 15-1

Use the law of sines when you have a given side opposite a given angle.

Example 15.1: In triangle ABC, $A = 48°$, $B = 35°$, and $a = 14.5$ in. Solve the triangle.

Solution: Make a sketch, roughly to scale, Fig. 15-2. By Eq. 60,

$$C = 180° - 48° - 35° = 97°$$

By the law of sines, Eq. 61,

$$\frac{14.5}{\sin 48°} = \frac{b}{\sin 35°} = \frac{c}{\sin 97°}$$

From which

$$b = \frac{14.5 \sin 35°}{\sin 48°} = 11.2 \text{ in.}$$

and

$$c = \frac{14.5 \sin 97°}{\sin 48°} = 19.4 \text{ in.}$$

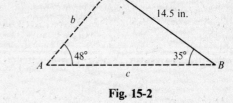

Fig. 15-2

Example 15.2: Solve triangle ABC, where $A = 33°$, $a = 55$ mm, and $c = 82$ mm.

Solution: Sketching the triangle, Fig. 15-3, we see that it's possible to construct it two different ways. This is the *ambiguous case*. From the law of sines, Eq. 61,

$$\frac{55}{\sin 33°} = \frac{82}{\sin C}$$

$$\sin C = \frac{82 \sin 33°}{55} = 0.812$$

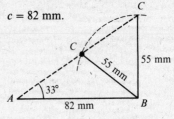

Fig. 15-3

421

As there are two possible values for C,

$$C = \arcsin{(0.812)} = 54°$$
$$C = 180° - 54° = 126°$$

and

For $C = 54°$,

$$B = 180° - 33° - 54° = 93°$$

and

$$\frac{b}{\sin 93°} = \frac{55}{\sin 33°}$$

$$b = \frac{55 \sin 93°}{\sin 33°} = 101 \text{ mm}$$

For $C = 126°$,

$$B = 180° - 33° - 126° = 21°$$

So,

$$\frac{b}{\sin 21°} = \frac{55}{\sin 33°}$$

$$b = \frac{55 \sin 21°}{\sin 33°} = 36 \text{ mm}$$

Common Error	Don't fail to sketch the triangle you are solving. It will provide a rough check of your calculations, as well as reveal ambiguities or inconsistencies in the given data.

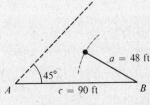

Fig. 15-4

Example 15.3:   Solve triangle ABC, where $A = 45°$, $c = 90$ ft,   and $a = 48$ ft.

Solution: A sketch of this data, Fig. 15-4, shows that side a is too short to complete the triangle.

15.3   LAW OF COSINES

In any triangle ABC, Fig. 15-1,

Law of Cosines	$a^2 = b^2 + c^2 - 2bc \cos A$ $b^2 = a^2 + c^2 - 2ac \cos B$ $c^2 = a^2 + b^2 - 2ab \cos C$	**62**

(See Problem 15.6 for the derivation.)

Use the law of cosines when (*a*)   two sides and the included angle are given or (*b*)   three sides are given.

Example 15.4:   Solve triangle ABC, where $a = 97.5$ m, $b = 61.4$ m, $C = 60.0°$.

Solution: Sketch the triangle, Fig. 15-5. By the law of cosines, Eq. 62.

$$c^2 = (97.5)^2 + (61.4)^2 - 2(97.5)(61.4) \cos 60.0° = 7290$$
$$c = 85.4 \text{ m}$$

By the law of sines, Eq. 61,

$$\frac{97.5}{\sin A} = \frac{85.4}{\sin 60.0°}$$

$$\sin A = \frac{97.5 \sin 60.0°}{85.4} = 0.989$$

So, $A = 81.4°$

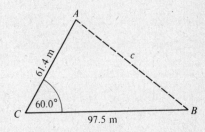

Fig. 15-5

By Eq. 60,

$$B = 180° - 81.4° - 60.0° = 38.6°$$

Common Error	The cosine of an obtuse angle is *negative*. Be sure to use the proper algebraic sign when applying the law of cosines to an obtuse angle.

Example 15.5:   Solve triangle ABC, where $b = 82.4$ ft, $c = 53.1$ ft, and $A = 134.1°$.

Solution: Make a sketch, Fig. 15-6. By the law of cosines,

$$a^2 = (82.4)^2 + (53.1)^2 - 2(82.4)(53.1) \cos 134.1°$$
$$= 9609 - 8751(-0.6959) = 9609 + 6090$$
$$= 15\,699$$
$$a = 125 \text{ ft}$$

By the law of sines,

$$\frac{82.4}{\sin B} = \frac{125}{\sin 134.1°}$$

$$\sin B = \frac{82.4 \sin 134.1°}{125} = 0.473$$

$$B = 28.3°$$

From Eq. 60,

$$C = 180° - 134.1° - 28.3° = 17.6°$$

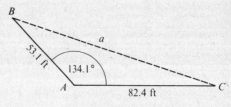

Fig. 15-6

Example 15.6:   Solve triangle ABC, where $a = 642$ ft, $b = 733$ ft, and $c = 1115$ ft.

Solution: A sketch is drawn, Fig. 15-7. By the law of cosines,

$$(642)^2 = (733)^2 + (1115)^2 - 2(733)(1115) \cos A$$

Transposing, $2(733)(1115) \cos A = (733)^2 + (1115)^2 - (642)^2$

or $1\,635\,000 \cos A = 1\,368\,000$

Dividing, $\cos A = 0.837$

$$A = 33.2°$$

By the law of sines,

$$\frac{733}{\sin B} = \frac{642}{\sin 33.2°}$$

$$\sin B = \frac{733 \sin 33.2°}{642} = 0.625$$

$$B = 38.7°$$

By Eq. 60,

$$C = 180° - 33.2° - 38.7° = 108.1°$$

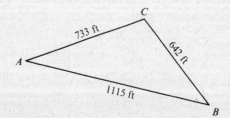

Fig. 15-7

Solved Problems

LAW OF SINES

15.1   Derive the law of sines for oblique triangle ABC, where all three angles are acute.

Solution

Draw altitude h to side AB, Fig. 15-8.   In right triangle ACD,

$$\sin A = \frac{h}{b} \qquad \text{or} \qquad h = b \sin A$$

In right triangle BCD,

$$\sin B = \frac{h}{a} \qquad \text{or} \qquad h = a \sin B$$

So, $h = b \sin A = a \sin B$

Dividing by $\sin A \sin B$,

$$\frac{a}{\sin A} = \frac{b}{\sin B}$$

Similarly, drawing altitude j to side AC,

$$j = a \sin C = c \sin A$$

or

$$\frac{a}{\sin A} = \frac{c}{\sin C}$$

So,
$$\boxed{\frac{a}{\sin A} = \frac{b}{\sin B} = \frac{c}{\sin C}} \qquad \mathbf{61}$$

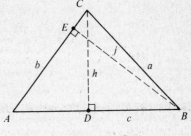

Fig. 15-8

15.2   Solve the oblique triangle ABC, Fig. 15-9, where $A = 35°$, $C = 75°$,   and $c = 2.0$ in.

Solution

By Eq. 60,

$$B = 180° - 35° - 75° = 70°$$

By Eq. 61, the law of sines,

$$\frac{2.0}{\sin 75°} = \frac{a}{\sin 35°} = \frac{b}{\sin 70°}$$

from which

$$a = \frac{2.0 \sin 35°}{\sin 75°} = 1.2 \text{ in.}$$

and

$$b = \frac{2.0 \sin 70°}{\sin 75°} = 1.9 \text{ in.}$$

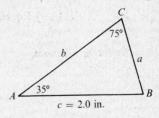

Fig. 15-9

15.3 Solve triangle ABC, Fig. 15-10, where $B = 55°$, $A = 60°$, and $b = 1.5$ cm.

Solution

By Eq. 60,
$$C = 180° - 55° - 60° = 65°$$

By Eq. 61, the law of sines,
$$\frac{1.5}{\sin 55°} = \frac{a}{\sin 60°} = \frac{c}{\sin 65°}$$

from which
$$a = \frac{1.5 \sin 60°}{\sin 55°} = 1.6 \text{ cm}$$

and
$$c = \frac{1.5 \sin 65°}{\sin 55°} = 1.7 \text{ cm}$$

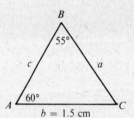

Fig. 15-10

15.4 Solve oblique triangle ABC, Fig. 15-11, where $B = 30°$, $b = 2.0$ in., and $c = 3.5$ in.

Solution

By Eq. 61, the law of sines
$$\frac{2.0}{\sin 30°} = \frac{3.5}{\sin C}$$
$$\sin C = \frac{3.5 \sin 30°}{2.0} = 0.875$$

As there are two possible values for C,
$$C = \arcsin(0.875) = 61°$$
and
$$C = 180° - 61° = 119°$$

For $C = 61°$, by Eq. 60,
$$A = 180° - 61° - 30° = 89°$$

By the law of sines,
$$\frac{a}{\sin 89°} = \frac{2.0}{\sin 30°}$$
$$a = \frac{2.0 \sin 89°}{\sin 30°} = 4.0 \text{ in.}$$

For $C = 119°$, by Eq. 60,
$$A = 180° - 30° - 119° = 31°$$

By the law of sines,
$$\frac{a}{\sin 31°} = \frac{2.0}{\sin 30°}$$
$$a = \frac{2.0 \sin 31°}{\sin 30°} = 2.1 \text{ in.}$$

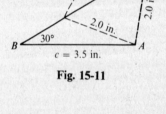

Fig. 15-11

15.5 Solve oblique triangle ABC, where $B = 50°$, $b = 15$ m, and $a = 30$ m.

Solution

The sketch, Fig. 15-12, immediately shows that the data are inconsistent, and the triangle cannot be solved.

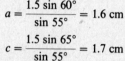

Fig. 15-12

LAW OF COSINES

15.6 Derive the law of cosines for oblique triangle ABC, where all three angles are acute.

Solution

Draw altitude h to side AC, Fig. 15-13. In right triangle ABD,

$$c^2 = h^2 + (AD)^2 = h^2 + (b - CD)^2$$

In right triangle BCD,

$$\cos C = \frac{CD}{a} \qquad CD = a \cos C$$

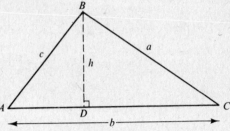

So, $c^2 = h^2 + (b - a \cos C)^2$

Squaring, $c^2 = h^2 + b^2 - 2ab \cos C + a^2 \cos^2 C$

but, applying the Pythagorean Theorem in triangle BCD,

$$h^2 = a^2 - (CD)^2 = a^2 - (a \cos C)^2$$

So, $c^2 = a^2 - a^2 \cos^2 C + b^2 - 2ab \cos C + a^2 \cos^2 C.$

Fig. 15-13

$$\boxed{c^2 = a^2 + b^2 - 2ab \cos C} \quad \mathbf{62}$$

15.7 Solve the oblique triangle ABC, Fig. 15-14, where $A = 37°$, $b = 1.7$ cm,   and $c = 3.4$ cm.

Solution

By Eq. 62, the law of cosines,

$$a^2 = (1.7)^2 + (3.4)^2 - 2(1.7)(3.4) \cos 37° = 5.22$$
$$a = 2.3 \text{ cm}$$

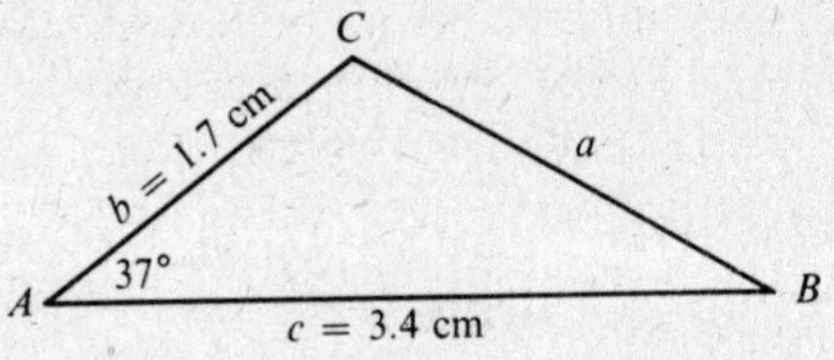

By Eq. 61, the law of sines,

$$\frac{2.3}{\sin 37°} = \frac{1.7}{\sin B}$$

So, $\sin B = \dfrac{1.7 \sin 37°}{2.3} = 0.445$

$$B = 26°$$

By Eq. 60,

$$C = 180° - 26° - 37° = 117°$$

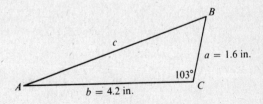

Fig. 15-14

15.8 Solve oblique triangle ABC, Fig. 15-15, where $C = 103°$, $a = 1.6$ in.,   and $b = 4.2$ in.

Solution

By the law of cosines,

$$c^2 = (1.6)^2 + (4.2)^2 - 2(1.6)(4.2) \cos 103° = 20.2 - 13.44(-0.225) = 23.2$$
$$c = 4.8 \text{ in.}$$

By the law of sines,

$$\frac{4.8}{\sin 103°} = \frac{1.6}{\sin A}$$

So, $\sin A = \dfrac{1.6 \sin 103°}{4.8} = 0.325$

$$A = 19.0°$$

By Eq. 60,

$$B = 180° - 103° - 19° = 58°$$

Fig. 15-15

15.9   Solve oblique triangle ABC, Fig. 15-16, where $a = 2.00$ m, $b = 1.80$ m,   and $c = 3.40$ m.

Solution

By the law of cosines,

$$(2.00)^2 = (1.80)^2 + (3.40)^2 - 2(1.80)(3.40) \cos A$$

So, $\cos A = \dfrac{(1.80)^2 + (3.40)^2 - (2.00)^2}{2(1.80)(3.40)} = 0.882$

$$A = 28.1°$$

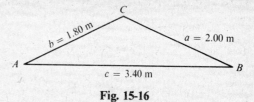

Fig. 15-16

By the law of sines,

$$\frac{2.00}{\sin 28.1°} = \frac{1.80}{\sin B}$$

$$\sin B = \frac{1.80 \sin 28.1°}{2.00} = 0.424$$

So, $B = 25.1$

By Eq. 60,

$$C = 180° - 25.1° - 28.1° = 126.8°$$

MEASURING INACCESSIBLE DISTANCES

Draw a triangle which has the inaccessible distance as one of its sides, and in which three other parts are known or can be found.   Use the law of sines or the law of cosines to solve the triangle.

15.10   In order to find the distance between towns A and B, Fig. 15-17, an Electronic Distance Measuring (EDM) instrument is placed on a hill from which both towns are visible.   The distances from each town to the EDM, as well as the angle between the two lines of sight, are recorded.   Find the distance a between the villages.

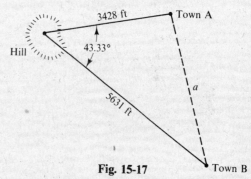

Solution

By the law of cosines, Eq. 62,

Fig. 15-17

$$a^2 = (3428)^2 + (5631)^2 - 2(3428)(5631) \cos 43.33° = 15\,377\,000$$

$$a = 3921 \text{ ft}$$

15.11   The long chord of a circular highway curve, Fig. 15-18, is $100\overline{0}$ ft and the central angle is 30.00°.   What is the radius R of the curve?

Solution

Since two sides of the triangle are equal to the radius R, the triangle is isosceles, and the base angles are equal.   By Eq. 60,

$$A + A + 30° = 180°$$

$$2A = 150°$$

$$A = 75.00°$$

Then by the law of sines, Eq. 61,

$$\frac{1000}{\sin 30°} = \frac{R}{\sin 75°}$$

$$R = \frac{\sin 75°(1000)}{\sin 30°} = 1932 \text{ ft}$$

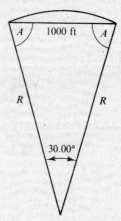

Fig. 15-18

15.12 Find the length of fence needed to enclose the triangular piece of land in Fig. 15-19.

Solution

By the law of sines,

$$\frac{500}{\sin 55°} = \frac{x}{\sin 45°}$$

$$x = \frac{500 \sin 45°}{\sin 55°} = 432 \text{ m}$$

By Eq. 60,

$$A = 180° - 55° - 45° = 80°$$

By Eq. 61,

$$\frac{500}{\sin 55°} = \frac{y}{\sin 80°}$$

$$y = \frac{500 \sin 80°}{\sin 55°} = 601 \text{ m}$$

Total fence length = 500 + 432 + 601 = 1533 m

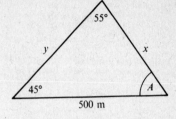

Fig. 15-19

15.13 To measure the length a of a pond, Fig. 15-20, a transit is placed at A and the angle CAB is measured, and the distances AC and AB are taped.   Find the length of the pond.

Solution

By the law of cosines,

$$a^2 = (517.31)^2 + (323.60)^2 - 2(517.31)(323.60) \cos 88° = 360\,642$$

$$a = 600.53 \text{ ft}$$

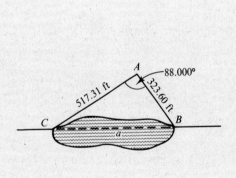

Fig. 15-20

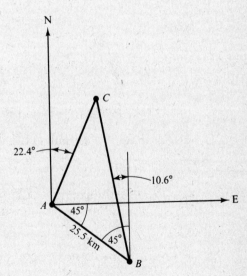

Fig. 15-21

15.14 From shore station A, a ship C is observed in the direction N22.4°E.   The same ship is observed to be in the direction N10.6°W from shore station B, located at a distance of 25.5 km SE of A. Find the distance of the ship from station A.

Solution

In triangle ABC, Fig. 15-21,

$$A = 135° - 22.4° = 112.6°$$

$$B = 45° - 10.6° = 34.4°$$

By Eq. 60,

$$C = 180° - 112.6° - 34.4° = 33.0°$$

By the law of sines,

$$\frac{AC}{\sin 34.4°} = \frac{25.5}{\sin 33.0°}$$

$$AC = \frac{25.5 \sin 34.4°}{\sin 33.0°} = 26.5 \text{ km}$$

15.15 A certain rocket tracking station, Fig. 15-22, consists of two telescopes A and B, 1.68 km apart, which lock onto the rocket C and continuously transmit the angles α and β to a computer.   Write an equation by which the distance AC may be computed.

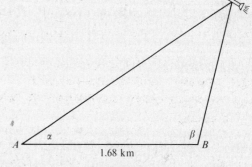

Solution

By Eq. 60,

Angle $ACB = 180° - \alpha - \beta$

By the law of sines,

$$\frac{AC}{\sin \beta} = \frac{1.68}{\sin (180° - \alpha - \beta)}$$

So,

$$AC = \frac{1.68 \sin \beta}{\sin (180 - \alpha - \beta)} \text{ km}$$

Fig. 15-22

15.16 Find the distance AC to the rocket in the previous problem when $\alpha = 57.8°$ and $\beta = 106.2°$.

Solution

Using the formula already derived,

$$AC = \frac{1.68 \sin 106.2°}{\sin (180° - 57.8° - 106.2°)} = \frac{1.68 \sin 106.2°}{\sin 16.0°} = 5.85 \text{ km}$$

15.17 An offshore oil drilling rig A is 2.85 mi, N42°15′W of a shore station C, and another rig B is 4.18 mi, N65°27′E of the same station, Fig. 15-23.   Find the distance from A to B.

Solution

Angle $ACB = 42°15′ + 65°27′$

$$= 107°42′ = 107.70°$$

By the law of cosines,

$$(AB)^2 = (2.85)^2 + (4.18)^2 - 2(2.85)(4.18) \cos 107.70°$$

$$= 32.8$$

$$AB = 5.73 \text{ mi}$$

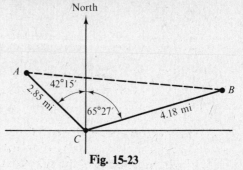

Fig. 15-23

MECHANISMS

15.18 In the slider crank mechanism of Fig. 15-24, find the distance x between the wrist pin W and the connecting rod center C, when $A = 25.5°$.

Solution

By the law of sines in triangle PWC,

$$\frac{\sin B}{11.2} = \frac{\sin 25.5°}{28.6}$$

So, $\sin B = \dfrac{11.2 \sin 25.5°}{28.6} = 0.169$

$$B = 9.71°$$

By Eq. 60,

$$\theta = 180° - 25.5° - 9.71° = 144.8°$$

By the law of sines,

$$\frac{x}{\sin 144.8°} = \frac{28.6}{\sin 25.5°}$$

$$x = \frac{28.6 \sin 144.8°}{\sin 25.5°}$$

$$= 38.3 \text{ cm}$$

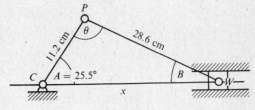

Fig. 15-24

15.19 In the four-bar linkage of Fig. 15-25, find angle ADC when angle BAD is $38.6°$.

Solution

Draw diagonal BD with length L. Applying the law of cosines to triangle ABD,

$$L^2 = (35.0)^2 + (92.0)^2 - 2(35.0)(92.0)\cos 38.6° = 4656$$

$$L = 68.2 \text{ cm}$$

Applying the law of sines in the same triangle,

$$\frac{\sin (ADB)}{35.0} = \frac{\sin 38.6°}{68.2}$$

$$\sin (ADB) = \frac{35.0 \sin 38.6°}{68.2} = 0.320$$

Angle $ADB = 18.7°$

Applying the law of cosines to triangle BCD,

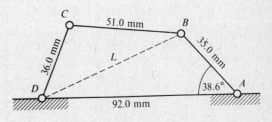

Fig. 15-25

$$(51.0)^2 = (36.0)^2 + (68.2)^2 - 2(36.0)(68.2)\cos (BDC)$$

$$4910 \cos (BDC) = 3346$$

$$\cos (BDC) = 0.681$$

$$\text{Angle } BDC = 47.0°$$

So, $\text{Angle } ADC = 47.0° + 18.7°$

$$= 65.7°$$

RESOLUTION OF VECTORS

15.20 Two forces of 25.6 N and 18.3 N are applied to a point on a body. The angle between the forces is 39.4°. Find the magnitude R of the resultant, and the angle θ it makes with the larger force.

Solution

Draw a parallelogram, Fig. 15-26, with the resultant as the diagonal. Then,

$$\text{Angle } ABC = 180° - 39.4° = 140.6°$$

Applying the law of cosines to triangle ABC,

$$R^2 = (18.3)^2 + (25.6)^2 - 2(18.3)(25.6) \cos 140.6° = 1714$$

$$R = 41.4 \text{ N}$$

By the law of sines,

$$\frac{18.3}{\sin \theta} = \frac{41.4}{\sin 140.6°}$$

$$\sin \theta = \frac{18.3 \sin 140.6°}{41.4} = 0.281$$

$$\theta = 16.3°$$

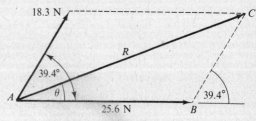

Fig. 15-26

15.21 A pilot wishes to fly in a northeasterly direction (exactly N45°E). The wind is from the west at 42.0 km/h, and the pilot's speed in still air is 285 km/h, Fig. 15-27. Find the heading H and the actual speed V.

Solution

In triangle ABC, $C = 45°$. By the law of sines,

$$\frac{42}{\sin A} = \frac{285}{\sin 45°}$$

$$\sin A = \frac{42 \sin 45°}{285} = 0.104$$

$$A = 5.98°$$

So,

$$H = 45.0° - 5.98° = 39.0°$$

By Eq. 60,

$$B = 180° - 45° - 5.98° = 129°$$

By the law of sines,

$$\frac{V}{\sin 129°} = \frac{285}{\sin 45°}$$

$$V = \frac{285 \sin 129°}{\sin 45°} = 313 \text{ km/h}$$

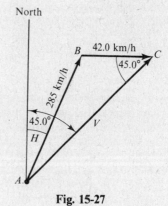

Fig. 15-27

15.22 Find the tension T in the cable, and the horizontal and vertical reactions at C for the derrick, Fig. 15-28, when angle A is 35.0°. Assume that the weight of the boom is 2000 lb concentrated at its center of gravity (the midspan of the boom).

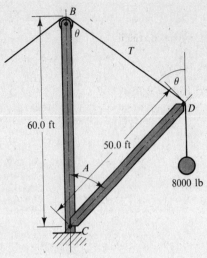

Fig. 15-28

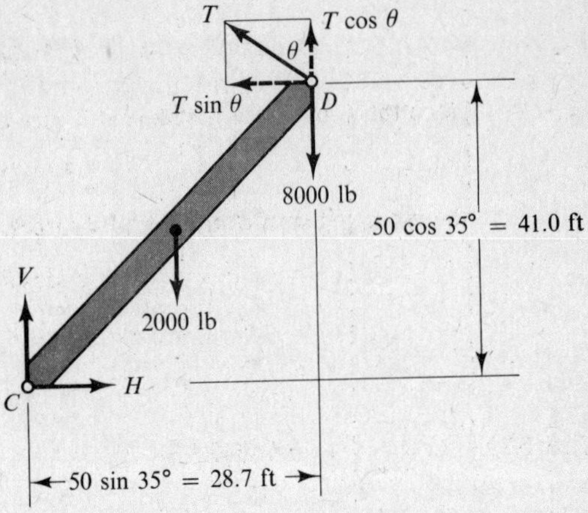

Fig. 15-29

Solution

In triangle BCD, by the law of cosines,

$$(BD)^2 = (50)^2 + (60)^2 - 2(50)(60) \cos 35° = 1185$$
$$BD = 34.4 \text{ ft}$$

By the law of sines,

$$\frac{50}{\sin \theta} = \frac{34.4}{\sin 35°}$$

$$\sin \theta = \frac{50 \sin 35°}{34.4} = 0.834$$

$$\theta = 56.5°$$

We draw the boom alone, with the forces acting upon it (*a free-body diagram*), Fig. 15-29. Let V and H represent the vertical and horizontal reactions at C, and $T \sin \theta$ and $T \cos \theta$ the vertical and horizontal components of the tension T. Since by Eq. 175 the sum of the moments of the forces about C must equal zero,

$$T \sin \theta(50 \cos A) + T \cos \theta(50 \sin A) - 8000(50 \sin A) - 2000 \frac{50 \sin A}{2} = 0$$

Therefore, $T \sin \theta(50 \cos A) + T \cos \theta(50 \sin A) = 8000(50 \sin A) + 2000 \dfrac{50 \sin A}{2}$

When $A = 35°$,

$$50 \sin 35° = 28.7$$

and

$$50 \cos 35° = 41.0$$

Also, $\theta = 56.5°$,   so,

$$\sin 56.5° = 0.834$$

and

$$\cos 56.5° = 0.552$$

Substituting, we get

$$0.834T(41.0) + 0.552T(28.7) = 8000(28.7) + 2000(14.3) = 258\,200$$

Solving for T,

$$T = \frac{258\,200}{0.834(41.0) + 0.552(28.7)} = 5160 \text{ lb}$$

Since the sum of the horizontal forces must be zero, by Eq. 173,

$$H = T \sin \theta = 5160 \sin 56.5° = 43\overline{0}0 \text{ lb}$$

and since the sum of the vertical forces must be zero, by Eq. 174,

$$V + T \cos \theta = 8000 + 2000$$

So, $$V = 10\,000 - 5160 \cos 56.5° = 7150 \text{ lb}$$

TRUSSES AND FRAMEWORKS

15.23   In the roof truss in Fig. 15-30, find the lengths of members AE, BE, DE, and BD.

Solution

In triangle ABE, angle $ABE = 180° - 41° = 139°$, and angle $AEB = 180° - 29° - 139° = 12°$. By the law of sines,

$$\frac{AE}{\sin 139°} = \frac{BE}{\sin 29°} = \frac{12.3}{\sin 12°}$$

So, $AE = \dfrac{12.3 \sin 139°}{\sin 12°} = 38.8 \text{ ft}$

and $BE = \dfrac{12.3 \sin 29°}{\sin 12°} = 28.7 \text{ ft}$

In right triangle BDE, by Eq. 73,

$$\sin 41° = \frac{DE}{28.7}$$

$$DE = 28.7 \sin 41° = 18.8 \text{ ft}$$

and by Eq. 74,

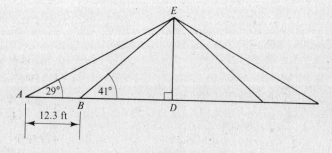

Fig. 15-30

$$\cos 41° = \frac{BD}{28.7}$$

$$BD = 28.7 \cos 41° = 21.7 \text{ ft}$$

15.24   The framework in Fig. 15-31 is symmetrical about the centerline.   Find all the missing sides and angles.

Solution

$$AB = \frac{12.4}{2} = 6.20 \text{ m}$$

and $\alpha = 90° - \dfrac{108.0°}{2} = 36.0°$

By the law of sines in triangle ABE, and since AB is one-half of AC,

$$\frac{4.14}{\sin 36.0°} = \frac{6.20}{\sin \beta}$$

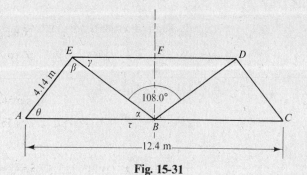

Fig. 15-31

So,

$$\sin \beta = \frac{6.20 \sin 36.0°}{4.14} = 0.880$$

$$\beta = 61.7°$$

By Eq. 60,

$$\theta = 180° - 36.0° - 61.7° = 82.3°$$

By the law of sines,

$$\frac{BE}{\sin 82.3°} = \frac{4.14}{\sin 36.0°}$$

$$BE = \frac{4.14 \sin 82.3°}{\sin 36.0°} = 6.98 \text{ m}$$

Since ED is parallel to AC, by Statement 56,

$$\gamma = \alpha = 36.0°$$

In right triangle BEF, by Eq. 74,

$$\cos \gamma = \frac{EF}{BE}$$

$$EF = BE \cos \gamma = 6.98 \cos 36.0° = 5.65 \text{ m}$$

So,

$$DE = 2(5.65) = 11.3 \text{ m}$$

Since the framework is symmetrical about the centerline,

$$CD = AE$$

$$BC = AB$$

$$BD = BE$$

and corresponding angles are equal.

15.25   An antenna mast, Fig. 15-32, is to be placed on sloping ground, with the cables making an angle of 40° with the top of the mast. Find the lengths x and y of the two cables shown.

Solution

Since a line parallel to the horizontal would make an angle of 20° with line BD and would be perpendicular to the vertical antenna mast, angle ABD must be

$$90° - 20° = 70°$$

So,   Angle $ABC = 180° - 70° = 110°$

Also, by Eq. 60,

Angle $ACB = 180° - 40° - 110° = 30°$

and   Angle $ADB = 180° - 40° - 70° = 70°$

Applying the law of sines to triangle ABC,

$$\frac{x}{\sin 110°} = \frac{155}{\sin 30°}$$

$$x = \frac{155 \sin 110°}{\sin 30°} = 291 \text{ ft}$$

Since triangle ABD is isosceles,

$$y = 155 \text{ ft}$$

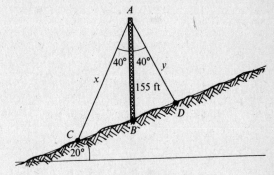

Fig. 15-32

Supplementary Problems

15.26 (a) Solve oblique triangle ABC, Fig. 15-33a, where $A = 40°$, $C = 80°$, $c = 2.7$ m.

      (b) Solve oblique triangle ABC, Fig. 15-33b, where $C = 60°$, $B = 37°$, $b = 1.8$ in.

      (c) Solve oblique triangle ABC, Fig. 15-33c, where $a = 3.1$ cm, $b = 1.7$ cm, $A = 48°$.

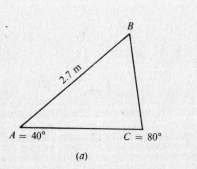
(a)

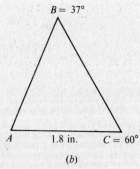
(b)

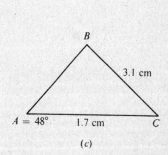
(c)

Fig. 15-33

15.27 (a) Solve oblique triangle ABC, Fig. 15-34a, where $A = 94°$, $c = 2.60$ in., $b = 3.90$ in.

      (b) Solve oblique triangle ABC, Fig. 15-34b, where $B = 75°$, $a = 7.30$ m, $c = 5.00$ m.

      (c) Solve oblique triangle ABC, Fig. 15-34c, where $C = 59°$, $a = 1.90$ in., $b = 2.30$ in.

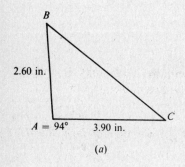
(a)

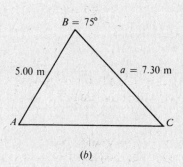
(b)

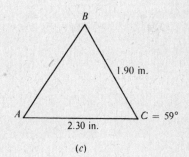
(c)

Fig. 15-34

15.28 Solve the triangles in Fig. 15-35 for any missing parts.

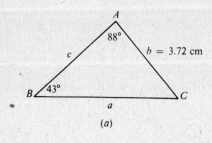
(a)

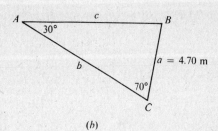
(b)

Fig. 15-35

15.29   Find all the missing parts in Fig. 15-36.

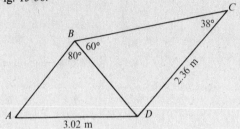

Fig. 15-36

15.30   Find all missing angles and lengths in the truss in Fig. 15-37.

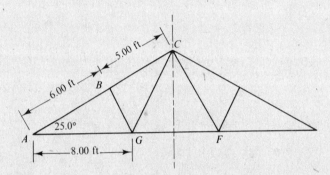

Fig. 15-37

Answers to Supplementary Problems

15.26   (a) $B = 60°$, $a = 1.8$ m, $b = 2.4$ m   (b) $A = 83°$, $a = 3.0$ in., $c = 2.6$ in.

(c) $C = 108°$, $B = 24°$, $c = 4.0$ cm

15.27   (a) $C = 32.4°$, $B = 53.6°$, $a = 4.84$ in.   (b) $A = 66.2°$, $C = 38.8°$, $b = 7.71$ m

(c) $A = 50.9°$, $B = 70.1°$, $c = 2.10$ in.

15.28   (a) $a = 5.45$ cm, $C = 49°$, $c = 4.12$ cm   (b) $c = 8.83$ m, $B = 80°$, $b = 9.26$ m

15.29 $BD = 1.68$ m, $\angle BDC = 82°$, $BC = 2.70$ m, $AB = 2.82$ m, $\angle BAD = 33.2°$, $\angle ADB = 66.8°$

15.30 $BG = 3.60$ ft, $CG = 5.05$ ft, $\angle CGF = 67.0°$, $\angle ABG = 110.2°$, $\angle BCG = 42.0°$, $\angle GCF = 46.0°$,
$\angle AGB = 44.8°$, $\angle BGC = 68.2°$, $GF = 3.95$ ft, $\angle GBC = 69.8°$

The right-hand side is symmetrical to the left-hand side; therefore, corresponding angles and sides are equal.

Appendixes

Appendix A

SI* Units of Measurement

Unit	Symbol	Relationship to standard
LENGTH		
†meter	m	
centimeter	cm	$m \times 10^{-2}$
millimeter	mm	$m \times 10^{-3}$
micrometer	μm	$m \times 10^{-6}$
kilometer	km	$m \times 10^{3}$
AREA		
square meter	m^2	
square millimeter	mm^2	$m^2 \times 10^{-6}$
**barn	b	$m^2 \times 10^{-28}$
**hectare	ha	$m^2 \times 10^{4}$
VOLUME		
cubic meter	m^3	
cubic millimeter	mm^3	$m^3 \times 10^{-9}$
‡liter	L or liter	$m^3 \times 10^{-3}$
milliliter	mL	$L \times 10^{-3} = m^3 \times 10^{-6}$
MASS		
†kilogram	kg	
gram	g	$kg \times 10^{-3}$
metric ton	t	$kg \times 10^{3}$
FORCE		
newton	N	
kilonewton	kN	$N \times 10^{3}$
meganewton	MN	$N \times 10^{6}$
PRESSURE		
pascal	Pa	
kilonewton per square meter	kPa	$Pa \times 10^{3}$

* From the French name, Le Système International d'Unités.

† An SI *base unit*, from which the others are derived.

** Units accepted for limited use.

‡ Used only for liquids and gases. The international symbol for liter is the lowercase "l," which can easily be confused with the numeral "1." Accordingly, the symbol "L" was recommended for United States use by the Department of Commerce in a *Federal Register* notice of December 10, 1976.

Unit	Symbol	Relationship to standard
ANGLES		
radian	rad	
degree	°	$\text{rad} \times \dfrac{180}{\pi}$
TIME		
†second	s	
millisecond	ms	$s \times 10^{-3}$
microsecond	μs	$s \times 10^{-6}$
nanosecond	ns	$s \times 10^{-9}$
picosecond	ps	$s \times 10^{-12}$
minute	min	$s \times 60$
hour	h	$s \times 3600$
day, month, year, century		
SPEED		
meter per second	m/s	
kilometer per hour	km/h	$\text{m/s} \times \dfrac{1}{3.6}$
ENERGY		
joule	J	$N \cdot m = kg \cdot m^2/s^2$
**kilowatthour	$kW \cdot h$	$J \times 3.6 \times 10^6$
POWER		
joule per second	J/s	
watt	W	$\text{J/s} \times 1$
kilowatt	kW	$\text{J/s} \times 10^3$
TEMPERATURE		
†kelvin	K	
degree Celsius	°C	$K - 273.15$
Temperature interval:		
Celsius degree	deg C	$1 \text{ deg C} = 1 \text{ K}$
ELECTRICAL		
Potential difference: volt	V	W/A
†Electric current: ampere	A	
Resistance: ohm	Ω	V/A
Conductance: siemens	S	A/V
Frequency: hertz	Hz	1/s
Power: watt	W	J/s

† An SI *base unit*, from which the others are derived.
** Units accepted for limited use.

Appendix B

SI Unit Prefixes

Amount	Multiples and submultiples	Prefixes	Symbols	Pronunciations	Means
1 000 000 000 000	10^{12}	tera	T	tĕr′à	One trillion times
1 000 000 000	10^9	giga	G	jĭ′ga	One billion times
1 000 000	10^6	mega	M*	mĕg′à	One million times
1 000	10^3	kilo	k*	kĭl′o	One thousand times
100	10^2	hecto	h	hĕk′tô	One hundred times
10	10	deka	da	dĕk′à	Ten times
0.1	10^{-1}	deci	d	dĕs′ĭ	One tenth of
0.01	10^{-2}	centi	c*	sĕn′tĭ	One hundredth of
0.001	10^{-3}	milli	m*	mĭl′ĭ	One thousandth of
0.000 001	10^{-6}	micro	μ*	mĭ′krô	One millionth of
0.000 000 001	10^{-9}	nano	n	năn′ô	One billionth of
0.000 000 000 001	10^{-12}	pico	p	pē′cô	One trillionth of
0.000 000 000 000 001	10^{-15}	femto	f	fĕm′tô	One quadrillionth of
0.000 000 000 000 000 001	10^{-18}	atto	a	ăt′tô	One quintillionth of

* Most commonly used.

Appendix C

Trigonometric Ratios

Angle θ	Sin θ	Cos θ	Tan θ	Angle θ	Sin θ	Cos θ	Tan θ
0°	0.0000	1.0000	0.0000	45°	0.7071	0.7071	1.0000
1°	0.0175	0.9998	0.0175	46°	0.7193	0.6947	1.036
2°	0.0349	0.9994	0.0349	47°	0.7314	0.6820	1.072
3°	0.0523	0.9986	0.0524	48°	0.7431	0.6691	1.111
4°	0.0698	0.9976	0.0699	49°	0.7547	0.6561	1.150
5°	0.0872	0.9962	0.0875	50°	0.7660	0.6428	1.192
6°	0.1045	0.9945	0.1051	51°	0.7771	0.6293	1.235
7°	0.1219	0.9925	0.1228	52°	0.7880	0.6157	1.280
8°	0.1392	0.9903	0.1405	53°	0.7986	0.6018	1.327
9°	0.1564	0.9877	0.1584	54°	0.8090	0.5878	1.376
10°	0.1736	0.9848	0.1763	55°	0.8192	0.5736	1.428
11°	0.1908	0.9816	0.1944	56°	0.8290	0.5592	1.483
12°	0.2079	0.9781	0.2126	57°	0.8387	0.5446	1.540
13°	0.2250	0.9744	0.2309	58°	0.8480	0.5299	1.600
14°	0.2419	0.9703	0.2493	59°	0.8572	0.5150	1.664
15°	0.2588	0.9659	0.2679	60°	0.8660	0.5000	1.732
16°	0.2756	0.9613	0.2867	61°	0.8746	0.4848	1.804
17°	0.2924	0.9563	0.3057	62°	0.8829	0.4695	1.881
18°	0.3090	0.9511	0.3249	63°	0.8910	0.4540	1.963
19°	0.3256	0.9455	0.3443	64°	0.8988	0.4384	2.050
20°	0.3420	0.9397	0.3640	65°	0.9063	0.4226	2.145
21°	0.3584	0.9336	0.3839	66°	0.9135	0.4067	2.246
22°	0.3746	0.9272	0.4040	67°	0.9205	0.3907	2.356
23°	0.3907	0.9205	0.4245	68°	0.9272	0.3746	2.475
24°	0.4067	0.9135	0.4452	69°	0.9336	0.3584	2.605
25°	0.4226	0.9063	0.4663	70°	0.9397	0.3420	2.747
26°	0.4384	0.8988	0.4877	71°	0.9455	0.3256	2.904
27°	0.4540	0.8910	0.5095	72°	0.9511	0.3090	3.078
28°	0.4695	0.8829	0.5317	73°	0.9563	0.2924	3.271
29°	0.4848	0.8746	0.5543	74°	0.9613	0.2756	3.487
30°	0.5000	0.8660	0.5774	75°	0.9659	0.2588	3.732
31°	0.5150	0.8572	0.6009	76°	0.9703	0.2419	4.011
32°	0.5299	0.8480	0.6249	77°	0.9744	0.2250	4.331
33°	0.5446	0.8387	0.6494	78°	0.9781	0.2079	4.705
34°	0.5592	0.8290	0.6745	79°	0.9816	0.1908	5.145
35°	0.5736	0.8192	0.7002	80°	0.9848	0.1736	5.671
36°	0.5878	0.8090	0.7265	81°	0.9877	0.1564	6.314
37°	0.6018	0.7986	0.7536	82°	0.9903	0.1392	7.115
38°	0.6157	0.7880	0.7813	83°	0.9925	0.1219	8.144
39°	0.6293	0.7771	0.8098	84°	0.9945	0.1045	9.514
40°	0.6428	0.7660	0.8391	85°	0.9962	0.0872	11.43
41°	0.6561	0.7547	0.8693	86°	0.9976	0.0698	14.30
42°	0.6691	0.7431	0.9004	87°	0.9986	0.0523	19.08
43°	0.6820	0.7314	0.9325	88°	0.9994	0.0349	28.64
44°	0.6947	0.7193	0.9657	89°	0.9998	0.0175	57.29
45°	0.7071	0.7071	1.0000	90°	1.0000	0.0000	———

Appendix D

Common Logarithms of Numbers

N	0	1	2	3	4	5	6	7	8	9
10	0000	0043	0086	0128	0170	0212	0253	0294	0334	0374
11	0414	0453	0492	0531	0569	0607	0645	0682	0719	0755
12	0792	0828	0864	0899	0934	0969	1004	1038	1072	1106
13	1139	1173	1206	1239	1271	1303	1335	1367	1399	1430
14	1461	1492	1523	1553	1584	1614	1644	1673	1703	1732
15	1761	1790	1818	1847	1875	1903	1931	1959	1987	2014
16	2041	2068	2095	2122	2148	2175	2201	2227	2253	2279
17	2304	2330	2355	2380	2405	2430	2455	2480	2504	2529
18	2553	2577	2601	2625	2648	2672	2695	2718	2742	2765
19	2788	2810	2833	2856	2878	2900	2923	2945	2967	2989
20	3010	3032	3054	3075	3096	3118	3139	3160	3181	3201
21	3222	3243	3263	3284	3304	3324	3345	3365	3385	3404
22	3424	3444	3464	3483	3502	3522	3541	3560	3579	3598
23	3617	3636	3655	3674	3692	3711	3729	3747	3766	3784
24	3802	3820	3838	3856	3874	3892	3909	3927	3945	3962
25	3979	3997	4014	4031	4048	4065	4082	4099	4116	4133
26	4150	4166	4183	4200	4216	4232	4249	4265	4281	4298
27	4314	4330	4346	4362	4378	4393	4409	4425	4440	4456
28	4472	4487	4502	4518	4533	4548	4564	4579	4594	4609
29	4624	4639	4654	4669	4683	4698	4713	4728	4742	4757
30	4771	4786	4800	4814	4829	4843	4857	4871	4886	4900
31	4914	4928	4942	4955	4969	4983	4997	5011	5024	5038
32	5051	5065	5079	5092	5105	5119	5132	5145	5159	5172
33	5185	5198	5211	5224	5237	5250	5263	5276	5289	5302
34	5315	5328	5340	5353	5366	5378	5391	5403	5416	5428
35	5441	5453	5465	5478	5490	5502	5514	5527	5539	5551
36	5563	5575	5587	5599	5611	5623	5635	5647	5658	5670
37	5682	5694	5705	5717	5729	5740	5752	5763	5775	5786
38	5798	5809	5821	5832	5843	5855	5866	5877	5888	5899
39	5911	5922	5933	5944	5955	5966	5977	5988	5999	6010
40	6021	6031	6042	6053	6064	6075	6085	6096	6107	6117
41	6128	6138	6149	6160	6170	6180	6191	6201	6212	6222
42	6232	6243	6253	6263	6274	6284	6294	6304	6314	6325
43	6335	6345	6355	6365	6375	6385	6395	6405	6415	6425
44	6435	6444	6454	6464	6474	6484	6493	6503	6513	6522
45	6532	6542	6551	6561	6571	6580	6590	6599	6609	6618
46	6628	6637	6646	6656	6665	6675	6684	6693	6702	6712
47	6721	6730	6739	6749	6758	6767	6776	6785	6794	6803
48	6812	6821	6830	6839	6848	6857	6866	6875	6884	6893
49	6902	6911	6920	6928	6937	6946	6955	6964	6972	6981
50	6990	6998	7007	7016	7024	7033	7042	7050	7059	7067
51	7076	7084	7093	7101	7110	7118	7126	7135	7143	7152
52	7160	7168	7177	7185	7193	7202	7210	7218	7226	7235
53	7243	7251	7259	7267	7275	7284	7292	7300	7308	7316
54	7324	7332	7340	7348	7356	7364	7372	7380	7388	7396
N	0	1	2	3	4	5	6	7	8	9

N	0	1	2	3	4	5	6	7	8	9
55	7404	7412	7419	7427	7435	7443	7451	7459	7466	7474
56	7482	7490	7497	7505	7513	7520	7528	7536	7543	7551
57	7559	7566	7574	7582	7589	7597	7604	7612	7619	7627
58	7634	7642	7649	7657	7664	7672	7679	7686	7694	7701
59	7709	7716	7723	7731	7738	7745	7752	7760	7767	7774
60	7782	7789	7796	7803	7810	7818	7825	7832	7839	7846
61	7853	7860	7868	7875	7882	7889	7896	7903	7910	7917
62	7924	7931	7938	7945	7952	7959	7966	7973	7980	7987
63	7993	8000	8007	8014	8021	8028	8035	8041	8048	8055
64	8062	8069	8075	8082	8089	8096	8102	8109	8116	8122
65	8129	8136	8142	8149	8156	8162	8169	8176	8182	8189
66	8195	8202	8209	8215	8222	8228	8235	8241	8248	8254
67	8261	8267	8274	8280	8287	8293	8299	8306	8312	8319
68	8325	8331	8338	.8344	8351	8357	8363	8370	8376	8382
69	8388	8395	8401	8407	8414	8420	8426	8432	8439	8445
70	8451	8457	8463	8470	8476	8482	8488	8494	8500	8506
71	8513	8519	8525	8531	8537	8543	8549	8555	8561	8567
72	8573	8579	8585	8591	8597	8603	8609	8615	8621	8627
73	8633	8639	8645	8651	8657	8663	8669	8675	8681	8686
74	8692	8698	8704	8710	8716	8722	8727	8733	8739	8745
75	8751	8756	8762	8768	8774	8779	8785	8791	8797	8802
76	8808	8814	8820	8825	8831	8837	8842	8848	8854	8859
77	8865	8871	8876	8882	8887	8893	8899	8904	8910	8915
78	8921	8927	8932	8938	8943	8949	8954	8960	8965	8971
79	8976	8982	8987	8993	8998	9004	9009	9015	9020	9025
80	9031	9036	9042	9047	9053	9058	9063	9069	9074	9079
81	9085	9090	9096	9101	9106	9112	9117	9122	9128	9133
82	9138	9143	9149	9154	9159	9165	9170	9175	9180	9186
83	9191	9196	9201	9206	9212	9217	9222	9227	9232	9238
84	9243	9248	9253	9253	9263	9269	9274	9279	9284	9289
85	9294	9299	9304	9309	9315	9320	9325	9330	9335	9340
86	9345	9350	9355	9360	9365	9370	9375	9380	9385	9390
87	9395	9400	9405	9410	9415	9420	9425	9430	9435	9440
88	9445	9450	9455	9460	9465	9469	9474	9479	9484	9489
89	9494	9499	9504	9509	9513	9518	9523	9528	9533	9538
90	9542	9547	9552	9557	9562	9566	9571	9576	9581	9586
91	9590	9595	9600	9605	9609	9614	9619	9624	9628	9633
92	9638	9643	9647	9652	9657	9661	9666	9671	9675	9680
93	9685	9689	9694	9699	9703	9708	9713	9717	9722	9727
94	9731	9736	9741	9745	9750	9754	9759	9763	9768	9773
95	9777	9782	9786	9791	9795	9800	9805	9809	9814	9818
96	9823	9827	9832	9836	9841	9845	9850	9854	9859	9863
97	9868	9872	9877	9881	9886	9890	9894	9899	9903	9908
98	9912	9917	9921	9926	9930	9934	9939	9943	9948	9952
99	9956	9961	9965	9969	9974	9978	9983	9987	9991	9996
N	0	1	2	3	4	5	6	7	8	9

Appendix E

Natural Logarithms

N	0	1	2	3	4	5	6	7	8	9
1.0	0.0 000	100	198	296	392	488	583	677	770	862
1.1	953	*044	*133	*222	*310	*398	*484	*570	*655	*740
1.2	0.1 823	906	989	*070	*151	*231	*311	*390	*469	*546
1.3	0.2 624	700	776	852	927	*001	*075	*148	*221	*293
1.4	0.3 365	436	507	577	646	716	784	853	920	988
1.5	0.4 055	121	187	253	318	383	447	511	574	637
1.6	700	762	824	886	947	*008	*068	*128	*188	*247
1.7	0.5 306	365	423	481	539	596	653	710	766	822
1.8	878	933	988	*043	*098	*152	*206	*259	*313	*366
1.9	0.6 419	471	523	575	627	678	729	780	831	881
2.0	931	981	*031	*080	*129	*178	*227	*275	*324	*372
2.1	0.7 419	467	514	561	608	655	701	747	793	839
2.2	885	930	975	*020	*065	*109	*154	*198	*242	*286
2.3	0.8 329	372	416	459	502	544	587	629	671	713
2.4	755	796	838	879	920	961	*002	*042	*083	*123
2.5	0.9 163	203	243	282	322	361	400	439	478	517
2.6	555	594	632	670	708	746	783	821	858	895
2.7	933	969	*006	*043	*080	*116	*152	*188	*225	*260
2.8	1.0 296	332	367	403	438	473	508	543	578	613
2.9	647	682	716	750	784	818	852	886	919	953
3.0	986	*019	*053	*086	*119	*151	*184	*217	*249	*282
3.1	1.1 314	346	378	410	442	474	506	537	569	600
3.2	632	663	694	725	756	787	817	848	878	909
3.3	939	969	*000	*030	*060	*090	*119	*149	*179	*208
3.4	1.2 238	267	296	326	355	384	413	442	470	499
3.5	528	556	585	613	641	669	698	726	754	782
3.6	809	837	865	892	920	947	975	*002	*029	*056
3.7	1.3 083	110	137	164	191	218	244	271	297	324
3.8	350	376	402	429	455	481	507	533	558	584
3.9	610	635	661	686	712	737	762	788	813	838
4.0	863	888	913	938	962	987	*012	*036	*061	*085
4.1	1.4 110	134	159	183	207	231	255	279	303	327
4.2	351	375	398	422	446	469	493	516	540	563
4.3	586	609	633	656	679	702	725	748	770	793
4.4	816	839	861	884	907	929	951	974	996	*019
4.5	1.5 041	063	085	107	129	151	173	195	217	239
4.6	261	282	304	326	347	369	390	412	433	454
4.7	476	497	518	539	560	581	602	623	644	665
4.8	686	707	728	748	769	790	810	831	851	872
4.9	892	913	933	953	974	994	*014	*034	*054	*074
5.0	1.6 094	114	134	154	174	194	214	233	253	273

Multiples of $\log_e 10$

$\log_e 10 = 2.3026$	$-\log_e 10 = 7.6974 - 10 = -2.3026$
$2 \log_e 10 = 4.6052$	$-2 \log_e 10 = 5.3948 - 10$
$3 \log_e 10 = 6.9078$	$-3 \log_e 10 = 3.0922 - 10$
$4 \log_e 10 = 9.2103$	$-4 \log_e 10 = 0.7897 - 10$
$5 \log_e 10 = 11.5129$	$-5 \log_e 10 = 9.4871 - 20$

N		0	1	2	3	4	5	6	7	8	9
5.1		292	312	332	351	371	390	409	429	448	467
5.2		487	506	525	544	563	582	601	620	639	658
5.3		677	696	715	734	752	771	790	808	827	845
5.4		864	882	901	919	938	956	974	993	*011	*029
5.5	1.7	047	066	084	102	120	138	156	174	192	210
5.6		228	246	263	281	299	317	334	352	370	387
5.7		405	422	440	457	475	492	509	527	544	561
5.8		579	596	613	630	647	664	681	699	716	733
5.9		750	766	783	800	817	834	851	867	884	901
6.0		918	934	951	967	984	*001	*017	*034	*050	*066
6.1	1.8	083	099	116	132	148	165	181	197	213	229
6.2		245	262	278	294	310	326	342	358	374	390
6.3		405	421	437	453	469	485	500	516	532	547
6.4		563	579	594	610	625	641	656	672	687	703
6.5		718	733	749	764	779	795	810	825	840	856
6.6		871	886	901	916	931	946	961	976	991	*006
6.7	1.9	021	036	051	066	081	095	110	125	140	155
6.8		169	184	199	213	228	242	257	272	286	301
6.9		315	330	344	359	373	387	402	416	430	445
7.0		459	473	488	502	516	530	544	559	573	587
7.1		601	615	629	643	657	671	685	699	713	727
7.2		741	755	769	782	796	810	824	838	851	865
7.3		879	892	906	920	933	947	961	974	988	*001
7.4	2.0	015	028	042	055	069	082	096	109	122	136
7.5		149	162	176	189	202	215	229	242	255	268
7.6		281	295	308	321	334	347	360	373	386	399
7.7		412	425	438	451	464	477	490	503	516	528
7.8		541	554	567	580	592	605	618	631	643	656
7.9		669	681	694	707	719	732	744	757	769	782
8.0		794	807	819	832	844	857	869	882	894	906
8.1		919	931	943	956	968	980	992	*005	*017	*029
8.2	2.1	041	054	066	080	090	102	114	126	138	150
8.3		163	175	187	199	211	223	235	247	258	270
8.4		282	294	306	318	330	342	353	365	377	389
8.5		401	412	424	436	448	460	471	483	494	506
8.6		518	529	541	552	564	576	587	599	610	622
8.7		633	645	656	668	679	691	702	713	725	736
8.8		748	759	770	782	793	804	815	827	838	849
8.9		861	872	883	894	905	917	928	939	950	961
9.0		972	983	994	*006	*017	*028	*039	*050	*061	*072
9.1	2.2	083	094	105	116	127	137	148	159	170	181
9.2		192	203	214	225	235	246	257	268	279	289
9.3		300	311	322	332	343	354	364	375	386	396
9.4		407	418	428	439	450	460	471	481	492	502
9.5		513	523	534	544	555	565	576	586	597	607
9.6		618	628	638	649	659	670	680	690	701	711
9.7		721	732	742	752	762	773	783	793	803	814
9.8		824	834	844	854	865	875	885	895	905	915
9.9		925	935	946	956	966	976	986	996	*006	*016
10.	2.3	026	036	046	056	066	076	086	096	106	115

Appendix F

Powers of e

n	e^n	e^{-n}	n	e^n	e^{-n}	n	e^n	e^{-n}
0.00	1.000	1.000	0.50	1.649	.607	1.0	2.718	.368
.01	1.010	0.990	.51	1.665	.600	.1	3.004	.333
.02	1.020	.980	.52	1.682	.595	.2	3.320	.301
.03	1.030	.970	.53	1.699	.589	.3	3.669	.273
.04	1.041	.961	.54	1.716	.583	.4	4.055	.247
0.05	1.051	.951	0.55	1.733	.577	1.5	4.482	.223
.06	1.062	.942	.56	1.751	.571	.6	4.953	.202
.07	1.073	.932	.57	1.768	.566	.7	5.474	.183
.08	1.083	.923	.58	1.786	.560	.8	6.050	.165
.09	1.094	.914	.59	1.804	.554	.9	6.686	.150
0.10	1.105	.905	0.60	1.822	.549	2.0	7.389	.135
.11	1.116	.896	.61	1.840	.543	.1	8.166	.122
.12	1.127	.887	.62	1.859	.538	.2	9.025	.111
.13	1.139	.878	.63	1.878	.533	.3	9.974	.100
.14	1.150	.869	.64	1.896	.527	.4	11.02	.0907
0.15	1.162	.861	0.65	1.916	.522	2.5	12.18	.0821
.16	1.174	.852	.66	1.935	.517	.6	13.46	.0743
.17	1.185	.844	.67	1.954	.512	.7	14.88	.0672
.18	1.197	.835	.68	1.974	.507	.8	16.44	.0608
.19	1.209	.827	.69	1.994	.502	.9	18.17	.0550
0.20	1.221	.819	0.70	2.014	.497	3.0	20.09	.0498
.21	1.234	.811	.71	2.034	.492	.1	22.20	.0450
.22	1.246	.803	.72	2.054	.487	.2	24.53	.0408
.23	1.259	.795	.73	2.075	.482	.3	27.11	.0369
.24	1.271	.787	.74	2.096	.477	.4	29.96	.0334
0.25	1.284	.779	0.75	2.117	.472	3.5	33.12	.0302
.26	1.297	.771	.76	2.138	.468	.6	36.60	.0273
.27	1.310	.763	.77	2.160	.463	.7	40.45	.0247
.28	1.323	.756	.78	2.181	.458	.8	44.70	.0224
.29	1.336	.748	.79	2.203	.454	.9	49.40	.0202
0.30	1.350	.741	0.80	2.226	.449	4.0	54.60	.0183
.31	1.363	.733	.81	2.248	.445	.1	60.34	.0166
.32	1.377	.726	.82	2.270	.440	.2	66.69	.0150
.33	1.391	.719	.83	2.293	.436	.3	73.70	.0136
.34	1.405	.712	.84	2.316	.432	.4	81.45	.0123
0.35	1.419	.705	0.85	2.340	.427	4.5	90.02	.0111
.36	1.433	.698	.86	2.363	.423	5.0	148.4	.00674
.37	1.448	.691	.87	2.387	.419	6.0	403.4	.00248
.38	1.462	.684	.88	2.411	.415	7.0	1097.	.000912
.39	1.477	.677	.89	2.435	.411			
0.40	1.492	.670	0.90	2.460	.407	8.0	2981.	.000335
.41	1.507	.664	.91	2.484	.403	9.0	8103.	.000123
.42	1.522	.657	.92	2.509	.399	10.0	22026.	.000045
.43	1.537	.651	.93	2.535	.395	$\pi/2$	4.810	.208
.44	1.553	.644	.94	2.560	.391	$2\pi/2$	23.14	.0432
0.45	1.568	.638	0.95	2.586	.387	$3\pi/2$	111.3	.00898
.46	1.584	.631	.96	2.612	.383	$4\pi/2$	535.5	.00187
.47	1.600	.625	.97	2.638	.379	$5\pi/2$	2576.	.000388
.48	1.616	.619	.98	2.664	.375	$6\pi/2$	12392.	.000081
.49	1.632	.613	.99	2.691	.372	$7\pi/2$	59610.	.000017
0.50	1.649	0.607	1.00	2.718	.368	$8\pi/2$	286751.	.000003

Greek Alphabet

A	α	alpha	N	ν	nu
B	β	beta	Ξ	ξ	xi
Γ	γ	gamma	O	o	omicron
Δ	δ	delta	Π	π	pi
E	ϵ	epsilon	P	ρ	rho
Z	ζ	zeta	Σ	σ	sigma
H	η	eta	T	τ	tau
Θ	θ	theta	Υ	υ	upsilon
I	ι	iota	Φ	ϕ	phi
K	κ	kappa	X	χ	chi
Λ	λ	lambda	Ψ	ψ	psi
M	μ	mu	Ω	ω	omega

Appendix H

Conversion Factors

Unit	equals	
LENGTH		
1 angstrom	1×10^{-10}	meter
	1×10^{-4}	micron
1 centimeter	10^{-2}	meter
	0.3937	inch
1 foot	12	inches
	0.3048	meter
1 inch	2.54×10^{8}	angstroms
	25.4	millimeters
	2.54	centimeters
1 kilometer	3281	feet
	0.5400	nautical mile
	0.6214	statute mile
	1094	yards
1 light-year	9.461×10^{12}	kilometers
	5.879×10^{12}	statute miles
1 meter	10^{10}	angstroms
	3.281	feet
	39.37	inches
	1.094	yards
1 micron	10^{4}	angstroms
	10^{-4}	centimeter
	10^{-6}	meter
1 nautical mile (International)	8.439	cables
	6076	feet
	1852	meters
	1.151	statute miles
1 statute mile	5280	feet
	8	furlongs
	1.609	kilometers
	0.8690	nautical mile
1 mil	10^{-3}	inch
	2.54×10^{-2}	millimeter
	25.4	micrometers
1 yard	3	feet
	0.9144	meter
ANGLES		
1 degree	60	minutes
	0.017 45	radian
	3600	seconds
	2.778×10^{-3}	revolution
1 minute of arc	0.016 67	degree
	2.909×10^{-4}	radian
	60	seconds

Unit	equals	
ANGLES (continued)		
1 radian	0.1592	revolution
	57.296	degrees
	3438	minutes
1 second of arc	2.778×10^{-4}	degree
	0.016 67	minute
AREA		
1 acre	4047	square meters
	43 560	square feet
1 are	0.024 71	acre
	1	square dekameter
	100	square meters
1 circular mil	10^{-6}	circular inch
	5.067×10^{-4}	square millimeter
	0.7854	square mil
1 hectare	2.471	acres
	100	ares
	10 000	square meters
1 square foot	144	square inches
	0.092 90	square meter
1 square inch	1.273×10^{6}	circular mils
	6.944×10^{-3}	square foot
	6.452	square centimeters
1 square kilometer	247.1	acres
1 square meter	10.76	square feet
1 square mile	640	acres
	2.788×10^{7}	square feet
	2.590	square kilometers
1 square mil	1.273	circular mils
	10^{-6}	square inch
VOLUME		
1 board-foot	144	cubic inches
1 bushel (U.S.)	1.244	cubic feet
	8	U.S. dry gallons
	35.24	liters
1 cord	12.8	cubic feet
	3.625	cubic meters
1 cubic foot	7.481	gallons (U.S. liquid)
	28.32	liters
1 cubic inch	0.016 39	liter
	16.39	milliliters

Unit	equals	
VOLUME (continued)		
1 cubic meter	35.31	cubic feet
	10^{-3}	cubic centimeter
1 cubic millimeter	6.102×10^{-5}	cubic inch
1 cubic yard	27	cubic feet
	0.7646	cubic meter
1 gallon (imperial)	277.4	cubic inches
	4.546	liters
1 gallon (U.S. liquid)	231	cubic inches
	3.785	liters
1 kiloliter	35.31	cubic feet
	1.000	cubic meter
	1.308	cubic yards
	220	imperial gallons
1 liter	10^3	cubic centimeters
	10^6	cubic millimeters
	10^{-3}	cubic meter
	61.02	cubic inches
	0.035 32	cubic foot
MASS		
1 grain	0.064 80	gram
1 gram	15.43	grains
	0.035 27	avoirdupois ounce
	2.204×10^{-3}	avoirdupois pound
1 kilogram	564.4	avoirdupois drams
	2.205	avoirdupois pounds
1 avoirdupois ounce	28.35	grams
	0.9115	troy ounce
1 avoirdupois pound	256	drams
	453.6	grams
	0.4536	kilogram
	16	ounces
1 metric ton	1000	kilograms
	2205	avoirdupois pounds
1 ton	2000	avoirdupois pounds
	907.2	kilograms
FORCE		
1 dyne	10^{-5}	newton
1 newton	10^5	dynes
	0.2248	pound-force
1 pound	4.448	newtons

Unit	equals	
VELOCITY		
1 foot/minute	0.3048	meter/minute
	5.08×10^{-3}	meter/second
	0.011 364	mile/hour
1 foot/second	1097	kilometers/hour
	18.29	meters/minute
	0.6818	mile/hour
1 kilometer/hour	3281	feet/hour
	54.68	feet/minute
	0.2778	meter/second
	0.6214	mile/hour
1 kilometer/minute	3281	feet/minute
	37.28	miles/hour
1 knot	6076	feet/hour
	101.3	feet/minute
	1.688	feet/second
	1.852	kilometers/hour
	30.87	meters/minute
	0.5144	meter/second
	1.151	miles/hour
1 meter/hour	3.281	feet/hour
	88	feet/minute
	1.466	feet/second
1 mile/hour	1.467	feet/second
	1.609	kilometers/hour
POWER		
1 British thermal unit/hour	0.2929	watt
1 Btu/pound	2.324	joules/gram
1 Btu/second	1.414	horsepower
	107.5	kilogrammeters/second
	1.054	kilowatts
	1054	watts
1 horsepower	42.44	Btu/minute
	550	footpounds/second
	746	kilowatts
1 kilogrammeter/second	9.807	watts
1 kilowatt	3414	Btu/hour
	737.6	footpounds/second
	1.341	horsepower
	10^3	joules/second
	3.671×10^5	kilogrammeters/hour
	999.8	international watt
1 watt	44.25	footpounds/minute
	1	joule/second

Unit	equals	
PRESSURE		
1 atmosphere	1.013	bars
	1033	grams/square centimeter
	14.70	pounds/square inch
	760	torrs
	101	kilopascals
1 bar	10^6	baryes
	10 197	kilogramsforce/square meter
	14.50	pounds-force/square inch
1 barye	10^{-6}	bar
1 inch of mercury	0.033 86	bar
	345.3	kilogramsforce/square meter
	70.73	pounds-force/square foot
1 pascal	1	newton/square meter
1 pound/square inch	0.068 03	atmosphere
	70.27	grams/square centimeter
ENERGY		
1 British thermal unit	1054	joules
	1054	wattseconds
1 foot-pound-force	1.356	joules
	0.1383	kilogramforce-meter
	1.356	newtonmeters
1 joule	0.7376	foot-pound-force
	0.1020	kilogramforce-meter
	1	wattsecond
1 kilogramforce-meter	9.288×10^{-3}	British thermal unit
	7.233	foot-pounds-force
	9.807	joules
	9.807	newtonmeters
	2.724×10^{-3}	watthour
1 kilowatthour	3410	British thermal units
	1.341	horsepowerhours
1 newtonmeter	0.1020	kilogramforce-meter
	0.7376	pound-force-foot
1 watthour	3.414	British thermal units
	2655	footpounds-force
	3600	joules
	367.1	kilogramforce-meters

Index

The italic numbers are *formula* numbers, *not* page numbers. They refer to the boxed and numbered formulas in the text and in the Summary of Formulas in the front of the book.

Catalog

If you are interested in a list of SCHAUM'S
OUTLINE SERIES send your name
and address, requesting your free catalog, to:

SCHAUM'S OUTLINE SERIES, Dept. C
McGRAW-HILL BOOK COMPANY
1221 Avenue of Americas
New York, N.Y. 10020